INTRODUCTORY RAMAN SPECTROSCOPY

INTRODUCTORY RAMAN SPECTROSCOPY

John R. Ferraro
Argonne National Laboratory
Argonne, Illinois

Kazuo Nakamoto
Marquette University
Milwaukee, Wisconsin

ACADEMIC PRESS, INC.
Harcourt Brace & Company, Publishers

Boston   San Diego   New York
London   Sydney   Tokyo   Toronto

This book is printed on acid-free paper. ∞

Coventry University

ACADEMIC PRESS, INC.
1250 Sixth Avenue, San Diego, CA 92101-4311

PO 5098

United Kingdom edition published by
ACADEMIC PRESS LIMITED
24–28 Oval Road, London NW1 7DX

Library of Congress Cataloging-in-Publication Data
Ferraro, John R., 1918–
    Introductory Raman spectroscopy / John R. Ferraro, Kazuo Nakamoto.
        p. cm.
    Includes bibliographical references and index.
    1. Raman spectroscopy.   I. Nakamoto, Kazuo, 1922–   . II. Title.
QC454.R36F47   1994
543′.08584--dc20   93-41279
                                                                                    CIP

International Standard Book Number: 0-12-253990-7

Printed in the United States of America
94 95 96 97 98 99   9 8 7 6 5 4 3 2 1

The authors wish to dedicate this book to their wives.

Contents

Contents

Preface

Raman spectroscopy has made remarkable progress in recent years. The synergism that has taken place with the advent of new detectors, Fourier-transform Raman and fiber optics has stimulated renewed interest in the technique. Its use in academia and especially in industry has grown rapidly.

A well-balanced Raman text on an introductory level, which explains basic theory, instrumentation and experimental techniques (including special techniques), and a wide variety of applications (particularly the newer ones) is not available. The authors have attempted to meet this deficiency by writing this book. This book is intended to serve as a guide for beginners.

One problem we had in writing this book concerned itself in how one defines 'introductory level.' We have made a sincere effort to write this book on our definition of this level, and have kept mathematics at a minimum, albeit giving a logical development of basic theory.

The book consists of Chapters 1 to 4, and appendices. The first chapter deals with basic theory of spectroscopy; the second chapter discusses instrumentation and experimental techniques; the third chapter deals with special techniques; Chapter 4 presents applications of Raman spectroscopy in structural chemistry, biochemistry, biology and medicine, solid-state chemistry and industry. The appendices consist of eight sections. As much as possible, the authors have attempted to include the latest developments.

Acknowledgements

The authors would like to express their thanks to Prof. Robert A. Condrate of Alfred University, Prof. Roman S. Czernuszewicz of the University of Houston, Dr. Victor A. Maroni of Argonne National Laboratory, and Prof. Masamichi Tsuboi of Iwaki-Meisei University of Japan who made many valuable suggestions. Special thanks are given to Roman S. Czernuszewicz for making drawings for Chapters 1 and 2. Our thanks and appreciation also go to Prof. Hiro-o Hamaguchi of Kanagawa Academy of Science and Technology of Japan and Prof. Akiko Hirakawa of the University of the Air of Japan who gave us permission to reproduce Raman spectra of typical solvents (Appendix 8). We would also like to thank Ms. Jane Ellis, Acquisition Editor for Academic Press, Inc., who invited us to write this book and for her encouragement and help throughout the project. Finally, this book could not have been written without the help of many colleagues who allowed us to reproduce figures for publication.

John R. Ferraro

1994   Kazuo Nakamoto

Chapter 1

Basic Theory

1.1 Historical Background of Raman Spectroscopy

In 1928, when Sir Chandrasekhra Venkata Raman discovered the phenomenon that bears his name, only crude instrumentation was available. Sir Raman used sunlight as the source and a telescope as the collector; the detector was his eyes. That such a feeble phenomenon as the Raman scattering was detected was indeed remarkable.

Gradually, improvements in the various components of Raman instrumentation took place. Early research was concentrated on the development of better excitation sources. Various lamps of elements were developed (e.g., helium, bismuth, lead, zinc) (1–3). These proved to be unsatisfactory because of low light intensities. Mercury sources were also developed. An early mercury lamp which had been used for other purposes in 1914 by Kerschbaum (1) was developed. In the 1930s mercury lamps suitable for Raman use were designed (2). Hibben (3) developed a mercury burner in 1939, and Spedding and Stamm (4) experimented with a cooled version in 1942. Further progress was made by Rank and McCartney (5) in 1948, who studied mercury burners and their backgrounds. Hilger Co. developed a commercial mercury excitation source system for the Raman instrument, which consisted of four lamps surrounding the Raman tube. Welsh *et al.* (6) introduced a mercury source in 1952, which became known as the Toronto Arc. The lamp consisted of a four-turn helix of Pyrex tubing

1

and was an improvement over the Hilger lamp. Improvements in lamps were made by Ham and Walsh (7), who described the use of microwave-powered helium, mercury, sodium, rubidium and potassium lamps. Stammreich (8–12) also examined the practicality of using helium, argon, rubidium and cesium lamps for colored materials. In 1962 laser sources were developed for use with Raman spectroscopy (13). Eventually, the Ar^+ (351.1–514.5 nm) and the Kr^+ (337.4–676.4 nm) lasers became available, and more recently the Nd-YAG laser (1,064 nm) has been used for Raman spectroscopy (see Chapter 2, Section 2.2).

Progress occurred in the detection systems for Raman measurements. Whereas original measurements were made using photographic plates with the cumbersome development of photographic plates, photoelectric Raman instrumentation was developed after World War II. The first photoelectric Raman instrument was reported in 1942 by Rank and Wiegand (14), who used a cooled cascade type RCA IP21 detector. The Heigl instrument appeared in 1950 and used a cooled RCA C-7073B photomultiplier. In 1953 Stamm and Salzman (15) reported the development of photoelectric Raman instrumentation using a cooled RCA IP21 photomultiplier tube. The Hilger E612 instrument (16) was also produced at this time, which could be used as a photographic or photoelectric instrument. In the photoelectric mode a photomultiplier was used as the detector. This was followed by the introduction of the Cary Model 81 Raman spectrometer (17). The source used was the 3 kW helical Hg arc of the Toronto type. The instrument employed a twin-grating, twin-slit double monochromator.

Developments in the optical train of Raman instrumentation took place in the early 1960s. It was discovered that a double monochromator removed stray light more efficiently than a single monochromator. Later, a triple monochromator was introduced, which was even more efficient in removing stray light. Holographic gratings appeared in 1968 (17), which added to the efficiency of the collection of Raman scattering in commercial Raman instruments.

These developments in Raman instrumentation brought commercial Raman instruments to the present state of the art of Raman measurements. Now, Raman spectra can also be obtained by Fourier transform (FT) spectroscopy. FT-Raman instruments are being sold by all Fourier transform infrared (FT-IR) instrument makers, either as interfaced units to the FT-IR spectrometer or as dedicated FT-Raman instruments.

1.2 Energy Units and Molecular Spectra

Figure 1-1 illustrates a wave of polarized electromagnetic radiation traveling in the z-direction. It consists of the electric component (x-direction) and

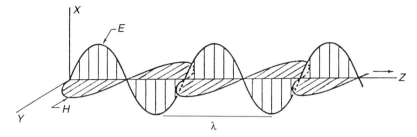

Figure 1-1   Plane-polarized electromagnetic radiation.

magnetic component (y-direction), which are perpendicular to each other. Hereafter, we will consider only the former since topics discussed in this book do not involve magnetic phenomena. The electric field strength (E) at a given time (t) is expressed by

$$E = E_0 \cos 2\pi\nu t, \qquad (1\text{-}1)$$

where E_0 is the amplitude and ν is the frequency of radiation as defined later.

The distance between two points of the same phase in successive waves is called the "wavelength," λ, which is measured in units such as Å (angstrom), nm (nanometer), mμ (millimicron), and cm (centimeter). The relationships between these units are:

$$1 \text{ Å} = 10^{-8} \text{ cm} = 10^{-1} \text{ nm} = 10^{-1} \text{ m}\mu. \qquad (1\text{-}2)$$

Thus, for example, 4,000 Å = 400 nm = 400 mμ.

The frequency, ν, is the number of waves in the distance light travels in one second. Thus,

$$\nu = \frac{c}{\lambda}, \qquad (1\text{-}3)$$

where c is the velocity of light (3×10^{10} cm/s). If λ is in the unit of centimeters, its dimension is (cm/s)/(cm) = 1/s. This "reciprocal second" unit is also called the "hertz" (Hz).

The third parameter, which is most common to vibrational spectroscopy, is the "wavenumber," $\tilde{\nu}$, defined by

$$\tilde{\nu} = \frac{\nu}{c}. \qquad (1\text{-}4)$$

The difference between ν and $\tilde{\nu}$ is obvious. It has the dimension of (1/s)/(cm/s) = 1/cm. By combining (1-3) and (1-4) we have

$$\tilde{\nu} = \frac{\nu}{c} = \frac{1}{\lambda} \text{ (cm}^{-1}). \qquad (1\text{-}5)$$

Table 1-1 Units Used in Spectroscopy*

10^{12}	tera	T
10^9	giga	G
10^6	mega	M
10^3	kilo	k
10^2	hecto	h
10^1	deca	da
10^{-1}	deci	d
10^{-2}	centi	c
10^{-3}	milli	m
10^{-6}	micro	μ
10^{-9}	nano	n
10^{-12}	pico	p
10^{-15}	femto	f
10^{-18}	atto	a

*Notations: T, G, M, k, h, da, μ, n—Greek; d, c, m—Latin; p—Spanish; f—Swedish; a—Danish.

Thus, 4,000 Å corresponds to 25×10^3 cm^{-1}, since

$$\tilde{v} = \frac{1}{\lambda \text{(cm)}} = \frac{1}{4 \times 10^3 \times 10^{-8}} = 25 \times 10^3 \ (\text{cm}^{-1}).$$

Table 1-1 lists units frequently used in spectroscopy. By combining (1-3) and (1-4), we obtain

$$v = \frac{c}{\lambda} = c\tilde{v}. \tag{1-6}$$

As shown earlier, the wavenumber ($\tilde{v}$) and frequency (v) are different parameters, yet these two terms are often used interchangeably. Thus, an expression such as "frequency shift of 30 cm^{-1}" is used conventionally by IR and Raman spectroscopists and we will follow this convention through this book.

If a molecule interacts with an electromagnetic field, a transfer of energy from the field to the molecule can occur only when Bohr's frequency condition is satisfied. Namely,

$$\Delta E = hv = h\frac{c}{\lambda} = hc\tilde{v}. \tag{1-7}$$

Here ΔE is the difference in energy between two quantized states, h is Planck's constant (6.62×10^{-27} erg s) and c is the velocity of light. Thus, $\tilde{v}$ is directly proportional to the energy of transition.

Suppose that

$$\Delta E = E_2 - E_1, \tag{1-8}$$

where E_2 and E_1 are the energies of the excited and ground states, respectively. Then, the molecule "absorbs" ΔE when it is excited from E_1 to E_2, and "emits" ΔE when it reverts from E_2 to E_1[1]:

$$\begin{array}{ll} \underline{\hspace{4cm}} \; E_2 & \qquad \underline{\hspace{4cm}} \; E_2 \\ \Delta E \uparrow \text{absorption} & \qquad \Delta E \downarrow \text{emission} \\ \underline{\hspace{4cm}} \; E_1 & \qquad \underline{\hspace{4cm}} \; E_1 \end{array}$$

Using the relationship given by Eq. (1-7), Eq. (1-8) is written as

$$\Delta E = E_2 - E_1 = hc\tilde{v}. \tag{1-9}$$

Since h and c are known constants, ΔE can be expressed in terms of various energy units. Thus, 1 cm^{-1} is equivalent to

$$\Delta E = [6.62 \times 10^{-27} \text{ (erg s)}][3 \times 10^{10} \text{ (cm/s)}][1(1/\text{cm})]$$

$$= 1.9 \times 10^{-16} \text{ (erg/molecule)}$$

$$= 1.99 \times 10^{-23} \text{ (joule/molecule)}$$

$$= 2.86 \text{ (cal/mole)}$$

$$= 1.24 \times 10^{-4} \text{ (eV/molecule)}.$$

In the preceding conversions, the following factors were used:

$$1 \text{ (erg/molecule)} = 2.39 \times 10^{-8} \text{ (cal/molecule)}$$

$$= 1 \times 10^{-7} \text{ (joule/molecule)}$$

$$= 6.2422 \times 10^{11} \text{ (eV/molecule)}$$

$$\text{Avogadro number, } N_0 = 6.025 \times 10^{23} \text{ (1/mole)}$$

$$1 \text{ (cal)} = 4.185 \text{ (joule)}$$

Figure 1-2 compares the order of energy expressed in terms of $\tilde{v}$ (cm^{-1}), λ (cm) and v (Hz).

As indicated in Fig. 1-2 and Table 1-2, the magnitude of ΔE is different depending upon the origin of the transition. In this book, we are mainly concerned with vibrational transitions which are observed in infrared (IR) or Raman spectra[2]. These transitions appear in the $10^4 \sim 10^2$ cm^{-1} region

---

[1] If a molecule loses ΔE via molecular collision, it is called a "radiationless transition."

[2] Pure rotational and rotational–vibrational transitions are also observed in IR and Raman spectra. Many excellent textbooks are available on these and other subjects (see general references given at the end of this chapter).

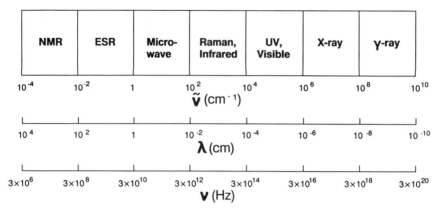

Figure 1-2 Energy units for various portions of electromagnetic spectrum.

Table 1-2 Spectral Regions and Their Origins

Spectroscopy	Range ($\tilde{\nu}$, cm^{-1})	Origin
γ-ray	10^{10}–10^8	Rearrangement of elementary particles in the nucleus
X-ray (ESCA, PES)	10^8–10^6	Transitions between energy levels of inner electrons of atoms and molecules
UV-Visible	10^6–10^4	Transitions between energy levels of valence electrons of atoms and molecules
Raman and infrared	10^4–10^2	Transitions between vibrational levels (change of configuration)
Microwave	10^2–1	Transitions between rotational levels (change of orientation)
Electron spin resonance (ESR)	1–10^{-2}	Transitions between electron spin levels in magnetic field
Nuclear magnetic resonance (NMR)	10^{-2}–10^{-4}	Transitions between nuclear spin levels in magnetic fields

and originate from vibrations of nuclei constituting the molecule. As will be shown later, Raman spectra are intimately related to electronic transitions. Thus, it is important to know the relationship between electronic and vibrational states. On the other hand, vibrational spectra of small molecules in the gaseous state exhibit rotational fine structures.[3] Thus, it is also

---

[3] In solution, rotational fine structures are not observed because molecular collisions (10^{-13} s) occur before one rotation is completed (10^{-11} s) and the levels of individual molecules are perturbed differently. In the solid state, molecular rotation does not occur because of intermolecular interaction.

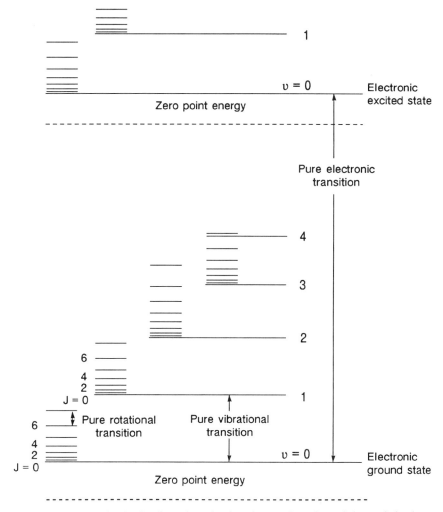

Figure 1-3 Energy levels of a diatomic molecule. (The actual spacings of electronic levels are much larger, and those of rotational levels much smaller, than those shown in the figure.)

important to know the relationship between vibrational and rotational states. Figure 1-3 illustrates the three types of transitions for a diatomic molecule.

1.3 Vibration of a Diatomic Molecule

Consider the vibration of a diatomic molecule in which two atoms are connected by a chemical bond.

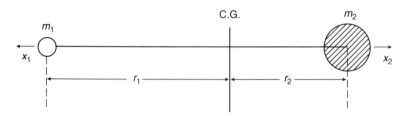

Here, m_1 and m_2 are the masses of atom 1 and 2, respectively, and r_1 and r_2 are the distances from the center of gravity (C.G.) to the atoms designated. Thus, $r_1 + r_2$ is the equilibrium distance, and x_1 and x_2 are the displacements of atoms 1 and 2, respectively, from their equilibrium positions. Then, the conservation of the center of gravity requires the relationships:

$$m_1 r_1 = m_2 r_2, \qquad (1\text{-}10)$$

$$m_1(r_1 + x_1) = m_2(r_2 + x_2). \qquad (1\text{-}11)$$

Combining these two equations, we obtain

$$x_1 = \left(\frac{m_2}{m_1}\right)x_2 \quad \text{or} \quad x_2 = \left(\frac{m_1}{m_2}\right)x_1. \qquad (1\text{-}12)$$

In the classical treatment, the chemical bond is regarded as a spring that obeys Hooke's law, where the restoring force, f, is expressed as

$$f = -K(x_1 + x_2). \qquad (1\text{-}13)$$

Here K is the force constant, and the minus sign indicates that the directions of the force and the displacement are opposite to each other. From (1-12) and (1-13), we obtain

$$f = -K\left(\frac{m_1 + m_2}{m_1}\right)x_2 = -K\left(\frac{m_1 + m_2}{m_2}\right)x_1. \qquad (1\text{-}14)$$

Newton's equation of motion ($f = ma$; $m = $ mass; $a = $ acceleration) is written for each atom as

$$m_1 \frac{d^2 x_1}{dt^2} = -K\left(\frac{m_1 + m_2}{m_2}\right)x_1, \qquad (1\text{-}15)$$

$$m_2 \frac{d^2 x_2}{dt^2} = -K\left(\frac{m_1 + m_2}{m_1}\right)x_2. \qquad (1\text{-}16)$$

By adding

$$(1\text{-}15) \times \left(\frac{m_2}{m_1 + m_2}\right) \quad \text{and} \quad (1\text{-}16) \times \left(\frac{m_1}{m_1 + m_2}\right),$$

we obtain

$$\frac{m_1 m_2}{m_1 + m_2}\left(\frac{d^2 x_1}{dt^2} + \frac{d^2 x_2}{dt^2}\right) = -K(x_1 + x_2). \tag{1-17}$$

Introducing the reduced mass (μ) and the displacement (q), (1-17) is written as

$$\mu\frac{d^2 q}{dt^2} = -Kq. \tag{1-18}$$

The solution of this differential equation is

$$q = q_0 \sin(2\pi v_0 t + \varphi), \tag{1-19}$$

where q_0 is the maximum displacement and φ is the phase constant, which depends on the initial conditions. v_0 is the classical vibrational frequency given by

$$v_0 = \frac{1}{2\pi}\sqrt{\frac{K}{\mu}}. \tag{1-20}$$

The potential energy (V) is defined by

$$dV = -f\,dq = Kq\,dq.$$

Thus, it is given by

$$V = \tfrac{1}{2}Kq^2 \tag{1-21}$$

$$= \tfrac{1}{2}Kq_0^2 \sin^2(2\pi v_0 t + \varphi)$$

$$= 2\pi^2 v_0^2 \mu q_0^2 \sin^2(2\pi v_0 t + \varphi).$$

The kinetic energy (T) is

$$T = \frac{1}{2}m_1\left(\frac{dx_1}{dt}\right)^2 + \frac{1}{2}m_2\left(\frac{dx_2}{dt}\right)^2$$

$$= \frac{1}{2}\mu\left(\frac{dq}{dt}\right)^2$$

$$= 2\pi^2 v_0^2 \mu q_0^2 \cos^2(2\pi v_0 t + \varphi). \tag{1-22}$$

Thus, the total energy (E) is

$$E = T + V$$

$$= 2\pi^2 v_0^2 \mu q_0^2 = \text{constant} \tag{1-23}$$

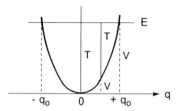

Figure 1-4   Potential energy diagram for a harmonic oscillator.

Figure 1-4 shows the plot of V as a function of q. This is a parabolic potential, $V = \frac{1}{2}Kq^2$, with $E = T$ at $q = 0$ and $E = V$ at $q = \pm q_0$. Such a vibrator is called a *harmonic oscillator*.

In quantum mechanics (18,19) the vibration of a diatomic molecule can be treated as a motion of a single particle having mass μ whose potential energy is expressed by (1-21). The Schrödinger equation for such a system is written as

$$\frac{d^2\psi}{dq^2} + \frac{8\pi^2\mu}{h^2}\left(E - \frac{1}{2}Kq^2\right)\psi = 0. \tag{1-24}$$

If (1-24) is solved with the condition that ψ must be single-valued, finite and continuous, the eigenvalues are

$$E_v = hv(v + \tfrac{1}{2}) = hc\tilde{v}(v + \tfrac{1}{2}), \tag{1-25}$$

with the frequency of vibration

$$v = \frac{1}{2\pi}\sqrt{\frac{K}{\mu}} \quad \text{or} \quad \tilde{v} = \frac{1}{2\pi c}\sqrt{\frac{K}{\mu}}. \tag{1-26}$$

Here, v is the vibrational quantum number, and it can have the values 0, 1, 2, 3,.... . The corresponding eigenfunctions are

$$\psi_v = \frac{(\alpha/\pi)^{1/4}}{\sqrt{2^v v!}}e^{-\alpha q^2/2}H_v(\sqrt{\alpha}q), \tag{1-27}$$

where

$$\alpha = 2\pi\sqrt{\mu K}/h = 4\pi^2\mu v/h \quad \text{and} \quad H_v(\sqrt{\alpha}q)$$

is a Hermite polynomial of the vth degree. Thus, the eigenvalues and the corresponding eigenfunctions are

$$v = 0, \quad E_0 = \tfrac{1}{2}hv, \quad \psi_0 = (\alpha/\pi)^{1/4}e^{-\alpha q^2/2}$$
$$v = 1, \quad E_1 = \tfrac{3}{2}hv, \quad \psi_1 = (\alpha/\pi)^{1/4}2^{1/2}qe^{-\alpha q^2/2}. \tag{1-28}$$
$$\vdots \qquad\qquad \vdots \qquad\qquad\quad \vdots$$

One should note that the quantum-mechanical frequency (1-26) is exactly the same as the classical frequency (1-20). However, several marked differences must be noted between the two treatments. First, classically, E is zero when q is zero. Quantum-mechanically, the lowest energy state ($v = 0$) has the energy of $\frac{1}{2}hv$ (zero point energy) (see Fig. 1-3) which results from Heisenberg's uncertainty principle. Secondly, the energy of a such a vibrator can change continuously in classical mechanics. In quantum mechanics, the energy can change only in units of hv. Thirdly, the vibration is confined within the parabola in classical mechanics since T becomes negative if $|q| > |q_0|$ (see Fig. 1-4). In quantum mechanics, the probability of finding q outside the parabola is not zero (tunnel effect) (Fig. 1-5).

In the case of a harmonic oscillator, the separation between the two successive vibrational levels is always the same (hv). This is not the case of an actual molecule whose potential is approximated by the Morse potential function shown by the solid curve in Fig. 1-6.

$$V = D_e(1 - e^{-\beta q})^2. \tag{1-29}$$

Here, D_e is the dissociation energy and β is a measure of the curvature at the bottom of the potential well. If the Schrödinger equation is solved with this potential, the eigenvalues are (18,19)

$$E_v = hc\omega_e(v + \tfrac{1}{2}) - hc\chi_e\omega_e(v + \tfrac{1}{2})^2 + \dots, \tag{1-30}$$

where ω_e is the wavenumber corrected for anharmonicity, and $\chi_e\omega_e$ indicates the magnitude of anharmonicity. Equation (1-30) shows that the energy levels of the anharmonic oscillator are no longer equidistant, and the separation decreases with increasing v as shown in Fig. 1-6. Thus far, anharmonicity

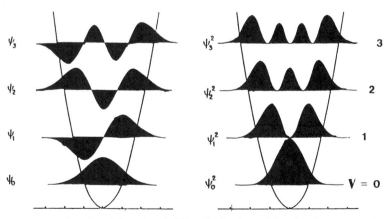

Figure 1-5 Wave functions (left) and probability distributions (right) of the harmonic oscillator.

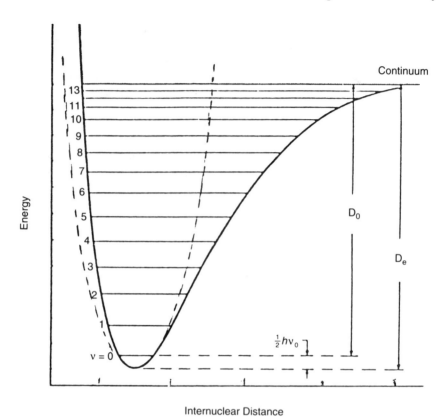

Figure 1-6 Potential energy curve for a diatomic molecule. Solid line indicates a Morse potential that approximates the actual potential. Broken line is a parabolic potential for a harmonic oscillator. D_e and D_0 are the theoretical and spectroscopic dissociation energies, respectively.

corrections have been made mostly on diatomic molecules (see Table 1-3), because of the complexity of calculations for large molecules.

According to quantum mechanics, only those transitions involving $\Delta v = \pm 1$ are allowed for a harmonic oscillator. If the vibration is anharmonic, however, transitions involving $\Delta v = \pm 2, \pm 3, \ldots$ (overtones) are also weakly allowed by selection rules. Among many $\Delta v = \pm 1$ transitions, that of $v = 0 \leftrightarrow 1$ (fundamental) appears most strongly both in IR and Raman spectra. This is expected from the Maxwell–Boltzmann distribution law, which states that the population ratio of the $v = 1$ and $v = 0$ states is given by

$$\frac{P_{v=1}}{P_{v=0}} = e^{-\Delta E/kT}, \qquad (1\text{-}31)$$

Table 1-3 Relationships among Vibrational Frequency, Reduced Mass and Force Constant

Molecule	Obs. $\tilde{\nu}$ (cm^{-1})	ω_e (cm^{-1})	μ (awu)	K (mdyn/A)
H_2	4,160	4,395	0.5041	5.73
HD	3,632	3,817	0.6719	5.77
D_2	2,994	3,118	1.0074	5.77
HF	3,962	4,139	0.9573	9.65
HCl	2,886	2,989	0.9799	5.16
HBr	2,558	2,650	0.9956	4.12
HI	2,233	2,310	1.002	3.12
F_2	892	—	9.5023	4.45
Cl_2	546	565	17.4814	3.19
Br_2	319	323	39.958	2.46
I_2	213	215	63.466	1.76
N_2	2,331	2,360	7.004	22.9
CO	2,145	2,170	6.8584	19.0
NO	1,877	1,904	7.4688	15.8
O_2	1,555	1,580	8.000	11.8

where ΔE is the energy difference between the two states, k is Boltzmann's constant (1.3807×10^{-16} erg/degree), and T is the absolute temperature. Since $\Delta E = hc\tilde{\nu}$, the ratio becomes smaller as $\tilde{\nu}$ becomes larger. At room temperature, 10^{-13} J/°

$$kT = 1.38 \times 10^{-6} \text{ (erg/degree) } 300 \text{ (degree)}$$
$$= 4.14 \times 10^{-14} \text{ (erg)}$$
$$= [4.14 \times 10^{-14} \text{ (erg)}]/[1.99 \times 10^{-16} \text{ (erg/cm}^{-1})]$$
$$= 208 \text{ (cm}^{-1}).$$

Thus, if $\tilde{\nu} = 4,160$ cm^{-1} (H_2 molecule), P $(v = 1)/P(v = 0) = 2.19 \times 10^{-9}$. Therefore, almost all of the molecules are at $v = 0$. On the other hand, if $\tilde{\nu} = 213$ cm^{-1} (I_2 molecule), this ratio becomes 0.36. Thus, about 27% of the I_2 molecules are at $v = 1$ state. In this case, the transition $v = 1 \rightarrow 2$ should be observed on the low-frequency side of the fundamental with much less intensity. Such a transition is called a "hot band" since it tends to appear at higher temperatures.

1.4 Origin of Raman Spectra

As stated in Section 1.1, vibrational transitions can be observed in either IR or Raman spectra. In the former, we measure the absorption of infrared light by the sample as a function of frequency. The molecule absorbs $\Delta E = h\nu$

IR

Raman

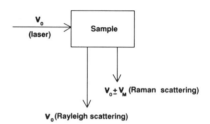

Figure 1-7   Differences in mechanism of Raman vs IR.

from the IR source at each vibrational transition. The intensity of IR absorption is governed by the Beer–Lambert law:

$$I = I_0 e^{-\varepsilon cd}. \tag{1-32}$$

Here, I_0 and I denote the intensities of the incident and transmitted beams, respectively, ε is the molecular absorption coefficient,[4] and c and d are the concentration of the sample and the cell length, respectively (Fig. 1-7). In IR spectroscopy, it is customary to plot the percentage transmission (T) versus wave number ($\tilde{\nu}$):

$$T(\%) = \frac{I}{I_0} \times 100. \tag{1-33}$$

It should be noted that T (%) is not proportional to c. For quantitative analysis, the absorbance (A) defined here should be used:

$$A = \log\frac{I_0}{I} = \varepsilon cd. \tag{1-34}$$

The origin of Raman spectra is markedly different from that of IR spectra. In Raman spectroscopy, the sample is irradiated by intense laser beams in the UV-visible region (ν_0), and the scattered light is usually observed in the direction perpendicular to the incident beam (Fig. 1-7; see also Chapter 2,

---

[4]ε has the dimension of l/moles cm when c and d are expressed in units of moles/liter and centimeters, respectively.

Section 2.7). The scattered light consists of two types: one, called *Rayleigh scattering*, is strong and has the same frequency as the incident beam (v_0), and the other, called *Raman scattering*, is very weak ($\sim 10^{-5}$ of the incident beam) and has frequencies $v_0 \pm v_m$, where v_m is a vibrational frequency of a molecule. The $v_0 - v_m$ and $v_0 + v_m$ lines are called the *Stokes* and *anti-Stokes* lines, respectively. Thus, in Raman spectroscopy, we measure the vibrational frequency (v_m) as a shift from the incident beam frequency (v_0).[5] In contrast to IR spectra, Raman spectra are measured in the UV-visible region where the excitation as well as Raman lines appear.

According to classical theory, Raman scattering can be explained as follows: The electric field strength (E) of the electromagnetic wave (laser beam) fluctuates with time (t) as shown by Eq. (1-1):

$$E = E_0 \cos 2\pi v_0 t, \tag{1-35}$$

where E_0 is the vibrational amplitude and v_0 is the frequency of the laser. If a diatomic molecule is irradiated by this light, an electric dipole moment P is induced:

$$P = \alpha E = \alpha E_0 \cos 2\pi v_0 t. \tag{1-36}$$

Here, α is a proportionality constant and is called *polarizability*. If the molecule is vibrating with a frequency v_m, the nuclear displacement q is written

$$q = q_0 \cos 2\pi v_m t, \tag{1-37}$$

where q_0 is the vibrational amplitude. For a small amplitude of vibration, α is a linear function of q. Thus, we can write

$$\alpha = \alpha_0 + \left(\frac{\partial \alpha}{\partial q}\right)_0 q_0 + \dots . \tag{1-38}$$

Here, α_0 is the polarizability at the equilibrium position, and $(\partial \alpha / \partial q)_0$ is the rate of change of α with respect to the change in q, evaluated at the equilibrium position.

Combining (1-36) with (1-37) and (1-38), we obtain

$$P = \alpha E_0 \cos 2\pi v_0 t$$

$$= \alpha_0 E_0 \cos 2\pi v_0 t + \left(\frac{\partial \alpha}{\partial q}\right)_0 q E_0 \cos 2\pi v_0 t$$

$$= \alpha_0 E_0 \cos 2\pi v_0 t + \left(\frac{\partial \alpha}{\partial q}\right)_0 q_0 E_0 \cos 2\pi v_0 t \cos 2\pi v_m t$$

---

[5] Although Raman spectra are normally observed for vibrational and rotational transitions, it is possible to observe Raman spectra of electronic transitions between ground states and low-energy excited states.

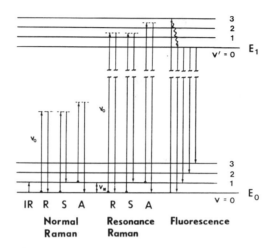

Figure 1-8 Comparison of energy levels for the normal Raman, resonance Raman, and fluorescence spectra.

$$= \alpha_0 E_0 \cos 2\pi v_0 t$$

$$+ \frac{1}{2}\left(\frac{\partial \alpha}{\partial q}\right)_0 q_0 E_0 [\cos\{2\pi(v_0 + v_m)t\} + \cos\{2\pi(v_0 - v_m)t\}]. \qquad (1\text{-}39)$$

According to classical theory, the first term represents an oscillating dipole that radiates light of frequency v_0 (Rayleigh scattering), while the second term corresponds to the Raman scattering of frequency $v_0 + v_m$ (anti-Stokes) and $v_0 - v_m$ (Stokes). If $(\partial \alpha / \partial q)_0$ is zero, the vibration is not Raman-active. Namely, to be Raman-active, the rate of change of polarizability (α) with the vibration must not be zero.

Figure 1-8 illustrates Raman scattering in terms of a simple diatomic energy level. In IR spectroscopy, we observe that $v = 0 \to 1$ transition at the electronic ground state. In normal Raman spectroscopy, the exciting line (v_0) is chosen so that its energy is far below the first electronic excited state. The dotted line indicates a "virtual state" to distinguish it from the real excited state. As stated in Section 1.2, the population of molecules at $v = 0$ is much larger than that at $v = 1$ (Maxwell–Boltzmann distribution law). Thus, the Stokes (S) lines are stronger than the anti-Stokes (A) lines under normal conditions. Since both give the same information, it is customary to measure only the Stokes side of the spectrum. Figure 1-9 shows that Raman spectrum of CCl_4.[6]

Resonance Raman (RR) scattering occurs when the exciting line is chosen

---

[6] A Raman spectrum is expressed as a plot, intensity vs. Raman shift ($\Delta \tilde{v} = \tilde{v}_0 \pm \tilde{v}$). However, $\Delta \tilde{v}$ is often written as $\tilde{v}$ for brevity.

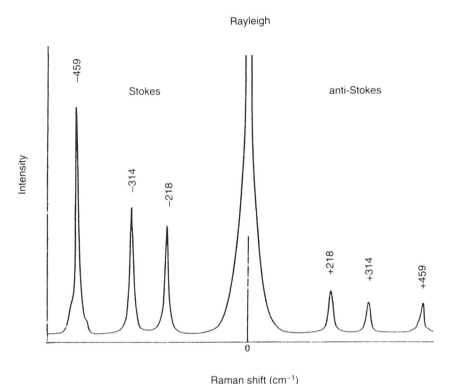

Figure 1-9   Raman spectrum of CCl_4 (488.0 nm excitation).

so that its energy intercepts the manifold of an electronic excited state. In the liquid and solid states, vibrational levels are broadened to produce a continuum. In the gaseous state, a continuum exists above a series of discrete levels. Excitation of these continua produces RR spectra that show extremely strong enhancement of Raman bands originating in this particular electronic transition. Because of its importance, RR spectroscopy will be discussed in detail in Section 1.15. The term "pre-resonance" is used when the exciting line is close in energy to the electronic excited state. Resonance fluorescence (RF) occurs when the molecule is excited to a discrete level of the electronic excited state (20). This has been observed for gaseous molecules such as I_2, Br_2. Finally, fluorescence spectra are observed when the excited state molecule decays to the lowest vibrational level via radiationless transitions and then emits radiation, as shown in Fig. 1-8. The lifetime of the excited state in RR is very short ($\sim 10^{-14}$ s), while those in RF and fluorescence are much longer ($\sim 10^{-8}$ to 10^{-5} s).

1.5 Factors Determining Vibrational Frequencies

According to Eq. (1-26), the vibrational frequency of a diatomic molecule is given by

$$\tilde{v} = \frac{1}{2\pi c}\sqrt{\frac{K}{\mu}}, \tag{1-40}$$

where K is the force constant and μ is the reduced mass. This equation shows that $\tilde{v}$ is proportional to $\sqrt{K}$ (force constant effect), but inversely proportional to $\sqrt{\mu}$ (mass effect). To calculate the force constant, it is convenient to rewrite the preceding equations as

$$K = 4\pi^2 c^2 \omega_e^2 \mu. \tag{1-41}$$

Here, the vibrational frequency (observed) has been replaced by ω_e (Eq. (1-30)) in order to obtain a more accurate force constant. Using the unit of millidynes/Å (mdyn/Å) or 10^5 (dynes/cm) for K, and the atomic weight unit (awu) for μ, Eq. (1-41) can be written as

$$K = 4(3.14)^2(3 \times 10^{10})^2 \left[\frac{\mu}{6{,}025 \times 10^{23}}\right]\omega_e^2$$

$$= (5.8883 \times 10^{-2})\mu\omega_e^2. \tag{1-42}$$

For $H^{35}Cl$, $\omega_e = 2{,}989$ cm^{-1} and μ is 0.9799. Then, its K is 5.16×10^5 (dynes/cm) or 5.16 (mdyn/Å). If such a calculation is made for a number of diatomic molecules, we obtain the results shown in Table 1-3. In all four series of compounds, the frequency decreases in going downward in the table. However, the origin of this downward shift is different in each case. In the $H_2 > HD > D_2$ series, it is due to the mass effect since the force constant is not affected by isotopic substitution. In the $HF > HCl > HBr > HI$ series, it is due to the force constant effect (the bond becomes weaker in the same order) since the reduced mass is almost constant. In the $F_2 > Cl_2 > Br_2 > I_2$ series, however, both effects are operative; the molecule becomes heavier and the bond becomes weaker in the same order. Finally, in the $N_2 > CO > NO > O_2$, series, the decreasing frequency is due to the force constant effect that is expected from chemical formulas, such as $N{\equiv}N$, and $O{=}O$, with CO and NO between them.

It should be noted, however, that a large force constant does not necessarily mean a stronger bond, since the force constant is the curvature of the potential well near the equilibrium position,

$$K = \left(\frac{d^2 V}{dq^2}\right)_{q \to 0}, \tag{1-43}$$

whereas the bond strength (dissociation energy) is measured by the depth of the potential well (Fig. 1-6). Thus, a large K means a sharp curvature near the bottom of the potential well, and does not directly imply a deep potential well. For example,

	HF		HCl		HBr		HI
K (mdyn/Å)	9.65	>	5.16	>	4.12	>	3.12
D_e (kcal/mole)	134.6	>	103.2	>	87.5	>	71.4

However,

	F_2		Cl_2		Br_2		I_2
K (mdyn/Å)	4.45	>	3.19	>	2.46	>	1.76
D_e (kcal/mole)	37.8	<	58.0	>	46.1	>	36.1

A rough parallel relationship is observed between the force constant and the dissociation energy when we plot these quantities for a large number of compounds.

1.6   Vibrations of Polyatomic Molecules

In diatomic molecules, the vibration occurs only along the chemical bond connecting the nuclei. In polyatomic molecules, the situation is complicated because all the nuclei perform their own harmonic oscillations. However, we can show that any of these complicated vibrations of a molecule can be expressed as a superposition of a number of "normal vibrations" that are completely independent of each other.

In order to visualize normal vibrations, let us consider a mechanical model of the CO_2 molecule shown in Fig. 1-10. Here, the C and O atoms are represented by three balls, weighing in proportion to their atomic weights,

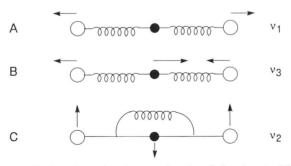

Figure 1-10   Atomic motions in normal modes of vibrations in CO_2.

that are connected by springs of a proper strength in proportion to their force constants. Suppose that the C—O bonds are stretched and released simultaneously as shown in Fig. 1-10a. Then, the balls move back and forth along the bond direction. This is one of the normal vibrations of this model and is called the symmetric (in-phase) stretching vibration. In the real CO_2 molecule, its frequency (v_1) is ca. 1,340 cm^{-1}. Next, we stretch one C—O bond and shrink the other, and release all the balls simultaneously (Fig. 1-10b). This is another normal vibration and is called the antisymmetric (out-of-phase) stretching vibration. In the CO_2 molecule, its frequency (v_3) is ca. 2,350 cm^{-1}. Finally, we consider the case where the three balls are moved in the perpendicular direction and released simultaneously (Fig. 1-10c). This is the third type of normal vibration called the (symmetric) bending vibration. In the CO_2 molecule, its frequency (v_2) is ca. 667 cm^{-1}.

Suppose that we strike this mechanical model with a hammer. Then, this model would perform an extremely complicated motion that has no similarity to the normal vibrations just mentioned. However, if this complicated motion is photographed with a stroboscopic camera with its frequency adjusted to that of the normal vibration, we would see that each normal vibration shown in Fig. 1-10 is performed faithfully. In real cases, the stroboscopic camera is replaced by an IR or Raman instrument that detects only the normal vibrations.

Since each atom can move in three directions (x,y,z), an N-atom molecule has $3N$ degrees of freedom of motion. However, the $3N$ includes six degrees of freedom originating from translational motions of the whole molecule in the three directions and rotational motions of the whole molecule about the three principal axes of rotation, which go through the center of gravity. Thus,

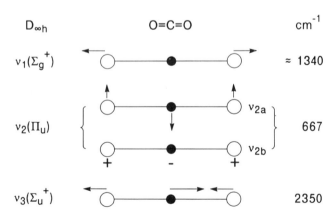

Figure 1-11 Normal modes of vibration in CO_2 (+ and − denote vibrations going upward and downward, respectively, in direction perpendicular to the paper plane).

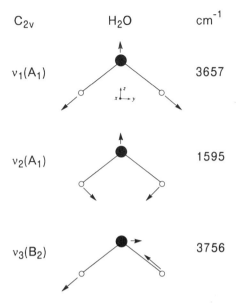

Figure 1-12   Normal modes of vibrations in H_2O.

the net vibrational degrees of freedom (number of normal vibrations) is $3N - 6$. In the case of linear molecules, it becomes $3N - 5$ since the rotation about the molecular axis does not exist. In the case of the CO_2 molecule, we have $3 \times 3 - 5 = 4$ normal vibrations shown in Fig. 1-11. It should be noted that v_{2a} and v_{2b} have the same frequency and are different only in the direction of vibration by 90°. Such a pair is called a set of doubly degenerate vibrations. Only two such vibrations are regarded as unique since similar vibrations in any other directions can be expressed as a linear combination of v_{2a} and v_{2b}. Figure 1-12 illustrates the three normal vibrations ($3 \times 3 - 6 = 3$) of the H_2O molecule.

Theoretical treatments of normal vibrations will be described in Section 1.20. Here, it is sufficient to say that we designate "normal coordinates" Q_1, Q_2 and Q_3 for the normal vibrations such as the v_1, v_2 and v_3, respectively, of Fig. 1-12, and that the relationship between a set of normal coordinates and a set of Cartesian coordinates ($q_1, q_2, \ldots$) is given by

$$q_1 = B_{11}Q_1 + B_{12}Q_2 + \ldots,$$
$$q_2 = B_{21}Q_1 + B_{22}Q_2 + \ldots, \tag{1-44}$$
$$\ldots,$$

so that the modes of normal vibrations can be expressed in terms of Cartesian coordinates if the B_{ij} terms are calculated.

1.7   Selection Rules for Infrared and Raman Spectra

To determine whether the vibration is active in the IR and Raman spectra, the selection rules must be applied to each normal vibration. Since the origins of IR and Raman spectra are markedly different (Section 1.4), their selection rules are also distinctively different. According to quantum mechanics (18,19) a vibration is IR-active if the dipole moment is changed during the vibration and is Raman-active if the polarizability is changed during the vibration.

The IR activity of small molecules can be determined by inspection of the mode of a normal vibration (normal mode). Obviously, the vibration of a homopolar diatomic molecule is not IR-active, whereas that of a heteropolar diatomic molecule is IR-active. As shown in Fig. 1-13, the dipole moment of the H_2O molecule is changed during each normal vibration. Thus, all these vibrations are IR-active. From inspection of Fig. 1-11, one can readily see that v_2 and v_3 of the CO_2 molecule are IR-active, whereas v_1 is not IR-active.

To discuss Raman activity, let us consider the nature of the polarizability (α) introduced in Section 1.4. When a molecule is placed in an electric field (laser beam), it suffers distortion since the positively charged nuclei are attracted toward the negative pole, and electrons toward the positive pole (Fig. 1-14). This charge separation produces an induced dipole moment (P) given by

$$P = \alpha E. \tag{1-45}$$

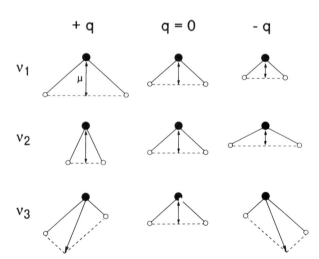

Figure 1-13   Change in dipole moment for H_2O molecule during each normal vibration.

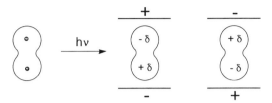

Figure 1-14 Polarization of a diatomic molecule in an electric field.

In actual molecules, such a simple relationship does not hold since both P and E are vectors consisting of three components in the x, y and z directions. Thus, Eq. (1-45) must be written as

$$P_x = \alpha_{xx}E_x + \alpha_{xy}E_y + \alpha_{xz}E_z,$$
$$P_y = \alpha_{yx}E_x + \alpha_{yy}E_y + \alpha_{yz}E_z, \qquad (1\text{-}46)$$
$$P_z = \alpha_{zx}E_x + \alpha_{zy}E_y + \alpha_{zz}E_z.$$

In matrix form, this is written as

$$\begin{bmatrix} P_x \\ P_y \\ P_z \end{bmatrix} = \begin{bmatrix} \alpha_{xx} & \alpha_{xy} & \alpha_{xz} \\ \alpha_{yx} & \alpha_{yy} & \alpha_{yz} \\ \alpha_{zx} & \alpha_{zy} & \alpha_{zz} \end{bmatrix} \begin{bmatrix} E_x \\ E_y \\ E_z \end{bmatrix}. \qquad (1\text{-}47)$$

The first matrix on the right-hand side is called the *polarizability tensor*. In normal Raman scattering, this tensor is symmetric; $\alpha_{xy} = \alpha_{yz}$, $\alpha_{xz} = \alpha_{zx}$ and $\alpha_{yz} = \alpha_{zy}$. According to quantum mechanics, the vibration is Raman-active if one of these components of the polarizability tensor is changed during the vibration.

In the case of small molecules, it is easy to see whether or not the polarizability changes during the vibration. Consider diatomic molecules such as H_2 or linear molecules such as CO_2. Their electron clouds have an elongated water melon like shape with circular cross-sections. In these molecules, the electrons are more polarizable (a larger α) along the chemical bond than in the direction perpendicular to it. If we plot α_i (α in the i-direction) from the center of gravity in all directions, we end up with a three-dimensional surface. Conventionally, we plot $1/\sqrt{\alpha_i}$ rather than α_i itself and call the resulting three-dimensional body a *polarizability ellipsoid*. Figure 1-15 shows the changes of such an ellipsoid during the vibrations of the CO_2 molecule.

In terms of the polarizability ellipsoid, the vibration is Raman-active if the *size*, *shape* or *orientation* changes during the normal vibration. In the v_1 vibration, the size of the ellipsoid is changing; the diagonal elements (α_{xx}, α_{yy} and α_{zz}) are changing simultaneously. Thus, it is Raman-active. Although

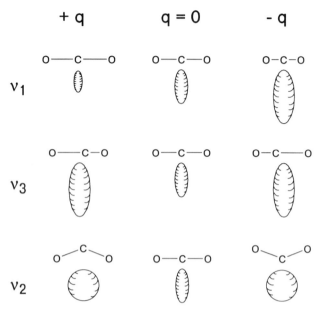

Figure 1-15 Changes in polarizability ellipsoids during vibration of CO_2 molecule.

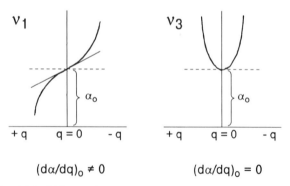

$$(d\alpha/dq)_0 \neq 0 \qquad\qquad (d\alpha/dq)_0 = 0$$

Figure 1-16 Difference between v_1 and v_3 vibrations in CO_2 molecule.

the size of the ellipsoid is changing during the v_3 vibration, the ellipsoids at two extreme displacements ($+q$ and $-q$) are exactly the same in this case. Thus, this vibration is not Raman-active if we consider a small displacement. The difference between the v_1 and v_3 is shown in Fig. 1-16. Note that the Raman activity is determined by $(d\alpha/dq)_0$ (slope near the equilibrium position). During the v_2 vibration, the shape of the ellipsoid is sphere like at two extreme configurations. However, the size and shape of the ellipsoid are

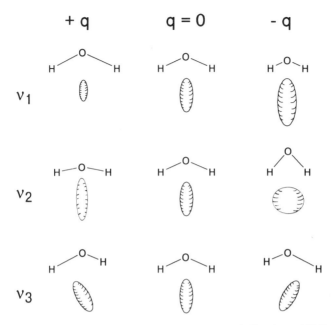

Figure 1-17 Changes in polarizability ellipsoid during normal vibrations of H_2O molecule.

exactly the same at $+q$ and $-q$. Thus, it is not Raman-active for the same reason as that of v_3. As these examples show, it is not necessary to figure out the exact size, shape or orientation of the ellipsoid to determine Raman activity.

Figure 1-17 illustrates the changes in the polarizability ellipsoid during the normal vibrations of the H_2O molecule. Its v_1 vibration is Raman-active, as is the v_1 vibration of CO_2. The v_2 vibration is also Raman-active because the *shape* of the ellipsoid is different at $+q$ and $-q$. In terms of the polarizability tensor, α_{xx}, α_{yy} and α_{zz} are all changing with different rates. Finally, the v_3 vibration is Raman-active because the *orientation* of the ellipsoid is changing during the vibration. This activity occurs because an off-diagonal element (α_{yz} in this case) is changing.

One should note that, in CO_2, the vibration that is symmetric with respect to the center of symmetry (v_1) is Raman-active but not IR-active, whereas those that are antisymmetric with respect to the center of symmetry (v_2 and v_3) are IR-active but not Raman-active. This condition is called the *mutual exclusion principle* and holds for any molecules having a center of symmetry.[7]

The preceding examples demonstrate that IR and Raman activities can be determined by inspection of the normal mode. Clearly, such a simple

[7] This principle holds even if a molecule has no atom at the center of symmetry (e.g., benzene).

approach is not applicable to large and complex molecules. As will be shown in Section 1.14, group theory provides elegant methods to determine IR and Raman activities of normal vibrations of such molecules.

1.8   Raman versus Infrared Spectroscopy

Although IR and Raman spectroscopies are similar in that both techniques provide information on vibrational frequencies, there are many advantages and disadvantages unique to each spectroscopy. Some of these are listed here.

1.   As stated in Section 1.7, selection rules are markedly different between IR and Raman spectroscopies. Thus, some vibrations are only Raman-active while others are only IR-active. Typical examples are found in molecules having a center of symmetry for which the mutual exclusion rule holds. In general, a vibration is IR-active, Raman-active, or active in both; however, totally symmetric vibrations are always Raman-active.

2.   Some vibrations are inherently weak in IR and strong in Raman spectra. Examples are the stretching vibrations of the C≡C, C=C, P=S, S—S and C—S bonds. In general, vibrations are strong in Raman if the bond is covalent, and strong in IR if the bond is ionic (O—H, N—H). For covalent bonds, the ratio of relative intensities of the C≡C, C=C and C—C bond stretching vibrations in Raman spectra is about 3:2:1.[8] Bending vibrations are generally weaker than stretching vibrations in Raman spectra.

3.   Measurements of depolarization ratios provide reliable information about the symmetry of a normal vibration in solution (Section 1.9). Such information can not be obtained from IR spectra of solutions where molecules are randomly orientated.

4.   Using the resonance Raman effect (Section 1.15), it is possible to selectively enhance vibrations of a particular chromophoric group in the molecule. This is particularly advantageous in vibrational studies of large biological molecules containing chromophoric groups (Sections 4.1 and 4.2.)

5.   Since the diameter of the laser beam is normally 1–2 mm, only a small sample area is needed to obtain Raman spectra. This is a great advantage over conventional IR spectroscopy when only a small quantity of the sample (such as isotopic chemicals) is available.

---

[8] In general, the intensity of Raman scattering increases as the $(d\alpha/dq)_0$ becomes larger.

6. Since water is a weak Raman scatterer, Raman spectra of samples in aqueous solution can be obtained without major interference from water vibrations. Thus, Raman spectroscopy is ideal for the studies of biological compounds in aqueous solution. In contrast, IR spectroscopy suffers from the strong absorption of water.

7. Raman spectra of hygroscopic and/or air-sensitive compounds can be obtained by placing the sample in sealed glass tubing. In IR spectroscopy, this is not possible since glass tubing absorbs IR radiation.

8. In Raman spectroscopy, the region from 4,000 to 50 cm^{-1} can be covered by a single recording. In contrast, gratings, beam splitters, filters and detectors must be changed to cover the same region by IR spectroscopy.

Some disadvantages of Raman spectroscopy are the following:

1. A powerful laser source is needed to observe weak Raman scattering. This may cause local heating and/or photodecomposition, especially in resonance Raman studies (Section 1.15) where the laser frequency is deliberately tuned in the absorption band of the molecule.

2. Some compounds fluoresce when irradiated by the laser beam.

3. It is more difficult to obtain rotational and rotation–vibration spectra with high resolution in Raman than in IR spectroscopy. This is because Raman spectra are observed in the UV-visible region where high resolving power is difficult to obtain.

4. The state of the art Raman system costs much more than a conventional FT-IR spectrophotometer.

Finally, it should be noted that vibrational (both IR and Raman) spectroscopy is unique in that it is applicable to the solid state as well as to the gaseous state and solution. In contrast, x-ray diffraction is applicable only to the crystalline state, whereas NMR spectroscopy is applicable largely to the sample in solution.

1.9 Depolarization Ratios

As stated in the preceding section, depolarization ratios of Raman bands provide valuable information about the symmetry of a vibration that is

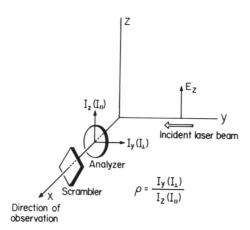

Figure 1-18 Irradiation of sample from the y-direction with plane polarized light, with the electronic vector in the z-direction.

indispensable in making band assignments. Figure 1-18 shows a coordinate system which is used for measurements of depolarization ratios. A molecule situated at the origin is irradiated from the y-direction with plane polarized light whose electric vector oscillates on the yz-plane (E_z). If one observes scattered radiation from the x-direction, and measure the intensities in the $y(I_y)$ and $z(I_z)$-directions using an analyzer, the depolarization ratio (ρ_p) measured by polarized light (p) is defined by

$$\rho_p = \frac{I_{\perp}(I_y)}{I_{\parallel}(I_z)}. \tag{1-48}$$

(For the use of the scrambler, see Chapter 2, Section 2.8.)

Suppose that a tetrahedral molecule such as CCl_4 is irradiated by plane polarized light (E_z). Then, the induced dipole (Section 1.7) also oscillates in the same yz-plane. If the molecule is performing the totally symmetric vibration, the polarizability ellipsoid is always sphere like; namely, the molecule is polarized equally in every direction. Under such a circumstance, $I_{\perp}(I_y) = 0$ since the oscillating dipole emitting the radiation is confined to the xz-plane. Thus, $\rho_p = 0$. Such a vibration is called *polarized* (abbreviated as p). In liquids and solutions, molecules take random orientations. Yet this conclusion holds since the polarizability ellipsoid is spherical throughout the totally symmetric vibration.

If the molecule is performing a non-totally symmetric vibration, the polarizability ellipsoid changes its shape from a sphere to an ellipsoid during the vibration. Then, the induced dipole would be largest along the direction of largest polarizability, namely along one of the minor axes of the ellipsoid.

Since these axes would be randomly oriented in liquids and solutions, the induced dipole moments would also be randomly oriented. In this case, the ρ_p is nonzero, and the vibration is called *depolarized* (abbreviated as *dp*). Theoretically, we can show (21) that

$$\rho_p = \frac{3g^S + 5g^a}{10g^0 + 4g^S},\tag{1-49}$$

where

$$g^0 = \tfrac{1}{3}(\alpha_{xx} + \alpha_{yy} + \alpha_{zz})^2,$$

$$g^S = \tfrac{1}{3}[(\alpha_{xx} - \alpha_{yy})^2 + (\alpha_{yy} - \alpha_{zz})^2 + (\alpha_{zz} - \alpha_{xx})^2]$$
$$+ \tfrac{1}{2}[(\alpha_{xy} + \alpha_{yx})^2 + (\alpha_{yz} + \alpha_{zy})^2 + (\alpha_{xz} + \alpha_{zx})^2],$$

$$g^a = \tfrac{1}{2}[(\alpha_{xy} - \alpha_{yz})^2 + (\alpha_{xz} - \alpha_{zx})^2 + (\alpha_{yz} - \alpha_{zy})^2].$$

In normal Raman scattering, $g^a = 0$ since the polarizability tensor is symmetric. Then, (1-49) becomes

$$\rho_p = \frac{3g^S}{10g^0 + 4g^S}\tag{1-50}$$

For totally symmetric vibrations, $g^0 > 0$ and $g^S \geqslant 0$. Thus, $0 \leqslant \rho_p < \tfrac{3}{4}$ (polarized). For non-totally symmetric vibrations, $g^0 = 0$ and $g^S > 0$. Then, $\rho_p = \tfrac{3}{4}$ (depolarized).

In resonance Raman scattering ($g^a \neq 0$), it is possible to have $\rho_p > \tfrac{3}{4}$. For example, if $\alpha_{xy} = -\alpha_{yx}$ and the remaining off-diagonal elements are zero, $g^0 = g^S = 0$ and $g^a \neq 0$. Then, (1-49) gives $\rho_p \to \infty$. This is called *anomalous* (or *inverse*) *polarization* (abbreviated as *ap* or *ip*). As will be shown in Section 1.15, resonance Raman spectra of metallopophyrins exhibit polarized (A_{1g}) and depolarized (B_{1g} and B_{2g}) vibrations as well as those of anomalous (or inverse) polarization (A_{2g}).

1.10   The Concept of Symmetry

The various experimental tools that are utilized today to solve structural problems in chemistry, such as Raman, infrared, NMR, magnetic measurements and the diffraction methods (electron, x-ray, and neutron), are based on symmetry considerations. Consequently, the symmetry concept as applied to molecules is thus very important.

Symmetry may be defined in a nonmathematical sense, where it is associated with beauty—with pleasing proportions or regularity in form, harmonious arrangement, or a regular repetition of certain characteristics (e.g., periodicity). In the mathematical or geometrical definition, symmetry refers to the correspondence of elements on opposite sides of a point, line, or plane, which we call the center, axis, or plane of symmetry (symmetry elements). It is the mathematical concept that is pursued in the following sections. The discussion in this section will define the symmetry elements in an isolated molecule (the point symmetry)—of which there are five. The number of ways by which symmetry elements can combine constitute a group, and these include the 32 crystallographic point groups when one considers a crystal. Theoretically, an infinite number of point groups can exist, since there are no restrictions on the order of rotational axes of an isolated molecule. However, in a practical sense, few molecules possess rotational axes C_n where $n > 6$. Each point group has a character table (see Appendix 1), and the features of these tables are discussed. The derivation of the selection rules for an isolated molecule is made with these considerations. If symmetry elements are combined with translations, one obtains operations or elements of symmetry that can define the symmetry of space as in a crystal. Two symmetry elements, the screw axis (rotation followed by a translation) and the glide plane (reflection followed by a translation), when added to the five point group symmetry elements, constitute the seven space symmetry elements. This final set of symmetry elements allows one to determine selection rules for the solid state.

Derivation of selection rules for a particular molecule illustrates the complementary nature of infrared and Raman spectra and the application of group theory to the determination of molecular structure.

1.11   Point Symmetry Elements

The spatial arrangement of the atoms in a molecule is called its equilibrium configuration or structure. This configuration is invariant under a certain set of geometric operations called a group. The molecule is oriented in a coordinate system (a right-hand xyz coordinate system is used throughout the discussion in this section). If by carrying out a certain geometric operation on the original configuration, the molecule is transformed into another configuration that is superimposable on the original (i.e., indistinguishable from it, although its orientation may be changed), the molecule is said to contain a symmetry element. The following symmetry elements can be cited.

1.11.1   IDENTITY (E)

The symmetry element that transforms the original equilibrium configuration into another one superimposable on the original without change in orientation, in such a manner that each atom goes into itself, is called the identity and is denoted by E or I (E from the German *Einheit* meaning "unit" or, loosely, "identical"). In practice, this operation means to leave the molecule unchanged.

1.11.2 ROTATION AXES (C_n)

If a molecule is rotated about an axis to a new configuration that is indistinguishable from the original one, the molecule is said to possess a rotational axis of symmetry. The rotation can be clockwise or counterclockwise, depending on the molecule. For example, the same configuration is obtained for the water molecule whether one rotates the molecule clockwise or counterclockwise. However, for the ammonia molecule, different configurations are obtained, depending on the direction around which the rotation is performed. The angle of rotation may be $2\pi/n$, or $360°/n$, where n can be 1, 2, 3, 4, 5, 6, ..., ∞. The order of the rotational axis is called n (sometimes p), and the notation C_n is used, where C denotes rotation. In cases where several axes of rotation exist, the highest order of rotation is chosen as the principal axis. Linear molecules have an infinitefold axes of symmetry (C_∞).

The selection of the axes in a coordinate system can be confusing. To avoid this, the following rules are used for the selection of the z axis of a molecule:

(1)   In molecules with only one rotational axis, this axis is taken as the z axis.
(2)   In molecules where several rotational axes exist, the highest-order axis is selected as the z axis.
(3)   If a molecule possesses several axes of the highest order, the axis passing through the greatest number of atoms is taken as the z axis.

For the selection of the x axis the following rules can be cited:

(1)   For a planar molecule where the z axis lies in this plane, the x axis can be selected to be normal to this plane.
(2)   In a planar molecule where the z axis is chosen to be perpendicular to the plane, the x axis must lie in the plane and is chosen to pass through the largest number of atoms in the molecule.
(3)   In nonplanar molecules the plane going through the largest number of atoms is located as if it were in the plane of the molecule and rule (1) or (2) is used. For complex molecules where a selection is difficult, one chooses the x and y axes arbitrarily.

1.11.3 PLANES OF SYMMETRY (σ)

If a plane divides the equilibrium configuration of a molecule into two parts that are mirror images of each other, then the plane is called a symmetry

plane. If a molecule has two such planes, which intersect in a line, this line
is an axis of rotation (see the previous section); the molecule is said to have
a vertical rotation axis C; and the two planes are referred to as vertical planes
of symmetry, denoted by σ_v. Another case involving two planes of symmetry
and their intersection arises when a molecule has more than one axis of
symmetry. For example, planes intersecting in an n-fold axis perpendicular
to n twofold axes, with each of the planes bisecting the angle between two
successive twofold axes, are called diagonal and are denoted by the symbol
σ_d. Figure 1-19a–c illustrates the symmetry elements of the AB_4 molecule

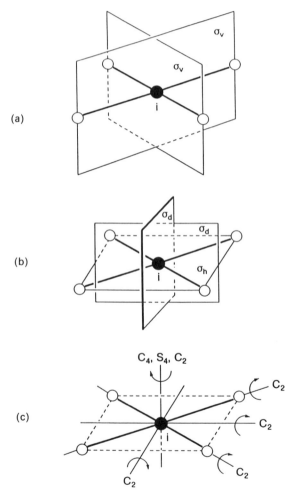

Figure 1-19 Symmetry elements for a planar AB_4 molecule (e.g., $PtCl_4^{2-}$ ion).

(e.g., $PtCl_4^{2-}$ ion). If a plane of symmetry is perpendicular to the rotational axis, it is called horizontal and is denoted by σ_h.

1.11.4 CENTER OF SYMMETRY (i)

If a straight line drawn from each atom of a molecule through a certain point meets an equivalent atom equidistant from the point, we call the point the center of symmetry of the molecule. The center of symmetry may or may not coincide with the position of an atom. The designation for the center of symmetry, or center of inversion, is i. If the center of symmetry is situated on an atom, the total number of atoms in the molecule is odd. If the center of symmetry is not on an atom, the number of atoms in the molecule is even. Figure 1-19b illustrates a center of symmetry and rotational axes for the planar AB_4 molecule.

1.11.5 ROTATION REFLECTION AXES (S_n)

If a molecule is rotated $360°/n$ about an axis and then reflected in a plane perpendicular to the axis, and if the operation produces a configuration indistinguishable from the original one, the molecule has the symmetry element of rotation–reflection, which is designated by S_n.

Table 1-4 lists the point symmetry elements and the corresponding symmetry operations. The notation used by spectroscopists and chemists, and used here, is the so-called Schoenflies system, which deals only with point groups. Crystallographers generally use the Hermann–Mauguin system, which applies to both point and space groups.

Table 1-4 Point Symmetry Elements and Symmetry Operations

	Symmetry Element	Symmetry Operation
1.	Identity (E or I)	Molecule unchanged
2.	Axis of rotation (C_n)	Rotation about axis by $2\pi/n$, $n = 1, 2, 3, 4,$ $5, 6, \ldots, \infty$ for an isolated molecule and $n = 1, 2, 3, 4$ and 6 for a crystal.
3.	Center of symmetry or center of inversion (i)	Inversion of all atoms through center.
4.	Plane (σ)	Reflection in the plane.
5.	Rotation reflection axis (S_n)	Rotation about axis by $2\pi/n$ followed by reflection in a plane perpendicular to the axis

Table 1-5 The 32 Crystallographic Point Groups[a]

Symbol	Plane σ	Axes of Symmetry $6(C_6)$	$4(C_4)$	$3(C_3)$	$2(C_2)$	Center i	Example
C_1	—	—	—	—	—	—	CH_3CHO
C_2	—	—	—	—	1	—	H_2O_2
C_3	—	—	—	1	—	—	$B(OH)_3$
C_4	—	—	1	—	—	—	$H_2S(s)$
C_6	—	1	—	—	—	—	—
C_s	1	—	—	—	—	—	$HCOCl$
C_{2s}	1	—	—	—	1	1	$trans\text{-}CHCl{=}CHCl$
C_{3h}	1	—	—	1	—	—	$C^+(NH_2)_3$
C_{4h}	1	—	1	—	—	1	$C_4H_4Cl_4$
C_{6h}	1	1	—	—	—	1	—
D_2	—	—	—	—	3	—	
D_3	—	—	—	1	3	—	$Co(H_2NCH_2CH_2NH_2)_3^{3+}$
D_4	—	—	1	—	4	—	cyclobutane
D_{2h}	3	—	—	—	3	1	C_2H_4
D_{3h}	4	—	—	1	3	—	BCl_3
D_{4h}	5	—	1	—	4	1	$PtCl_4^{2-}$

Point group							Example
D_{6h}	7	1	—	—	6	1	C_6H_6
$S_2(C_i)$	—	—	—	—	1	1	FClHC–CHClF
S_4	—	—	—	1	—	—	$C_{10}H_4F_4$
S_6	2	—	—	—	3	—	—
D_{2d}	3	—	—	1	3	1	$CH_2{=}C{=}CH_2$
D_{3d}	2	—	—	—	1	—	Cyclohexane
C_{2v}	3	—	1	—	—	—	$H_2O(g)$
C_{3v}	3	—	—	1	—	1	$NH_3(g)$
C_{4v}	4	1	—	—	—	—	IF_5
C_{6v}	6	—	—	—	—	—	—
T	—	—	4	—	3	—	$NH_3(s)$
O	3	3	4	—	6	—	—
T_h	3	—	4	—	3	1	$CO_2(s)$
O_h	9	3	4	3	6	1	SF_6
T_d	6	—	4	—	3	—	CH_4

[a]Special Point Groups:

$C_{\infty v}$—One C_∞, any C_p, infinite number of σ_v—e.g., HCN

$D_{\infty h}$—Infinite-fold axes (C_∞), infinite number $C_2 \perp$ to C_∞, infinite number of planes through C_∞, plane $\perp C_\infty$ axis, i—e.g., CO_2

I_h—Six fivefold, 10 threefold, 15 twofold axes, i, numerous planes (σ) and rotation–reflection axes—e.g., C_{60}

(a) Point Groups

It can be shown that a group consists of mathematical elements (symmetry elements or operations), and if the operation is taken to be performing one symmetry operation after another in succession, and the result of these operations is equivalent to a single symmetry operation in the set, then the set will be a mathematical group. The postulates for a complete set of elements $A, B, C, \ldots$ are as follows:

(1) For every pair of elements A and B, there exists a binary operation that yields the product AB belonging to the set.
(2) This binary product is associative, which implies that $A(BC) = (AB)C$.
(3) There exists an identity element E such that for every A, $AE = EA = A$.
(4) There is an inverse A^{-1} for each element A such that $AA^{-1} = A^{-1}A = E$.

For molecules it would seem that the point symmetry elements can combine in an unlimited way. However, only certain combinations occur. In the mathematical sense, the sets of all its symmetry elements for a molecule that adhere to the preceding postulates constitute a point group. If one considers an isolated molecule, rotation axes having $n = 1, 2, 3, 4, 5, 6$ to ∞ are possible. In crystals n is limited to $n = 1, 2, 3, 4,$ and 6 because of the *space-filling requirement*. Table 1-5 lists the symmetry elements of the 32 point groups.

(b) Rules for Classifying Molecules into their Proper Point Group

The method for the classification of molecules into different point groups suggested by Zeldin[22] is outlined in Table 1-6. The method can be described as follows:

(1) Determine whether the molecule belongs to a special group such as $\mathbf{D}_{\infty h}$, $\mathbf{C}_{\infty v}$, $\mathbf{T}_d$, $\mathbf{O}_h$, or $\mathbf{I}_h$. If the molecule is linear, it will be either $\mathbf{D}_{\infty h}$ or $\mathbf{C}_{\infty v}$. If the molecule has an infinite number of twofold axes perpendicular to the C_∞ axis, it will fall into point group $\mathbf{D}_{\infty h}$. If not, it is $\mathbf{C}_{\infty v}$.
(2) If the molecule is not linear, it may belong to a point group of extremely high symmetry such as $\mathbf{T}_d$, $\mathbf{O}_h$, or $\mathbf{I}_h$.
(3) If (1) or (2) is not found to be the case, look for a proper axis of rotation of the highest order in the molecule. If none is found, the molecule is of low symmetry, falling into point group $\mathbf{C}_3$, $\mathbf{C}_i$, or $\mathbf{C}_1$. The presence in the molecule of a plane of symmetry or an inversion center will distinguish among these point groups.
(4) If C_n axes exist, select the one of highest order. If the molecule also has an S_{2n} axis, with or without an inversion center, the point group is $\mathbf{S}_n$.
(5) If no S_n exists look for a set of n twofold axes lying perpendicular to

Table 1-6 Method of Classifying Molecules into Point Groups (22)

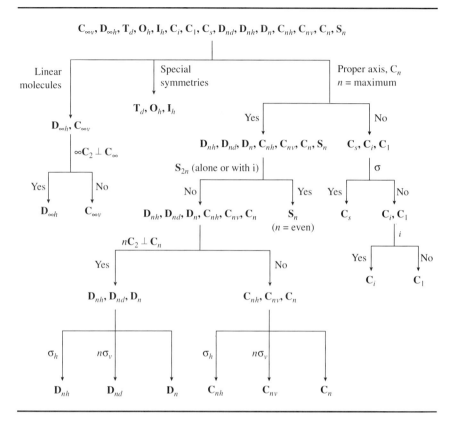

the major C_n axis. If no such set is found, the molecule belongs to $\mathbf{C}_{nh}$, $\mathbf{C}_{nv}$ or $\mathbf{C}_n$. If in a σ_h plane exists, the molecule is of $\mathbf{C}_{nh}$ symmetry even if other planes of symmetry are present. If no σ_h plane exists and a σ_v plane is found, the molecule is of $\mathbf{C}_{nv}$ symmetry. If no planes exist, it is of C_n symmetry.

(6) If in (5) nC_2 axes perpendicular to C_n axes are found, the molecule belongs to the $\mathbf{D}_{nh}$, $\mathbf{D}_{nd}$, or $\mathbf{D}_n$ point group. These can be differentiated by the presence (or absence) of symmetry planes (σ_h, σ_v, or no σ, respectively).

Several examples will be considered to illustrate the classification of molecules into point groups. Consider, for instance, the bent triatomic molecule of type AB_2 (H_2O) shown in Fig. 1-20. Following the rules and Table 1-6, it can be determined that the molecule is not of a special symmetry.

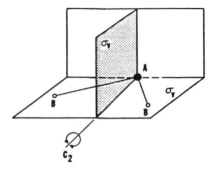

Figure 1-20 Symmetry elements for a bent AB_2 molecule. (Reproduced with permission from Ref. 57.)

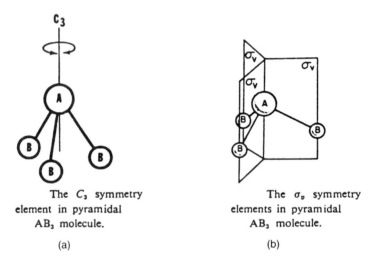

The C_3 symmetry
element in pyramidal
AB_3 molecule.

(a)

The σ_v symmetry
elements in pyramidal
AB_3 molecule.

(b)

Figure 1-21 Symmetry elements for a pyramidal AB_3 molecule: (a) C_3 element; (b) σ_v elements. (Reproduced with permission from Ref. 57.)

It does have a C_2 axis of rotation but no S_4 axis. There are no $nC_2 \perp C_n$, and therefore the molecule is either $\mathbf{C}_{2h}$, $\mathbf{C}_{2v}$, or $\mathbf{C}_2$. The molecule possesses two vertical planes of symmetry bot no σ_h plane, and therefore belongs to the $\mathbf{C}_{2v}$ point group.

Now consider the pyramidal molecule of type AB_3 (NH_3) shown in Fig. 1-21. This molecule also is not of a special symmetry. It has a C_3 axis of rotation but no S_6 axis. There are no nC_n axes perpendicular to the C_3 axis, and therefore the molecule belongs to the C classification. Since three vertical

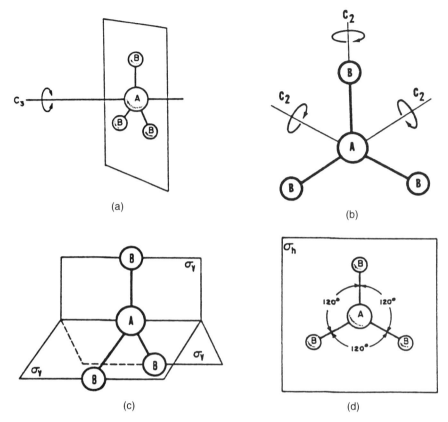

Figure 1-22 Symmetry elements for a planar AB_3 molecule: (a) C_3, (b) C_2, (c) σ_v, and (d) σ_h. (Reproduced with permission from Ref. 57.)

planes of symmetry are found but no σ_h plane, the molecule can be classified into $\mathbf{C}_{3v}$.

Next, consider the planar AB_3 molecule (BF_3) shown in Fig. 1-22. This molecule has no special symmetry. It has a C_3 axis of rotation without a collinear S_6 axis. It has three C_2 axes perpendicular to the C_3 axis, and therefore falls into the D classification. It has a σ_h plane of symmetry perpendicular to the C_3 axis and three σ_v planes of symmetry. However, the σ_h plane predominates and the molecule is of $\mathbf{D}_{3h}$ symmetry.

The next example is the hexagonal planar molecule of type A_6 or A_6B_6 (benzene) shown in Fig. 1-23. The molecule is not of a special symmetry. It has a center of symmetry and a C_6 axis of symmetry. No S_2 axis exists. Since six C_2 axes perpendicular to the C_6 axis are found, this molecule also falls

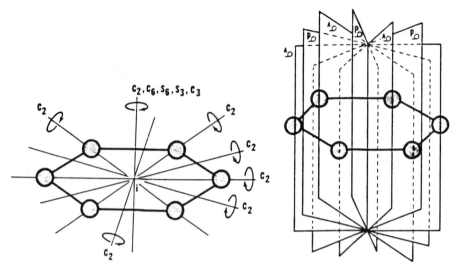

Figure 1-23 Symmetry elements for a planar hexagonal A_6B_6 molecule. (Reproduced with permission from Ref. 57.)

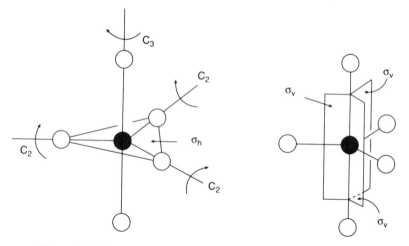

Figure 1-24 Symmetry elements for a trigonal bipyramidal AB_5 molecule.

into the **D** classification. Since it has a horizontal plane of symmetry perpendicular to the C_6 axis, the molecule belongs to the $\mathbf{D}_{6h}$ point group.

As the last example, consider the AB_5 trigonal bipyramid (e.g., gaseous PCl_5) shown in Fig. 1-24. This molecule does not belong to a special symmetry. The axis of highest order is C_3. There is no S_6 collinear with C_3. There are three C_2 axes perpendicular to the C_3 axis, and therefore the molecule belongs to one of the **D** groups. Since it possesses a σ_h plane perpendicular to the C_3 axis, the proper classification is $\mathbf{D_{3h}}$. Figure 1-25a, b, c shows symmetry elements for several other common point groups.

1.12   The Character Table

Prior to interpreting the character table, it is necessary to explain the terms *reducible* and *irreducible representations*. We can illustrate these concepts using the NH_3 molecule as an example. Ammonia belongs to the point group C_{3v} and has six elements of symmetry. These are E (identity), two C_3 axes (threefold axes of rotation) and three σ_v planes (vertical planes of symmetry) as shown in Fig. 1-21. If one performs operations corresponding to these symmetry elements on the three equivalent NH bonds, the results can be expressed mathematically by using 3×3 matrices.[9]

Consider the C_3^+ (clockwise rotation by 120°) operation, shown with its changes:

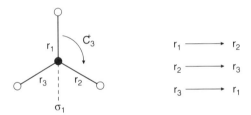

Using matrix language, this is expressed as

$$C_3^+ \begin{bmatrix} r_1 \\ r_2 \\ r_3 \end{bmatrix} = \begin{bmatrix} 0 & 1 & 0 \\ 0 & 0 & 1 \\ 1 & 0 & 0 \end{bmatrix} \begin{bmatrix} r_1 \\ r_2 \\ r_3 \end{bmatrix}. \tag{1-51}$$

The square matrix on the right-hand side is called a *representation* for the symmetry operation, C_3^+. For the σ_1 operation, we obtain

$$\sigma_1 \begin{bmatrix} r_1 \\ r_2 \\ r_3 \end{bmatrix} = \begin{bmatrix} 1 & 0 & 0 \\ 0 & 0 & 1 \\ 0 & 1 & 0 \end{bmatrix} \begin{bmatrix} r_1 \\ r_2 \\ r_3 \end{bmatrix}. \tag{1-52}$$

[9] The reader should consult introductory textbooks on matrix theory.

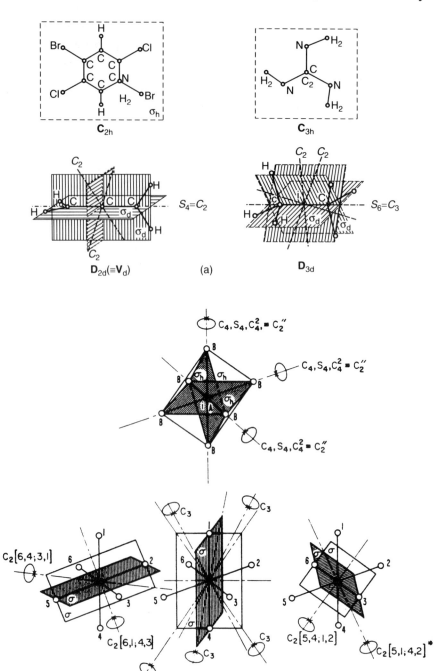

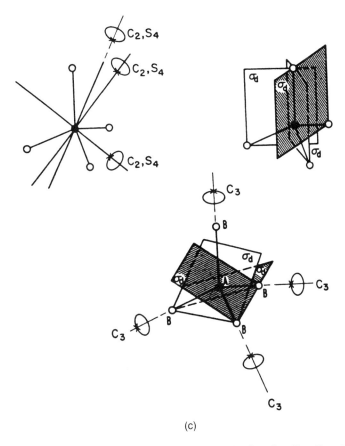

(c)

Figure 1-25 Symmetry elements for other point groups. (a) $\mathbf{C}_{2h}$, $\mathbf{C}_{3h}$, $\mathbf{D}_{2d}$, $\mathbf{D}_{3d}$; (b) $\mathbf{T}_d$; and (c) $\mathbf{O}_h$. (Reproduced with permission from Ref. 57.)

Similarly, for the E operation, we obtain

$$E\begin{bmatrix} r_1 \\ r_2 \\ r_3 \end{bmatrix} = \begin{bmatrix} 1 & 0 & 0 \\ 0 & 1 & 0 \\ 0 & 0 & 1 \end{bmatrix}\begin{bmatrix} r_1 \\ r_2 \\ r_3 \end{bmatrix}. \tag{1-53}$$

These representations are called *reducible representations* since they can be block-diagonalized in the form

$$\begin{bmatrix} 1 & 0 & 0 \\ 0 & A & B \\ 0 & C & D \end{bmatrix} \tag{1-54}$$

via similarity transformation.[10] If such simplification is no longer possible, the resulting representations are called *irreducible representations*. In the present case, the irreducible representations thus obtained are

for C_3^+:
$$
\begin{bmatrix} 0 & 1 & 0 \\ 0 & 0 & 1 \\ 1 & 0 & 0 \end{bmatrix} \rightarrow
\begin{bmatrix} 1 & 0 & 0 \\ 0 & -\dfrac{1}{2} & \dfrac{\sqrt{3}}{2} \\ 0 & -\dfrac{\sqrt{3}}{2} & -\dfrac{1}{2} \end{bmatrix},
\tag{1-55}
$$

for σ_1:
$$
\begin{bmatrix} 1 & 0 & 0 \\ 0 & 0 & 1 \\ 0 & 1 & 0 \end{bmatrix} \rightarrow
\begin{bmatrix} 1 & 0 & 0 \\ 0 & 1 & 0 \\ 0 & 0 & -1 \end{bmatrix}.
\tag{1-56}
$$

The reducible representation for E (1-53) is already diagonalized.

The sum of the diagonal elements of a matrix is called the character (χ) of the matrix. Hereafter, we use the term character rather than the representation since there is a one-to-one correspondence between them and since mathematical manipulation with χ is simpler than with the representation. The characters of the reducible representations for the E, C_3^+ and σ_1 operations are $3, 0$ and 1, respectively. The characters for C_3^- (counterclockwise rotation by $120°$) is the same as that of C_3^+, and those for σ_2 and σ_3 are the same as that of σ_1. By grouping symmetry operations of the same character ("class"), we obtain

C_{3v}	E	$2C_3$	$3\sigma_v$
	3	0	1

As seen in (1-53), (1-55) and (1-56), this set of reducible representations can be resolved into a sum of characters of the irreducible representations:

A_1	1	1	1
E	2	-1	0
$A_1 + E$	3	0	1

The characters of irreducible representations are listed in the *character*

---

[10] In general, similarity transformation is expressed as $S^{-1}R(K)s$ where $R(K)$ is a reducible representation for the symmetry operation, $R(K)$, S is a matrix of the same dimension, and S^{-1} is its reciprocal (or inverse), defined by the relationship, $S^{-1}S = SS^{-1} = E$, where E is a unit matrix such as (1-53). In the present case, the S matrix is obtained by writing a U matrix (Section 1.20) for the pyramidal XY_3 molecule.

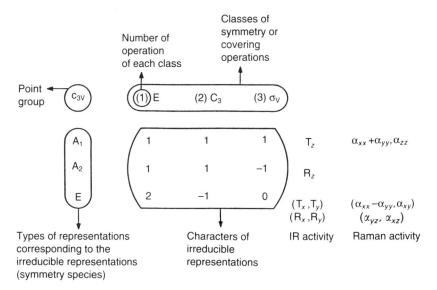

Figure 1-26 Diagrammatic interpretation of the character table for the C_{3v} point group.

table of each point group (Appendix 1). Figure 1-26 shows the character table for the point group C_{3v}. It is seen that there are three sets of characters corresponding to the A_1, A_2 and E species. In practical terms, the above result indicates that the three N—H bond stretching vibrations of the NH_3 molecule can be classified into one A_1 and one E (doubly degenerate) vibration. Thus, the character tables are important in classifying normal vibrations according to their symmetry properties (Section 1.13).

The last two columns of the character table provide information about IR and Raman activities of normal vibrations. One column lists the symmetry species of translational motions along the x, y and z axes (T_x, T_y and T_z) and rotational motions around the x, y and z axes (R_x, R_y and R_z). The last column lists the symmetry species of the six components of polarizability. As will be discussed in Section 1.14, the vibration is IR-active if it belongs to a symmetry species that contains any T components and is Raman-active if it belongs to a symmetry species that contains any α components. Pairs of these components are listed in parentheses when they belong to degenerate species (E and F).

TYPES OF SPECIES OF IRREDUCIBLE REPRESENTATIONS

(a) Nonlinear Molecules

A species is designated by the letter A if the transformation of the molecule is symmetric ($+1$) with respect to the rotation about the principal axis of symmetry. In NH_3, this axis is C_3, and, as can be seen, A_1 is totally symmetric, being labeled with positive 1's for all symmetry classes. A species that is symmetric with respect to the rotation, but is antisymmetric with respect to a rotation about the C_2 axis perpendicular to the principal axis or the vertical plane of reflection, is designated by the symbol A_2.

If a species of vibration belongs to the antisymmetric (-1) representation, it is designated by the letter B. If it is symmetric with respect to the rotation about the C_2 axis perpendicular to the principle axis of symmetry or to the vertical plane of reflection, it is a B_1 vibration, and if it is antisymmetric, it is a B_2 vibration. The letter E designates a twofold degenerate[11] vibration and the letter F denotes[12] a triply degenerate vibration. The character under the class of identity gives the degeneracy of the vibration, 1 for singly degenerate, 2 for doubly degenerate, and 3 for triply degenerate. For point groups containing a σ_h operation, primes (e.g., A') and double primes (e.g., A'') are used. The single prime indicates symmetry and the double prime antisymmetry with respect to σ_h. In molecules with a center of symmetry i, the symbols g and u are used, g standing for the German word *gerade* (which means even) and u for *ungerade* (or uneven). The symbol g goes with the species that transforms symmetrically with respect to i, and the symbol u goes with the species that transforms antisymmetrically with respect to i.

(b) Linear Molecules

Different symbols are used for linear molecules belonging to the point groups $\mathbf{C}_{\infty v}$ and $\mathbf{D}_{\infty h}$, namely Greek letters identical with the designations used for the electronic states of any diatomic molecules. The symbols σ or Σ are used for species symmetric with respect to the principal axis. A superscript plus sign (σ^+ or Σ^+) is used for species that are symmetric, and a superscript minus sign (σ^- or Σ^-) for species that are antisymmetric with respect to a plane of symmetry through the molecular axis. The symbols π, Δ, and ψ are used for degenerate vibrations, with the degree of degeneracy increasing in this order. This is illustrated in Table 1-7.

(c) Molecules of Highest Symmetry

Although it has been generally believed that molecules would never be found in icosahedral ($\mathbf{I}_h$) symmetry (23,24), today we know that such is not the case.

---

[11] The bending vibration of CO_2 is an example of a degenerate vibration. The frequency and character of the vibrations are the same, but they occur perpendicular to one another.

[12] Some texts use the symbol T for the triply degenerate vibration.

Table 1-7 Character Table for the $\mathbf{C}_{\infty v}$ Point Group

$\mathbf{C}_{\infty v}$	E	$2C_{\infty}^{\phi}$	$2C_{\infty}^{2\phi}$	$2C_{\infty}^{3\phi}$	$\cdots$	$\infty \sigma_v$
Σ^+	$+1$	$+1$	$+1$	$+1$	$\cdots$	$+1$
Σ^-	$+1$	$+1$	$+1$	$+1$	$\cdots$	-1
Π	$+2$	$2\cos\phi$	$2\cos 2\phi$	$2\cos 3\phi$	$\cdots$	0
Δ	$+2$	$2\cos 2\phi$	$2\cos 2\cdot 2\phi$	$2\cos 3\cdot 2\phi$	$\cdots$	0
ψ	$+2$	$2\cos 3\phi$	$2\cos 2\cdot 3\phi$	$2\cos 3\cdot 3\phi$	$\cdots$	0
$\cdots$	$\cdots$	$\cdots$	$\cdots$	$\cdots$	$\cdots$	$\cdots$

There are at least three molecules possessing icosahedral symmetry. These are the borohydride anion (25) $B_{12}H_{12}^{2-}$; dodecahedrane, $C_{20}H_{20}$ (26), and the buckminsterfullerene or buckyball C_{60} cluster (27) (see Chapter 4, Section 4.3.6). The $\mathbf{I}_h$ symmetry contains fivefold to twofold axes of rotation as well as a center of symmetry. As a result, additional species of vibrations such as G and H appear in the $\mathbf{I}_h$ character table (28). These correspond to four-dimensional and five-dimensional representations. G is a quadruple and H is a pentagonal degenerate vibration. The character table for the $\mathbf{I}_h$ point group is shown in Appendix 1.

1.13 Classification of Normal Vibrations by Symmetry

The $3N - 6(5)$ normal vibrations of an N-atom molecule can be classified into symmetry species of a point group according to their symmetry properties. As an example, consider the displacements of individual atoms of the H_2O molecule ($\mathbf{C}_{2v}$) using the Cartesian coordinates shown below:

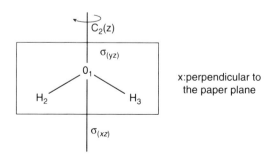

The symmetry operations to be considered are given in the table.

C_{2v}	$C_2(z)$	$\sigma(xz)$	$\sigma(yz)$
A_1	$+1$	$+1$	$+1$
A_2	$+1$	-1	-1
B_1	-1	$+1$	-1
B_2	-1	-1	$+1$

It is not necessary to consider $\sigma(yz)$ since $C_2 \times \sigma(xz) = \sigma(yz)$.

First, the symmetry species of the three displacements of the oxygen atom are readily determined as shown below:

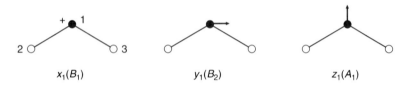

Here the $+$ sign denotes the out-of-plane displacement in the $+x$ direction.

Since the two hydrogen atoms are equivalent, we consider symmetry species of six linear combinations of their displacements:

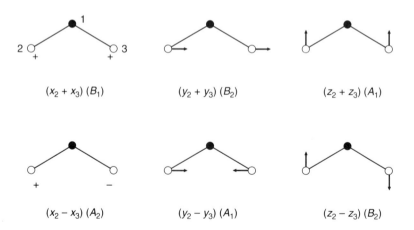

Since these nine displacements include three translational (T) and three rotational (R) motions of the whole molecule, we must subtract them from our calculations. It is readily seen that T_x, T_y and T_z belong to the B_1, B_2 and A_1, whereas R_x, R_y and R_z belong to the B_2, B_1 and A_2 species,

Table 1-8 Number of Normal Vibrations of H_2O Molecule

C_{2v}	Number of Coordinates			Translation and Rotation	Number of Vibrations
	O	H	Total		
A_1	1	2	3	T_z	$3-1=2$
A_2	0	1	1	R_z	$1-1=0$
B_1	1	1	2	T_x, R_y	$2-2=0$
B_2	1	2	3	T_y, R_x	$3-2=1$

respectively. Table 1-8 summarizes these results. Thus, we find that two vibrations belong to the A_1 and one vibration belongs to the B_2 species. The approximate vibrational modes of the two A_1 type vibrations can be derived by combining z_1, $(z_2 + z_3)$ and $(y_2 - y_3)$ as follows:

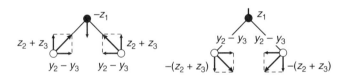

The vibrational mode of the B_2 vibration is obtained by combining $-y_1$, $(y_2 + y_3)$ and $(z_2 - z_3)$ as follows:

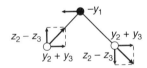

More accurate mode descriptions can be made if we consider the masses of individual atoms, bond distances, bond angles and force constants (Section 1.20). It is clear that these three vibrations correspond to the v_1, v_2 and v_3, respectively, of Fig. 1-12.

More generally, the number of normal vibrations in each species can be calculated by using Herzberg's formulas (23) given in Appendix 2. In the case of the C_{2v} point group, they are expressed as:

$$A_1: \quad 3m + 2m_{xz} + 2m_{yz} + m_0 - 1,$$

$$A_2: \quad 3m + m_{xz} + m_{yz} - 1,$$

$$B_1: \quad 3m + 2m_{xz} + m_{yz} + m_0 - 2,$$

$$B_2: \quad 3m + m_{xz} + 2m_{yz} + m_0 - 2,$$

and N (total number of atoms) is given by $4m + 2m_{xz} + 2m_{yz} + m_0$. In the foregoing, the parameters are defined as follows:

m: number of sets of nuclei not on any symmetry elements.

m_0: number of nuclei on all symmetry elements.

m_{xy}, m_{yz}, m_{xz}: number of sets of nuclei lying on the xy, yz and xz planes, respectively, but not on any other axis going through these planes.

For H_2O, $m = 0$, $m_0 = 1$ (oxygen atom), $m_{xy} = 0$, $m_{yz} = 1$ (hydrogen atom), $m_{xz} = 0$ and $N = 3$. Thus, the numbers of normal vibrations are 2, 0, 0 and 1 for the A_1, A_2, B_1 and B_2 species, respectively.

The preceding general equations can be obtained from our calculations on H_2O. As shown earlier, the displacements of the oxygen atom (m_0) are distributed into $1A_1$, $1B_1$, and $1B_2$. The displacements of the two hydrogen atoms (m_{yz}) are distributed among $2A_1$, $1A_2$, $1B_1$ and $2B_2$. Although m_{xz} is zero for H_2O, it is not zero for other C_{2v} molecules. For example, it is 1 for CH_2Cl_2:

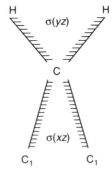

It is easily seen that m_{xz} is distributed into $2A_1$, $1A_2$, $2B_1$ and $1B_2$. m is zero in the case of H_2O. However, it is not zero for other molecules of C_{2v} symmetry. For example, consider an imaginary conformation of urea in which the four hydrogen atoms take the positions shown below:

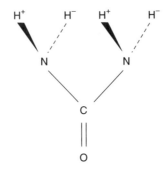

Here, H^+ and H^- are mirror images of each other with respect to the molecular plane (the whole molecule is nearly planar in the real molecule). Then, $m = 1$. The displacements of $4m$ atoms are expressed by $3 \times 4m$ coordinates, which are distributed equally into the four species ($3m$ for each). The summation of these calculations leads to the general equations given earlier.

As another example, consider the NH_3 molecule of $\mathbf{C}_{3v}$ symmetry. Using the table given in Appendix 2, we find that $m_0 = 1$ (nitrogen atom), $m = 0$ and $m_v = 1$ (hydrogen atom). Thus, the number of normal vibrations in the A_1, A_2 and E species are 2, 0, and 2, respectively.

Although not described earlier, classification of normal vibrations can be made by applying group theoretical treatments to individual molecules (29). The latter is useful for confirming the results obtained from Herzberg's tables.

1.14 Symmetry Selection Rules

As shown in Section 1.7, IR and Raman activities for small molecules can be determined by inspection of their normal modes. Clearly, it is difficult to apply such an approach to large and complex molecules. This problem can be solved by using the group theoretical consideration described next.

According to quantum mechanics (18,19), the selection rule for the IR spectrum is determined by the integrals

$$[\mu_x]_{v',v''} = \int \psi_{v'}^*(Q_a)\mu_x\psi_{v''}(Q_a)\,dQ_a,$$

$$[\mu_y]_{v',v''} = \int \psi_{v'}^*(Q_a)\mu_y\psi_{v''}(Q_a)\,dQ_a \qquad (1\text{-}57)$$

$$[\mu_z]_{v',v''} = \int \psi_{v'}^*(Q_a)\mu_z\psi_{v''}(Q_a)\,dQ_a.$$

Here, μ_x, μ_y and μ_z are the x, y and z components of the dipole moment at the electronic ground state, respectively. $\psi_{v'}$ and $\psi_{v''}$ are vibrational wavefunctions where v' and v'' are the vibrational quantum numbers before and after the transition, respectively. Q_a is the *normal coordinate* of the normal vibration, a. If one of these integrals is nonzero, this vibration is infrared-active. If all three integrals are zero, it is infrared-inactive.

Using (1-57) as an example, let us determine whether such an integral is zero or nonzero. For this purpose, we first expand μ_x in terms of the normal coordinate, Q_a:

$$\mu_x = (\mu_x)_0 + \left(\frac{\partial \mu_x}{\partial Q_a}\right)Q_a + \cdots .$$

Then, (1-57) can be rewritten as

$$[\mu_x]_{v',v''} = (\mu_x)_0 \int \psi_v^*(Q_a)\psi_{v''}(Q_a)\, dQ_a$$

$$+ \left(\frac{\partial \mu_x}{\partial Q_a}\right) \int \psi_v^*(Q_a)Q_a\psi_{v''}(Q_a)\, dQ_a + \dots. \qquad (1\text{-}58)$$

The integral in the first term vanishes because of the orthogonality of $\psi_{v'}$ and $\psi_{v''}$ (except for $v' = v''$, no transition). In order for the second term to be nonzero,

$$\left(\frac{\partial \mu_x}{\partial Q_a}\right) \neq 0 \quad \text{and} \quad \int \psi_v^*(Q_a)Q_a\psi_{v''}(Q_a)\, dQ_a \neq 0.$$

The latter is nonzero only when $\Delta v = \pm 1$ (harmonic oscillator, Section 1.3). In order to understand the significance of the former, we approximate the second term in (1-58) as

$$\left(\frac{\partial \mu_x}{\partial Q_a}\right) \int \psi_v^*(Q_a)Q_a\psi_{v''}(Q_a)\, dQ_a \approx e \int \psi_{v'}(Q_a)x\psi_{v''}(Q_a)\, dQ_a. \qquad (1\text{-}59)$$

Here, $\mu_x = \Sigma_i e_i x_i$, where e_i is the charge on the ith electron or nucleus and x_i is the x component of its position.

Consider the fundamental vibration in which the transition occurs from $v' = 0$ to $v'' = 1$. From Eq. (1-28), it is obvious that ψ_0 is invariant under any symmetry operation since it contains the Q_a^2 term. On the other hand, the symmetry of ψ_1 is the same as that of Q_a since it contains a $Qe^{-Q^2/2}$ term. In general, any integral such as

$$\int f_A f_B f_C\, d\tau$$

does not vanish if the representation of the direct product, $f_A f_B f_C$, contains the totally symmetric representation.[13] In the present case, ψ_0 is totally symmetric. Then the integral such as (1-59) is nonzero if the representation of the product $x\psi_1$ is totally symmetric. This is possible only when x and ψ_1 belong to the same symmetry species.

The symmetry species of $\mu_x(x)$, $\mu_y(y)$ and $\mu_z(z)$ are listed in the character tables of the respective point groups (Appendix 1).[14] As shown in the preceding section, the symmetry species of normal vibrations can be found by using Herzberg's tables (Appendix 2). Thus, IR activity is readily determined by

---

[13] The proof of this theorem is given in Ref. 28.
[14] The symmetry property of μ_x (x-component of the dipole moment) is the same as that of T_x (translational motion along the x-axis).

the inspection of character tables; *the vibration is IR-active if the component(s) of the dipole moment belong(s) to the same symmetry species as that of the vibration.*

As an example, consider IR activity of the six normal vibrations of the NH_3 molecule, which are classified into $2A_1$ and $2E$ species of C_{3v} point group. The character table shows that μ_z belongs to the A_1 and the pair of (μ_x, μ_y) belongs to the E species. Thus, all six normal vibrations are IR-active.

The selection rule for Raman spectrum is determined by the integrals

$$[\alpha_{xx}]_{v',v''} = \int \psi_{v'}^*(Q_a)\alpha_{xx}\psi_{v''}(Q_a) \, dQ_a, \tag{1-60}$$

$$[\alpha_{xy}]_{v',v''} = \int \psi_{v'}^*(Q_a)\alpha_{xy}\psi_{v''}(Q_a) \, dQ_a, \tag{1-61}$$

$$\vdots \qquad \qquad \vdots$$

Here α_{xx}, α_{yy}, α_{zz}, α_{xy}, α_{yz} and α_{xz} are the components of the polarizability tensor discussed in Section 1.7. If one of these six integrals is nonzero, this vibration is Raman-active. If all the integrals are zero, it is Raman-inactive.

Symmetry selection rules for Raman spectrum can be derived by using a procedure similar to that for the IR spectrum. One should note, however, that the symmetry property of α_{xy}, for example, is determined by the product, $\mu_x\mu_y(xy)$ (19). The symmetry species of six components of polarizability are readily found in character tables. In point group C_{3v}, for example, $\alpha_{xx} + \alpha_{yy}$ and α_{zz} belong to the A_1 species while two pairs, $(\alpha_{xx} - \alpha_{yy}, \alpha_{xy})$ and $(\alpha_{yz}, \alpha_{xz})$, belong to the E species.[15] Thus, all six normal vibrations of the NH_3 molecule ($2A_1$ and $2E$) are Raman-active. More generally, *the vibration is Raman-active if the component(s) of the polarizability belong(s) to the same symmetry species as that of the vibration.*

As another example, consider an octahedral XY_6-type molecule of O_h symmetry. Using Herzberg's table, its 15 ($3 \times 7 - 6$) vibrations are classified into $A_{1g} + E_g + 2F_{1u} + F_{2g} + F_{2u}$. It is readily seen from the character table that only F_{1u} vibrations are IR-active, while A_{1g}, E_g and F_{2g} vibrations are Raman-active.

1.15 Resonance Raman Spectra

As stated in Section 1.4, resonance Raman (RR) scattering occurs when the sample is irradiated with an exciting line whose energy corresponds to that

---

[15]Instead of α_{xx} and α_{yy}, their linear combinations, $\alpha_{xx}+\alpha_{yy}$ and $\alpha_{xx}-\alpha_{yy}$ are used for mathematical convenience.

of the electronic transition of a particular chromophoric group in a molecule. Under these conditions, the intensities of Raman bands originating in this chromophore are selectively enhanced by a factor of 10^3 to 10^5. This selectivity is important not only for identifying vibrations of this particular chromophore in a complex spectrum, but also for locating its electronic transitions in an absorption spectrum.

Theoretically, the intensity of a Raman band observed at $v_0 - v_{mn}$ is given by (30):

$$I_{mn} = \text{constant} \cdot I_0 \cdot (v_0 - v_{mn})^4 \sum_{p\sigma} |(\alpha_{p\sigma})_{mn}|^2. \tag{1-62}$$

Here, m and n denote the initial and final states, respectively, of the electronic ground state. Although not explicit in Eq. (1-62), e represents an electronic excited state (Fig. 1-27) involved in Raman scattering. I_0 is the intensity of the incident laser beam of frequency v_0. The $(v_0 - v_{mn})^4$ term expresses the v^4 rule to be discussed in Chapter 2, Section 2.6. Finally $(\alpha_{p\sigma})_{mn}$ represents the change in polarizability α caused by the $m \to e \to n$ transition, and ρ and σ are x, y and z components of the polarizability tensor (Section 1.7). This term can be rewritten as (30)

$$(\alpha_{p\sigma})_{mn} = \frac{1}{h} \sum_e \left(\frac{M_{me}M_{en}}{v_{em} - v_0 + i\Gamma_e} + \frac{M_{me}M_{en}}{v_{en} + v_0 + i\Gamma_e} \right), \tag{1-63}$$

where v_{em} and v_{en} are the frequencies corresponding to the energy differences between the states subscribed and h is Planck's constant. M_{me}, etc., are the

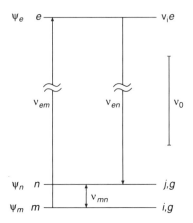

Figure 1-27   Energy level diagram for resonance Raman transition.

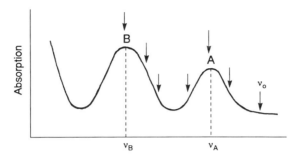

Figure 1-28 Absorption spectrum of a compound containing two chromophoric groups (A and B).

electric transition moments, such as

$$M_{me} = \int \Psi_m^* \mu_\sigma \Psi_e \, d\tau. \qquad (1\text{-}64)$$

Here, Ψ_m and Ψ_e are total wavefunctions of the m and e states, respectively, and μ_σ is the σ component of the electric dipole moment. Γ_e is the band width of the eth state, and the $i\Gamma_e$ term is called the damping constant. In normal Raman scattering, v_0 is chosen so that $v_0 \ll v_{em}$. Namely, the energy of the incident beam is much smaller than that of an electronic transition. Under these conditions, the Raman intensity is proportional to $(v_0 - v_{mn})^4$. As v_0 approaches v_{em}, the denominator of the first term in the brackets of Eq. (1-63) becomes very small. Hence, this term ("resonance term") becomes so large that the intensity of the Raman band at $v_0 - v_{mn}$ increases enormously. This phenomenon is called *resonance Raman (RR) scattering*.

Suppose that a compound contains two chromophoric groups that exhibit electronic bands at v_A and v_B as shown in Fig. 1-28. Then, vibrations of chromophore A are resonance-enhanced when v_0 is chosen near v_A, and those of chromophore B are resonance-enhanced when v_0 is chosen near v_B. For example, heme proteins such as hemoglobin and cytochromes (Chapter 4, Section 4.2) exhibit porphyrin core π–π^* transitions in the 400–600 nm region and peptide chain transitions below 250 nm. Thus, the porphyrin core and peptide chain vibrations can be selectively enhanced by choosing exciting lines, in the regions of these electronic absorptions.

More specific information can be obtained by expressing the total wavefunction as the product of the electronic and vibrational wavefunctions. (See the right-hand side labeling in Fig. 1-27). The results are (30, 31)

$$(\alpha_{p\sigma})_{mn} \cong A + B. \qquad (1\text{-}65)$$

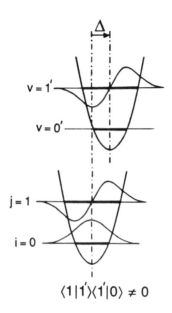

$$\langle 1|1'\rangle\langle 1'|0\rangle \neq 0$$

Figure 1-29 Shift of equilibrium position caused by a totally symmetric vibration.

The A-term is written as

$$A \cong M_e^2 \frac{1}{h} \sum_v \frac{\langle j|v\rangle\langle v|i\rangle}{\nu_{vi} - \nu_0 + i\Gamma_v}. \qquad (1\text{-}66)^{16}$$

Here M_e is the pure electronic transition moment for the resonant excited state e, of which v is a vibrational level of band width Γ_v, ν_{vi} is the transition frequency from the ground state vibrational level (i) to the excited vibrational level (v). The A-term becomes larger as the denominator becomes smaller (resonance condition) and as M_e becomes larger (stronger electronic absorption). The numerator contains the product of two overlap integrals of vibrational wavefunctions (*Franck–Condon overlap*) involving the i, j and v states. Because of the orthogonality of vibrational wavefunctions (Section 1.2), either one of the integrals becomes zero unless the equilibrium position is shifted upon electronic excitation (Fig. 1-29). Since this occurs only for totally symmetric vibrations, the A-term enhancement can be seen only for totally symmetric modes.

The A-term resonance has been observed for a number of compounds. Figure 1-30 shows the RR spectra of TiI_4 obtained by 514.5 nm excitation (32). An overtone (nv) series up to $n = 12$ was observed in this case. (Also

---

[16] Only the resonance term is shown in (1-66) as well as in (1-67).

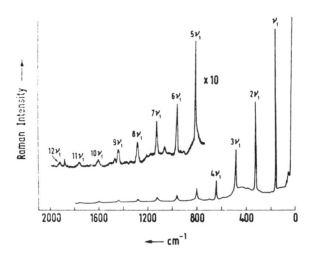

Figure 1-30 Resonance Raman spectrum of solid TiI_4 (514.5 nm excitation). $v_1 = 160.8$ cm^{-1}. (Reproduced with permission from Ref. 32. Copyright 1973 American Chemical Society.)

see Fig. 3-11 and Fig. 4-11.) The appearance of such a series can be explained if one calculates Franck–Condon overlaps under rigorous resonance conditions (33). Among the totally symmetric modes, the mode that leads to the excited state configuration is most strongly resonance-enhanced (34). For example, the NH_3 molecule is pyramidal in the ground state and planar at the excited state (216.8 nm above the ground state). Thus, the symmetric bending mode near 950 cm^{-1} is enhanced 10 times more than the symmetric stretching mode near 3,300 cm^{-1} when the exciting line is changed from 514.5 to 351.1 nm.

The B-term involves two electronic excited states (e and s) and provides a mechanism for resonance-enhancement of non-totally symmetric vibrations. The B-term can be expressed as

$$B \cong M_e M'_e \frac{1}{h} \sum_v \frac{\langle j|Q|v\rangle\langle v|i\rangle + \langle j|v\rangle\langle v|Q|i\rangle}{v_{vi} - v_0 + i\Gamma_v}, \tag{1-67}$$

$$M'_e = \mu_s \langle s|\partial H/\partial Q|e\rangle / \langle v_s - v_e\rangle, \tag{1-68}$$

where v_s and μ_s are the frequency and transition dipole moment of the second excited state(s). Although the denominator in the B-term is the same as that of the A-term, its numerator contains Q-dependent vibrational overlap integrals as well as Franck–Condon overlap integrals. Here, Q is the normal coordinate of a particular vibration. Thus, it does not vanish even when the equilibrium position is not shifted by electronic excitation (non-totally symmetric vibration). The most important term in Eq. (1-67) is the vibronic coupling operator $\langle s|\partial H/\partial Q|e\rangle$, where H is the electronic Hamiltonian. As will be shown later, this term becomes nonzero if normal coordinates of proper symmetry are chosen. Thus, it is possible to resonance-enhance these vibrations via the B-term. A typical example of B-term resonance is found in RR spectra of heme proteins and their model compounds (35). As shown in Fig. 1-31, metalloporphyrins such as Ni(OEP) exhibit two electronic transitions; Q_0 (α) and B (or Soret) bands together with a vibronic sideband, Q_1 (or β) in the 600–350 nm region. According to MO calculations on the porphyrin core of $\mathbf{D}_{4h}$ symmetry, the $a_{1u} \rightarrow e_g^*$ and $a_{2u} \rightarrow e_g^*$ transitions have similar energies and the same excited state symmetry, since $a_{1u} \times e_g = a_{2u} \times e_g = E_u$. This results in a strong configuration interaction to produce two separate states, Q_0 and B (see the inset in Fig. 1-31). The transition dipole moments nearly cancel each other to produce a weak Q_0 band in the visible region, but add up to produce a strong B band in the UV region.

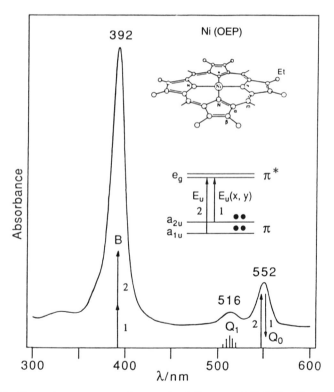

Figure 1-31 Absorption spectrum and energy level diagram of Ni (OEP). (Reproduced with permission from Ref. 35. Copyright © 1988 John Wiley & Sons, Inc.)

As stated earlier, the vibronic coupling operator, $\langle s|\partial H/\partial Q|e\rangle$, determines which normal vibration is resonance-enhanced via the B-term. This term becomes nonzero only when the normal vibration has a proper symmetry. As discussed in Section 1.14, an integral such as

$$\int f_A f_B f_C \, d\tau \tag{1-69}$$

is nonzero if the direct product of the irreducible representations of f_A, f_B and f_C contains the totally symmetric component. Since both $e(Q_0)$ and $s(B)$ states are of E_u symmetry, only the A_{1g}, B_{1g}, B_{2g} and A_{2g} vibrations,

$$E_u \times E_u = A_{1g} + B_{1g} + B_{2g} + A_{2g}, \tag{1-70}$$

can be resonance-enhanced via the B-term.[17] Figure 1-32 shows the RR spectra of Ni(OEP) obtained by B, Q_1 and Q_0 excitation (36). The former is dominated by totally symmetric vibrations, whereas the latter two are dominated by B_{1g}, B_{2g} and A_{2g} vibrations since A_{1g} modes are not effective in mixing the B and Q_0 states because of the high symmetry of the porphyrin core. Distinction of these modes can be made by measuring the depolarization ratios (Section 1.9), which should be $\frac{3}{4}$ (depolarized) for B_{1g} and B_{2g}, and $\frac{3}{4} \sim \infty$ (inverse polarization) for A_{2g} vibrations. It should be noted that the RR spectrum obtained by Soret (B) excitation is dominated by A_{1g} vibrations, since resonance enhancement occurs through the A-term discussed earlier.

1.16   Space Group Symmetry

1.16.1   SYMMETRY ELEMENTS FOR A MOLECULE IN THE CONDENSED STATE

If one takes into account symmetry elements combined with translations, one obtains operations or elements that can be used to define the symmetry of space. Here, a translation is defined as the superposition of atoms or molecules from one site onto the same atoms or molecules in another site without the use of a rotation. The symmetry element called the *screw axis* involves an operation combining a translation with a rotation. The symmetry element called the glide plane involves an operation combining a reflection with a translation. These operations can be used to describe the symmetry of homogeneous spatial materials such as crystals and polymers, as well as space itself if one takes the identity element to be the set of all translations. This condition implies that these translations in a crystal are those that

[17] The general method for resolving direct products into symmetry species is found in textbooks of group theory[24,28]. Appendix 3 summarizes the general rules for obtaining the symmetry species of direct products.

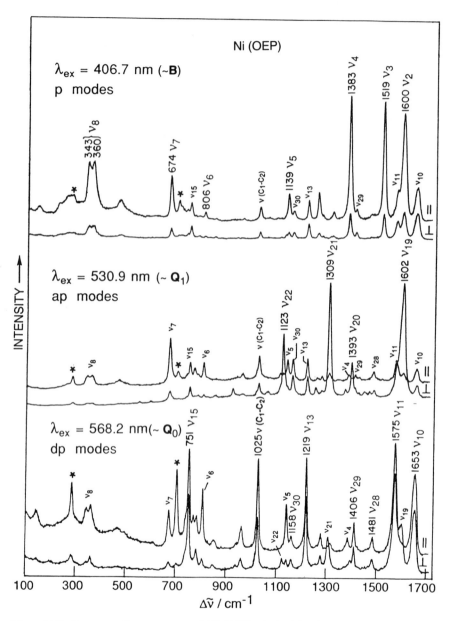

Figure 1-32 Resonance Raman spectra of Ni (OEP) obtained by three different excitations. (Reproduced with permission from Ref. 36. Copyright 1990 American Chemical Society.)

describe the lattice in contrast to those that are fractional translations associated with some rotation (screw motions and glides). A crystal may be considered to be made up of a large number of blocks of the same size and shape. One such block is defined as a unit cell. The unit cell must be capable of repetition in space without leaving any gap. The unit cell may be a primitive or nonprimitive unit cell. The distinction between these will be made later in this section.

The screw axis and the glide plane are further defined as follows:

Screw axis (n_p)

Rotation followed by a translation;

n = order of axis;

p/n = fraction of the unit cell over which translation occurs;

$n = 2, 3, 4$ or 6;

$p = 1, 2, 3, \ldots, n - 1$.

For example, 2_1 = twofold screw axis, translation one-half the distance of the unit cell; and 3_1 = threefold screw axis, translation one-third the distance of the unit cell.

Figure 1-33 shows the operation of a twofold screw axis, in which the translation is $\frac{1}{2}$ of the unit cell (fractional translation).

Glide Plane

Reflection followed by a translation.

Figure 1-34 shows the operation of a glide plane.

The new symmetry elements that are added to the point symmetry elements are the screw axis and glide plane. As a result, 230 different combinations of symmetry elements become possible in what are called space groups. For a development of space groups see Burns and Glazer (38). Table 1-9 shows the distribution of the space groups among the seven crystal systems. Some

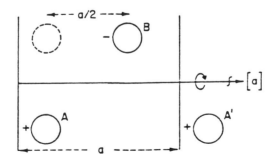

Figure 1-33 The operation of a twofold screw axis. (Reproduced with permission from Ref. 37.)

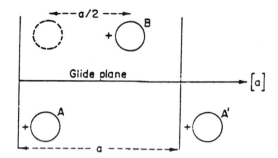

Figure 1-34 The operation of a glide plane. (Reproduced with permission from Ref. 37.)

Table 1-9 Distribution of Space Groups among the Seven
Crystal Systems

Crystal System	Number of Space Groups
Triclinic	2
Monoclinic	13
Orthorhombic	59
Trigonal	25
Hexagonal	27
Tetragonal	68
Cubic	36

of the space groups are never found in actual crystals, and about one-half of the known crystals belong to the 13 space groups of the monoclinic system. Figure 1-35 shows the lattices of the seven major crystal systems (one triclinic, four orthorhombic, two monoclinic, two tetragonal, three cubic, one hexagonal, one trigonal).

1.16.2 Space Group

Just as the point group collects all of the point symmetry elements, the space group is seen collecting all of the space symmetry elements in crystals involving translation. For the space group selection rules, it is necessary to work on crystals for which space groups are known from x-ray studies, or where sufficient information is available to make a choice of structure. Alternatively, the structure may be assumed, and the space group selection rules can serve as a test of this assumption. In deriving space group selection rules, one must deal with a primitive unit cell. A primitive unit cell is the

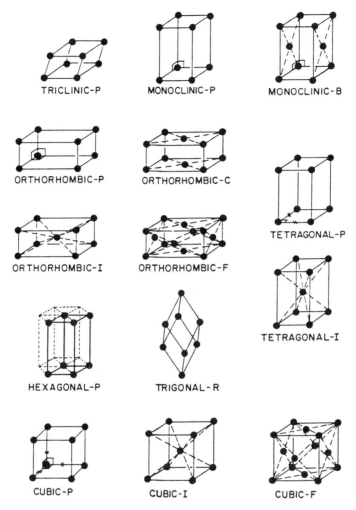

Figure 1-35 The seven crystal systems corresponding to 14 Bravais lattices. (Reproduced with permission from Ref. 38.)

smallest unit in a crystal which, by a series of translations, would build up the whole crystal.

The Hermann–Mauguin notation is generally used by crystalographers to describe the space group. Tables exist to convert this notation to the Schoenflies notation. The first symbol is a capital letter and indicates whether the lattice is primitive. The next symbol refers to the principal axis, whether it is rotation, inversion, or screw, e.g.,

$P2_1$ = primitive lattice with a two-fold axis of rotation, translation one-half the distance of the unit cell

$C2$ = nonprimitive centered lattice with a twofold axis of rotation.

A mirror plane is symbolized as m, and a glide plane by c, e.g.,

$P2_1/m$ m = mirror plane perpendicular to principal axis;

$C2/c$ c = glide plane perpendicular to principal axis.

Table 1-10 shows the space group symbolism used.

As previously mentioned, the primitive unit cell is the smallest unit of a crystal that reproduces itself by translations. Figure 1-36 illustrates the difference between a primitive and a centered or nonprimitive cell. The primitive cell can be defined by the lines a and c. Alternatively, we could have defined it by the lines a' and c'. Choosing the cell defined by the lines a'' and c'' gives us a nonprimitive cell or centered cell, which has twice the volume and two repeat units. Table 1-11 illustrates the symbolism used for the various types of lattices and records the number of repeat units in the cell for a primitive and a nonprimitive lattice. The spectroscopist is concerned with the primitive (Bravais) unit cell in dealing with lattice vibrations. For factor group selection rules, it is necessary to convert the number of molecules per crystallographic unit cell Z to Z', discussed later, which is the number

Table 1-10   Space Group Symbolism

First symbol refers to the Bravais lattice
   P = primitive lattice
   C = centered lattice
   F = face-centered lattice
   I = body-centered lattice
   R = rhombohedral (unit cell can be primitive or nonprimitive; see notes to Table 1-11)
   Principal axis of ration given number n = order
      e.g., 2 = twofold axis of rotation
   For screw axis p/n = fraction of primitive lattice over which translation parallel to screw axis occurs.
      e.g., $P2_1$ = primitive lattice with a twofold axis of rotation, translation; one-half unit cell
   Mirror plane = m
   Glide planes = symbols a, b, c along (a), (b), (c) axes
      symbol $n = (b+c)/2$ or $(a+b)/2$
      symbol $d = (a+b)/4$ or $(b+c)/4$ or $(a+c)/4$
      e.g., $P2_1/m$; m = mirror plane perpendicular to principal axis
         $C2/c$; c = glide plane axis perpendicular to principal axis

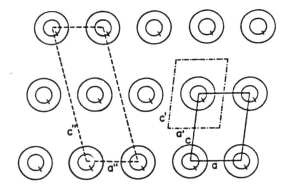

Figure 1-36 Differentiation between primitive and centered unit cells. (Reproduced with permission from Ref. 37.)

Table 1-11 Primitive and Centered Lattices

Type	Symbol	Number of Repeat Units in Cell
Primitive	P	1
Rhombohedral[a]	R	3 or 1[b]
Body-centered	I	2
Side-centered	A, B, or C	2
Face-centered	F	4

[a] Also called trigonal.

[b] There are cases in which the number of repeat units in the crystallographic cell may be three or one. For the cases where it is three, Z will be divisible by three. For example, for TiS, $\mathbf{D}_{3d}^5$—$R\overline{3}m$ (No. 166), $Z = 9$, and therefore $Z' = 9/3 = 3$. However, for Cr_2O_3, $\mathbf{D}_{3d}^6$—$R\overline{3}c$ (No. 167), $Z = 2$, and $Z' = 2/1 = 2$. Thus, in the latter crystal the cell can be considered to be primitive.

of molecules per primitive cell. For example,

$$Z' = \text{number of molecules in the primitive cell}$$

$$= \frac{Z(\text{number of molecules in crystallographic cell})}{\text{repeat units in cell}}.$$

If $Z = 4$ for an F-type lattice, then

$$Z' = 4/4 = 1.$$

In the site symmetry compilation for the 230 space groups given in Appendix 4, the data are for a primitive cell and can be used directly.

1.16.3   FACTOR GROUP

It is necessary to define a factor group and to describe how it relates to a space group. In a crystal, one primitive cell or unit cell can be carried into another primitive cell or unit cell by a translation. The number of translations of unit cells then would seem to be infinite since a crystal is composed of many such units. If, however, one considers only one translation and consequently only two unit cells, and defines the translation that takes a point in one unit cell to an equivalent point in the other unit cell as the identity, one can define a finite group, which is called a *factor group* of the space group.

The factor groups are isomorphic (one-to-one correspondence) with the 32 point groups and, consequently, the character table of the factor group can be obtained from the corresponding isomorphic point group.

1.16.4   SITE GROUP

It also becomes necessary to define a site group. A unit cell of a crystal is composed of points (molecules or ions) located at particular positions in the cell. It turns out, however, that the points can only be located at certain positions in the lattice that are called sites, that is, they can only be located on one of the symmetry elements of the factor group and thus remain invariant under that operation independent of translation. The point has fewer symmetry elements than the parent factor group and belongs to what is called a "site group," which is a subgroup of the factor group. [A subgroup (S) contains a set of symmetry elements that are also part of a parent group (G).] In general, factor groups can have a variety of different sites possible, that is, many subgroups can be formed from the factor group. Also, a number of distinct sites in the Bravais unit cell with the same site group are possible.

1.17   Normal Vibrations in a Crystal

In order to discuss the selection rules for crystalline lattices it is necessary to consider elementary theory of solid vibrations. The treatment essentially follows that of Mitra (39). A crystal can be regarded as a mechanical system of nN particles, where n is the number of particles (atoms) per unit cell and N is the number of primitive cells contained in the crystal. Since N is very large, a crystal has a huge number of vibrations. However, the observed spectrum is relatively simple because, as shown later, only where equivalent atoms in primitive unit cells are moving in phase are they observed in the IR or Raman spectrum. In order to describe the vibrational spectrum of such a solid, a frequency distribution or a distribution relationship is necexsary. The development that follows is for a simple one-dimensional crystalline diatomic linear lattice. See also Turrell (40).

Consider a simple one-dimensional infinite chain, consisting of alternating masses M and m separated by a distance a with a force constant f:

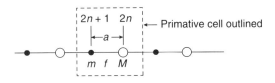

The two particles are located at the even- and odd-numbered lattice points $2n$ and $2n + 1$, respectively. The displacements u_{2n} and u_{2n+1} of the even and odd particles are given by the equations of motion

$$M\ddot{u}_{2n} = f(u_{2n+1} + u_{2n-1} - 2u_{2n}),$$
$$m\ddot{u}_{2n+1} = f(u_{2n+2} + u_{2n} - 2u_{2n+1}). \tag{1-71}$$

Assuming the following solutions for u_{2n} and u_{2n+1}:

$$u_{2n} = y_1 \exp i(2\pi vt + 2nka), \tag{1-72}$$

$$u_{2n+1} = y_2 \exp i[2\pi vt + (2n + 1)ka], \tag{1-73}$$

and substituting the values of u_{2n} and u_{2n+1} in Eq. (1-71) one obtains two equations for the amplitudes y_1 and y_2. Here k is the wave vector and corresponds to the phase differences for each successive cell. A solution for these equations exists, and the secular determinant is illustrated as follows:

$$\begin{vmatrix} 2f - 4\pi^2 v^2 M & -2f \cos ka \\ -2f \cos ka & 2f - 4\pi^2 v^2 m \end{vmatrix} = 0. \tag{1-74}$$

A dispersion formula results, based on frequency dependency on masses, force constant and distance between the two masses, such as

$$v^2 = \frac{1}{4\pi^2} \left[\frac{f}{\mu} \pm \left(\frac{f^2}{\mu^2} - \frac{4f^2 \sin^2 ka}{Mm} \right)^{\frac{1}{2}} \right], \tag{1-75}$$

where μ is the reduced mass. The finite length of the lattice restricts the values of k in the range $-\pi/2a \leqslant k \leqslant \pi/2a$. The region between these limits of k is called the first Brillouin zone. There are two solutions for v, since in Eq. (1-75) v depends on the positive or negative signs and these correspond to the optical and acoustical branches, respectively. The two roots are

$$v = 1/2\pi(2f/\mu)^{1/2} \qquad \text{(optical branch)},$$
$$v = 1/2\pi[2f(M + m)]^{1/2}ka \qquad \text{(acoustic branch)}. \qquad (1\text{-}76)$$

The positive roots of Eq. (1:75) are plotted in the positive half of the Brillouin zone as shown in Fig. 1-37a. It may be observed that the upper curve, which is called the optical branch, represents frequencies occurring in the optical spectral region (infrared or Raman). The low curve passes through $v = 0$ and is termed the acoustical branch (so-called because the frequencies occur in the sonic or ultrasonic region). Various wave motions are associated with both the optical and acoustical branches illustrated in Fig. 1-37b. Figure 1-37b-1 illustrates the wave motion of the optical branch at $K = 0$ and at point Q in Fig. 1-37a. Here the two atoms vibrate rigidly against each other. Figure 1-37b-2 shows the wave motion at point S on the optical branch. Figure 1-37b-3 demonstrates the wave motion at point R on the optical branch, where the light atoms are moving back and forth against each other

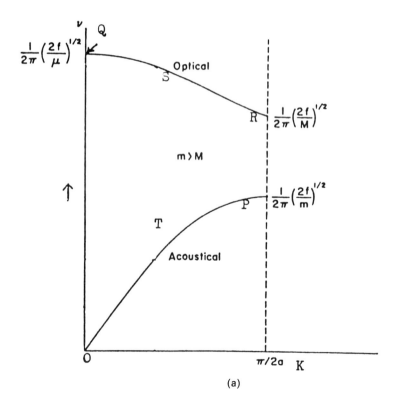

(a)

with the heavy atoms being fixed. In Fig. 1-37b-4 at point O $(K = 0)$ of the acoustic branch, the wave motion involves a translation at the entire lattice. Figure 1-37b-5 shows the wave motion at point T on the acoustic branch. Figure 1-37b-6 shows the wave motion at point P of the acoustical branch, where only the heavy atoms vibrate back and forth against each other and the light atoms are stationary.

The optical spectral region consists of internal vibrations (discussed in Section 1.13) and lattice vibrations (external). The fundamental modes of

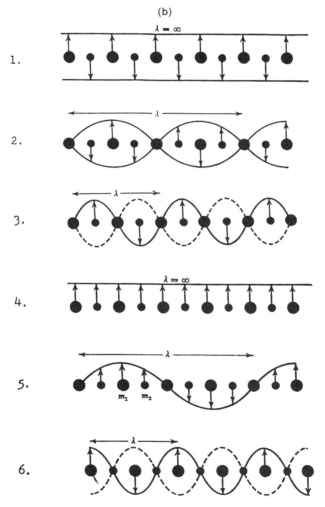

Figure 1-37 (a) Dispersion relation for the optical and acoustic branches in solids. (b) Wave motion in an infinite diatomic lattice. (Reproduced with permission from Ref. 41.)

vibration that show infrared and/or Raman activities are located in the center Brillouin zone where $k = 0$, and for a diatomic linear lattice, are the long-wave limit. The lattice (external) modes are weak in energy and are found at lower frequencies (far infrared region). These modes are further classified as translations and rotations (or librations), and occur in ionic or molecular crystals. Acoustical and optical modes are often termed "phonon" modes because they involve wave motions in a crystal lattice chain (as demonstrated in Fig. 1-37b) that are quantized in energy.

If the primitive cell contains σ molecules, each of which contains ρ atoms, then the number of acoustic modes is 3, and that of optical modes is $(3\sigma\rho - 3)$. The latter is classified into $(3\rho - 6)\sigma$ internal modes and $(6\sigma - 3)$ lattice modes. Analysis of these optical modes will be carried out in the following section.

Further discussion of solid vibrations of three-dimensional lattices is beyond the scope of this text. The reader may refer to Turrell (40) or other solid state texts (41).

1.18   Selection Rules for Solids (Factor Group)

By simply extending the methods used for the point group selection rules, one can obtain selection rules for molecules involving rotation–translation and reflection–translation. Two approaches are available. The older method is the Bhagavantum–Ventkatarayudu (BV) method (42), and necessitates the availability of the structure of the material being studied. The other method is that of Halford–Hornig (HH) (43–45) and considers only the local symmetry of a solid and the number of molecules in the unit cell and is simpler to work with. This method is also called the correlation method and depends on the proper selection of the site symmetry in the unit cell.

1.18.1   UNAMBIGUOUS CHOICE OF SITE SYMMETRY IN THE UNIT CELL

For cases where an unambiguous choice of site symmetry cannot be made, the use of Wyckoff's tables of crystallographic data (46) can prove helpful. Wyckoff's tables consist of the site correlations for some space groups. In instances where there is some doubt as to which site correlates (Appendix 5) with an axis of rotation, e.g., $[C_2(x), C_2(y), C_2(z)]$, or with a plane of symmetry $[\sigma(xy), \sigma(yz), \sigma(zx)]$, the proper site can be chosen. For example, consider orthorhombic $PuBr_3$, which has a $\mathbf{D}_{2h}^{17}$ ($Cmcm$) space group and a crystallographic unit cell with $Z = 4$. For a C-type lattice there are two repeat units in the cell, and therefore

$$Z' = \text{number of molecules in the cell}$$

$$= \frac{Z(\text{number of molecules in crystallographic cell})}{\text{repeat units in cell}} = \frac{4}{2} = 2.$$

From Appendix 4, we can observe that for D_{2h}^{17} (space group 63) the following site symmetries are possible: $2C_{2h}(2); C_{2v}(2); C_i(4); C_2(4); 2C_s(4); C_1(8)$. With the number of molecules in the unit cell equal to two, we must place two Pu^{3+} ions on a set of particular sites and six Br^- on other sets of sites. We observe that two site symmetries are available for the two Pu^{3+} ions—either C_{2h} or C_{2v}, each having two equivalent sites per set to place the metal ions. An unambiguous choice cannot be made with the data available. For the six Br^- ions, no site symmetry has six equivalent sites available. Thus, we must conclude that the six Br^- ions must be nonequivalent, and some are on one site and others on another site. At this point one must consult the Wyckoff tables (see Appendix 5) on published crystallographic data, and when this is done, we find the notation tabulated here.

Ion	Site Position
$2Pu^{3+}$	c
$2Br^-$	c
$4Br^-$	f

We can deduce the Wyckoff nomenclature of the site positions from the site symmetries by listing the site positions in alphabetical order, as shown in the next table.

Site in Appendix 3	Alphabetical Order	Wykoff's Alphabetical Ordering of Site Position	Ion Site
$2C_{2h}(2)$	$C_{2h}(2)$	a	
	$C_{2h}(2)$	b	
$C_{2v}(2)$	$C_{2v}(2)$	c	$2Pu^{3+}(c)$ $2Br^-(c)$
$C_i(4)$	$C_i(4)$	d	
$C_2(4)$	$C_2(4)$	e	
$2C_s(4)$	$C_s(4)$	f	$4Br^-(f)$
	$C_s(4)$	g	
$C_1(8)$	$C_1(8)$	h	

We can place the two Pu^{3+} ions on a c site (C_{2v}), two Br^- ions on a c site

(C_{2v}), and four Br^- must be on an f site (C_s). If we examine the correlation tables in Appendix 6, we observe that three correlations are possible for a D_{2h} space group with a site symmetry of C_{2v}. Similarly, three correlations are possible for the site symmetry C_s. Each correlation is based on a different rotational axis or reflection plane being involved. For example:

	$C_2(z)$	$C_2(y)$	$C_2(x)$	$\sigma(xy)$	$\sigma(zx)$	$\sigma(yz)$
D_{2h}	C_{2v}	C_{2v}	C_{2v}	C_s	C_s	C_s
Site Correlation		c	a, b, e	g		f

One must decide which site group to use. Appendix 5 can be used to determine the proper site. For each space group, the correlation to go with each site is included. Knowing the site symmetry as given by the Wyckoff tables, one can determine which site correlation to use. For this example, the c-site position for a C_{2v} site is correlated with C_{2v} involving a C_2 rotation around the y axis, and the f-site position for a C_s site is correlated with C_s, involving a reflection plane in the yz plane. In this manner an unambiguous choice of the site symmetry for the Pu^{3+} and Br^- ions is made. This method of obtaining the proper site symmetry is possible whenever the Wyckoff tables contain the molecule of interest. If the information is not available in the Wyckoff tables, then one must resort to a study of the actual crystallographic structure of the crystalline material, if it is available.

Although only two equivalent sites per set are available for C_{2v} symmetry, it is possible to place the two Pu^{3+} and two Br^- ions in a C_{2v} site, since the number of such sites is infinite. When the site symmetry is C_p, C_{pv}, or C_s and $p = 1, 2, 3$, etc., the number of sites is infinite. This point should be kept in mind when using Appendix 5. Figure 1-38 demonstrates the packing diagram of $PuBr_3$.

1.18.2   EXAMPLES OF THE HALFORD-HORNIG SITE GROUP METHOD

In this section, we shall attempt to illustrate the HH method using several examples.

To derive factor group (space group) selection rules, it is necessary to utilize x-ray data for a molecule from a literature source or from Wyckoff's (46) *Crystal Structures*. The factor group and site symmetries of the ion, molecule, or atoms must be available, as well as the number of molecules per unit cell reduced to a primitive unit cell.

(a)   LaCl₃ Solid

Let us consider the $LaCl_3$ crystal. The unit cell of $LaCl_3$ is seen in Fig. 1-39. The data available from Wyckoff indicate a space group #176, $C_{6h}^2-P6_3/m$.

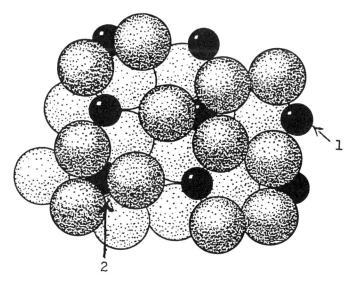

Figure 1-38 Packing diagram for PuBr$_3$. (Reproduced with permission from Ref. 47. Copyright © 1972 John Wiley & Sons, Inc.)

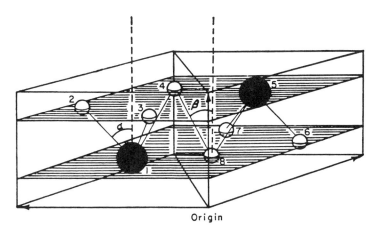

Origin

Figure 1-39 Unit cell for LaCl$_3$. The large circles represent lanthanum ions while the small circles represent chlorine ions.

The unit cell is 2(Z'). The two La atoms sit on a $\mathbf{C}_{3h}$ site, and the six chlorine atoms are on a $\mathbf{C}_s$ site (see Appendix 4). Since the Hermann–Mauguin nomenclature cites that the unit cell is primitive (Pb6$_3$/m) we need not reduce it. For the two La atoms there are six degrees of freedom $(3n, Z') = 3 \times 1 \times 2 = 6$. The six Cl atoms possess 18 degrees of freedom $(3n, Z') = 3 \times 3 \times 2 =$

Table 1-12 Correlation Table for La^{3+} in LaCl$_3$

Degrees of Freedom (DOF)		Site Group	Factor Group	Modes	
T	R	C_{3h}	C_{6h}^2	T	R
0	0	A' ⟨	A_g (R_z)	0	0
			B_u	0	0
4	0	$(T_x, T_y)E'$ ⟨	E_{2g}	1	0
			E_{1u} (T_x, T_y)	1	0
2	0	$(T_z)A''$ ⟨	A_u (T_z)	1	0
			B_g	1	0
0	0	E'' ⟨	E_{2u}	0	0
			E_{1g} (R_x, R_y)	0	0

18. Since all vibrational modes can be considered external modes, we need only correlate the site group to factor group. For the La atoms we can initiate a correlation chart using the correlation tables (Appendix 6) (47). See Table 1-12. For a derivation of the correlation tables, see Ref. 38.

The six degrees of freedom (DOF) for the La atoms are placed where the site group indicates translation vectors. For example, the E' species in the C_{3h} site has T_x, T_y (two vectors). Therefore, the two La atoms with four DOF are placed with E' species. Likewise, the remaining two DOF are placed with A'' species. We need not consider site rotations since there can be no site rotations for single atoms. Examining the correlation tables (see Appendix 6), we can assign the four E' species to one E_{2g} and one E_{1u} in the C_{6h}^2 factor group, since "doubly degenerate" counts for two. Likewise, we can assign the two A'' species in C_{3h} sites to A_u and B_g in the factor group C_{6h}^2. Thus, two La atoms have E_{2g}, E_{1u}, B_g, and A_u as the active modes totaling the six DOF.

Similarly, one can calculate the correlation chart for the six Cl atoms as illustrated in Table 1-13. For the six Cl atoms, activity is demonstrated for $2A_g$, $2B_u$, $2E_{2g}$, $2E_{1u}$, A_u, B_g, E_{2u} and E_{1g} totaling 18 DOF. Summarizing the modes for LaCl$_3$; we obtain

For 2 La: $\Gamma_{T'} = E_{2g} + E_{1u} + A_u + B_g$,

For 6 Cl: $\Gamma_{T'} = 2A_g + 2E_{2g} + E_{1g} + B_g + 2B_u + 2E_{1u} + E_{2u} + A_u$,

Table 1-13 Correlation Table For Cl^- in $LaCl_3$

Degrees of Freedom (DOF)		Site Group	Factor Group	Modes	
T	R	C_s	C_{6h}^2	T	R
12	0	$(T_x, T_y)A'$	$A_g (R_z)$	2	0
			B_u	2	0
			E_{2g}	2	0
			$E_{1u} (T_x, T_y)$	2	0
6	0	$(T_z)A''$	$A_u (T_z)$	1	0
			B_g	1	0
			E_{2u}	1	0
			$E_{1g} (R_x, R_y)$	1	0

where T' equals the total lattice or external modes of vibration. Including the three acoustic modes ($A_u + E_{1u}$), the total of 24 modes for $LaCl_3$ is distributed as follows:

$$\Gamma_{n_i} = 3E_{2g} + E_{1g} + 2B_g + 2A_g + 3E_{1u} + 2B_u + E_{2u} + 2A_u.$$

For a C_{6h}^2 factor group, the vibrations B_{1g}, B_u and E_{2u} are inactive (see C_{6h}^2 character table in Appendix 1). Subtracting off the inactive modes, the three acoustic, total modes are

$$3E_{2g} + E_{1g} + 2A_g + 2E_{1u} + A_u$$
$$\text{R} \quad \text{R} \quad \text{R} \quad \text{IR} \quad \text{IR}$$

The $3E_{2g}$, E_{1g} and $2A_g$ modes would be Raman-active, and the $2E_u + A_u$ modes would be IR-active. The summary for $LaCl_3$ is

$$\begin{array}{cccc} \text{R} & \text{IR} & \text{C} & \text{P} \\ 10 & 5 & 0 & 1 \end{array}$$

At $K = 0$, acoustic modes have zero frequency and are not observed in the Raman or IR experiments. They may be observed by performing slow neutron scattering experiments.

As shown in the following section, one can apply the same procedures for an organic molecule such as cyclopropane, C_3H_6. For such molecules one correlates for the molecular point group → site group → factor group to obtain the internal modes, and the site group → factor group for the external modes. This would be the procedure if one is dealing with covalent organic compounds with internal modes of vibrations as well as external modes.

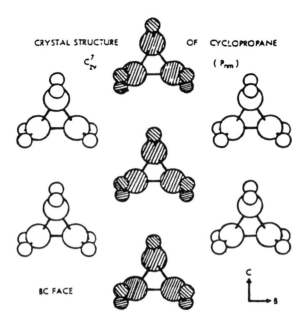

CRYSTAL STRUCTURE   OF   CYCLOPROPANE

C^7_{2v}

(P_{nm})

BC FACE

C

B

Figure 1-40 Proposed crystal structure of cyclopropane. The shaded molecules are not in the same plane as the unshaded ones and are inclined oppositely. (Reproduced with permission from Ref. 39.)

(b) Solid Cyclopropane, C_3H_6

Cyclopropane belongs to the C^7_{2v}–$Pmn2_1$ space group (No. 31) with $Z' = 2$. Figure 1-40 shows the structure of cyclopropane. The molecular point group is $\mathbf{D}_{3h}$. The site group $\mathbf{C}_s$ is a subgroup of both the $\mathbf{C}^7_{2v}$ and $\mathbf{D}_{3h}$ groups. The proper choice of $\mathbf{C}_s$ is obtained from Appendix 5 and is found to be $\mathbf{C}_s(\sigma_{yz})$. A total of $3n \cdot Z' = 3 \cdot 9 \cdot 2 = 54$ modes are expected, of which $(3n - 6)Z' = 42$ are internal modes. There are, therefore, $54 - 42 = 12$ external modes. For organic molecules such as cyclopropane, it is necessary to correlate the molecular point group to the site group and factor group to obtain the internal modes. For cyclopropane the correlation follows:

$$\text{Point Group} \rightarrow \text{Site Group} \rightarrow \text{Factor Group}.$$

$$\mathbf{D}_{3h} \qquad\qquad \mathbf{C}_s(\sigma_{yz}) \qquad\qquad \mathbf{C}^7_{2v}$$

The external modes are determined as for the $LaCl_3$ case by correlating the site group → factor group.

Internal modes for C_3H_6:

$$\Gamma_{n_i} = 12A_1 + 9A_2 + 9B_1 + 12B_2$$

External modes for C_3H_6:

$$\Gamma_{T+T'} = A_2 + B_1 + 2A_1 + 2B_2 \qquad \text{Total translations + acoustics}$$

$$\Gamma_T \quad = A_1 + B_1 + B_2 \qquad\qquad\quad \text{Translations}$$

$$\Gamma_{T'} \quad = A_2 + A_1 + B_2 \qquad\qquad\quad \text{Acoustics}$$

$$\Gamma_{R'} \quad = A_1 + 2B_1 + 2A_2 + B_2 \qquad\quad \text{Rotations}$$

Summary for C_3H_6:
A total of $3nZ' = 3{\cdot}9{\cdot}2 = 54$ modes are expected:

$$\Gamma_{n_i} = 12A_1 + 9A_2 + 9B_1 + 12B_2 \qquad \text{Internal modes}$$

$$\Gamma_T = \quad A_1 + \qquad\quad B_1 + \quad B_2 \qquad \text{Translations}$$

$$\Gamma_{T'} = \quad A_1 + \quad A_2 + \qquad\quad B_2 \qquad \text{Acoustics}$$

$$\Gamma_{R'} = \quad A_1 + \quad 2A_2 + \quad 2B_1 + \quad B_2 \qquad \text{Rotations}$$

$$\Gamma_n = 15A_1 + 12A_2 + 12B_1 + 15B_2 \qquad \text{Total}$$

$$\text{Activity} \quad \text{(IR,R)} \quad \text{(R)} \qquad \text{(IR,R)} \quad \text{(IR,R)}$$

Of these, 39 are infrared-active and 51 are Raman-active, and $A_1 + B_1 + B_2$ are acoustic modes and are not observed.

Correlation tables for cyclopropane internal and external modes are tabulated in Tables 1-14 and 1-15.

We have illustrated the methods to obtain solid state selection rules. It should be mentioned that tables for factor group or point group analyses have been prepared by Adams and Newton (48, 49) where one can read the

Table 1-14 Correlation Table for Cyclopropane Internal Modes

Molecular Point Group $\mathbf{D}_{3h}{}^a$	Site Group $C_3(\sigma_{yz})$	Factor Group C_{2v}^7
$3A_1'$		
$1A_1''$		
$(R_z)1A_2'$	$12A'$	$12A_1(T_z)$
$(T_z)2A_2''$		$9A_2$
$(T_x, T_y)4E'$	$9A''$	$9B_1(T_x)$
$(R_x, R_y)3E''$		$12B_2(T_y)$

^{a}Number of species for molecular point group determined from Appendix 2.

Table 1-15 Correlation Table for Cyclopropane External Modes

Degrees of Freedom (DOF)		Site Group $C_s(\sigma_{yz})$	Factor Group C_{2v}^7	Modes	
T	R			T	R
4	2	$(T_x, T_y)(R_z)A'$	$A_1(T_z)$	2	1
			$A_2(R_z)$	1	2
2	4	$(T_z)(R_x, R_y)A''$	$B_1(T_x)(R_y)$	1	2
			$B_2(T_y)(R_x)$	2	1

number and type of species allowed directly from the table. Although useful, the approach neglects the procedures as how to obtain results in the tables. For further examples of the correlation method, see Refs. 50–53, and the Correlation Theory Bibliography.

In general, a vibrational band in the free state splits into several bands as a result of solid intermolecular interactions in the unit cell. The number of split components can be predicted by the factor group analysis discussed earlier. Such splitting is termed factor group splitting or Davydov splitting, and the magnitude of this splitting is determined by the strength of the intermolecular interaction and the number of molecules in the unit cell interacting. In molecular crystals this splitting is in the range of 0–10 cm^{-1}.

1.19 Polarized Raman Spectra of Single Crystals

Porto illustrated the importance of polarized Raman spectra in obtaining the symmetry properties of normal vibrations, and their assignments. If one examines the character table for the $\mathbf{D}_{3d}$ point group (see Appendix 1), one may observe that the A_{1g} and E_g vibrations are Raman-active. The A_{1g} modes become Raman-active if any of the polarizability components α_{xx}, α_{yy}, α_{zz} change during irradiation. Let us consider a single crystal of calcite, which exists in a $\mathbf{D}_{3d}^6$ space group ($Z = 2$), and having $1A_{1g}$ and $4E_g$ vibrations (a total of nine) as determined by factor group analysis. If we irradiate this crystal in the y direction (see Fig. 1-41) and observe the Raman scattering in the x direction and polarized scattering in the z direction ($y(zz)x$ in the Damen–Porto nomenclature); we should observe only the A_{1g} species. We can understand how this occurs by simplifying Eq. (1-46) and obtaining

$$P_x = \alpha_{zz}E_z,$$

since $E_x = E_y = 0$ and $P_x = P_y = 0$. The polarizability component α_{zz} belongs

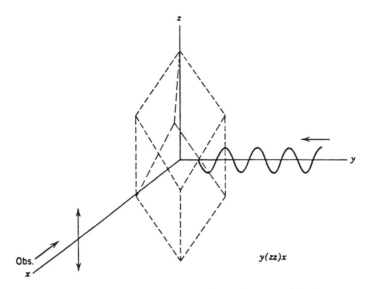

Figure 1-41 Schematic representation of experimental conditions used for the measurement of depolarization ratios of calcite crystal. (Reproduced with permission from Ref. 29. Copyright © 1986, John Wiley & Sons, Inc.)

to the A_{1g} species, and only A_{1g} modes should be observed under the conditions of $y(zz)x$. Figure 1-42a illustrates the Raman spectrum obtained at $y(zz)x$ radiation (53). Only the 1,088 cm^{-1} vibration appears, and this belongs to the A_{1g} species. Under the radiation conditions $z(xx)y$, both A_{1g} and E_g species should appear. Figure 1-42b shows four of five vibrations appearing (the 1,434 cm^{-1} band is not shown). The 1,088 cm^{-1} band is the A_{1g} mode, while the others are the E_g modes. The assignments for calcite are therefore made using the polarized Raman technique. These assignments may be confirmed by irradiating the calcite crystal under the $y(xy)x$ and $x(zx)y$ conditions (Fig. 1-42c,d), where only the E_g modes appear.

This example illustrates the usefulness of the polarized Raman technique. For a further discussion on the analysis of calcite, see Nakamoto (29), Nakagawa and Walter (54) and Mitra (39) for data on gypsum.

1.20   Normal Coordinate Analysis

As shown in Section 1.3, force constants of diatomic molecules can be calculated by using Eq. (1-20). In the case of polyatomic molecules, force constants can be calculated via normal coordinate analysis (NCA), which is much more involved than simple application of Eq. (1-20). Its complete

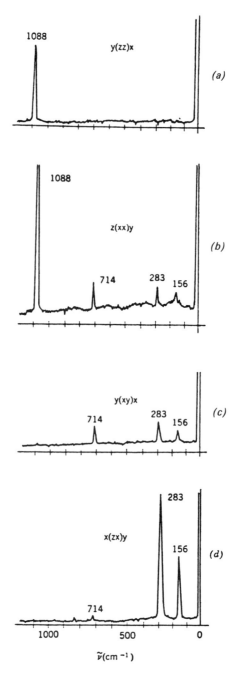

Figure 1-42   Polarized Raman spectra of calcite. (Reproduced with permission from Ref. 53.)

description requires complex and lengthy mathematical treatments that are beyond the scope of this book. Here, we give only the outline of NCA using the H_2O molecule as an example. For complete description of NCA, the reader should consult reference books cited at the end of this chapter (29, 55–57).

1.20.1 INTERNAL COORDINATES

The kinetic and potential energies of a polyatomic molecule can be expressed in terms of Cartesian coordinates (Δx, Δy, Δz) or internal coordinates such as increments of bond length (Δr) and bond angles ($\Delta \alpha$). In the former case, $3N$ coordinates are required for an N-atom molecule. Figure 1-43 shows the nine Cartesian coordinates of the H_2O molecule. Since the number of normal vibrations is $3(3 \times 3 - 6)$, this set of Cartesian coordinates includes six extra coordinates. On the other hand, only three coordinates (Δr_1, Δr_2 and $\Delta \alpha$) shown in Fig. 1-43 are necessary to express the energies in terms of internal coordinates. Furthermore, the latter has the advantage that the force constants obtained have clearer physical meaning than those obtained by using Cartesian coordinates since they represent force constants for particular bond stretching and angle bending. Thus, internal coordinates are commonly used for NCA. If it is necessary to use more than $3N - 6(5)$ internal coordinates, such a set includes extra (redundant) coordinates that can be eliminated during the process of calculation. Using the general formulas developed by Decius (58), one can calculate the types and numbers of internal coordinates for a given molecule, as follows.

The number of bond stretching coordinates is given by

$$n_r = b, \tag{1-77}$$

where b is the number of bonds disregarding type, which for H_2O is two. The number of angle bending coordinates is given by

$$n_\alpha = 4b - 3a + a_1, \tag{1-78}$$

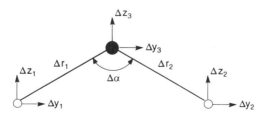

Figure 1-43 The nine Cartesian and three internal coordinates for H_2O. The three x coordinates are not shown since they are in the direction perpendicular to the paper plane.

where a is the number of atoms in the molecule, and a_1 is the number of bonds meeting at the central atom. For H_2O, n_α is equal to one.

1.20.2 SYMMETRY COORDINATES

If a molecule contains equivalent coordinates due to its symmetry properties, it is possible to simplify the calculation (*vide infra*) by using symmetry coordinates rather than internal coordinates. In the case of H_2O, they are

$$R_1 \sim (\Delta r_1 + \Delta r_2) \tag{1-79}$$

$$R_2 \sim \Delta\alpha, \tag{1-80}$$

$$R_3 \sim (\Delta r_1 - \Delta r_2). \tag{1-81}$$

R_1 and R_2 correspond to the two A_1, whereas R_3 corresponds to the B_2 vibration (see Section 2.13). Selection of symmetry coordinates can be facilitated by use of the method of Nielsen and Berryman (59). However, the preceding symmetry coordinates must be normalized so that

$$\sum_k (U_{jk})^2 = 1, \tag{1-82}$$

where U_{jk} is the coefficient of the kth internal coordinate in the jth symmetry coordinate. For R_1, $(U_{11})^2 + (U_{12})^2 = 1$. This gives $U_{11} = U_{12} = 1/\sqrt{2}$. Thus,

$$R_1 = (1/\sqrt{2})(\Delta r_1 + \Delta r_2).$$

Similarly

$$R_3 = (1/\sqrt{2})(\Delta r_1 - \Delta r_2), \tag{1-83}$$

$$R = \Delta\alpha.$$

Next, a set of symmetry coordinates must satisfy the orthogonality condition:

$$\sum_k (U_{jk})(U_{lk}) = 0. \tag{1-84}$$

For R_1 and R_2, $(1/\sqrt{2})(0)$ $+ (1/\sqrt{2})(0)$ $+ (0)(1) = 0$.

For R_1 and R_3, $(1/\sqrt{2})(1/\sqrt{2}) + (1/\sqrt{2})(-1/\sqrt{2}) + (0)(0) = 0$.

For R_2 and R_3, $(0)(1/\sqrt{2})$ $+ (0)(-1/\sqrt{2})$ $+ (1)(0) = 0$.

Thus, R_1, R_2 and R_3 shown in Eqs. (1-83) are orthogonal to each other.

It is necessary to determine if the preceding symmetry coordinates transform according to the character table of the point group C_{2v} (Appendix 1). By applying each symmetry operation, we find that R_1 and R_2 transform as A_1 species, while R_3 transforms as B_2 species. For example,

$$E(R_1) = 1, \quad C_2(R_1) = \quad 1, \quad \sigma_v(xz)(R_1) = \quad 1 \quad \text{and} \quad \sigma_v(yz)(R_1) = 1;$$

$$E(R_3) = 1, \quad C_2(R_3) = -1, \quad \sigma_v(xz)(R_3) = -1 \quad \text{and} \quad \sigma_v(yz)(R_3) = 1.$$

Using matrix notation, the relationship between the internal and symmetry coordinates is written as

$$\begin{bmatrix} R_1(A_1) \\ R_2(A_1) \\ R_3(B_2) \end{bmatrix} = \begin{bmatrix} \dfrac{1}{\sqrt{2}} & \dfrac{1}{\sqrt{2}} & 0 \\ 0 & 0 & 1 \\ \dfrac{1}{\sqrt{2}} & \dfrac{-1}{\sqrt{2}} & 0 \end{bmatrix} \begin{bmatrix} \Delta r_1 \\ \Delta r_2 \\ \Delta \alpha \end{bmatrix}, \tag{1-85}$$

where the first matrix on the right is called the U-matrix.

1.20.3   POTENTIAL ENERGY—F-MATRIX

The next step is to express the potential energy in terms of the F-matrix, which consists of a set of force constants. In the case of H_2O, it is written as

$$2V = f_{11}(\Delta r_1)^2 + f_{11}(\Delta r_2)^2 + f_{33}r^2(\Delta \alpha)^2 + 2f_{12}(\Delta r_1)(\Delta r_2)$$
$$+ 2f_{13}r(\Delta r_1)(\Delta \alpha) + 2f_{13}r(\Delta r_2)(\Delta \alpha) \tag{1-86}$$

Here, f_{11}, f_{12}, f_{13} and f_{33} are the stretching, stretching–stretching interaction, stretching–bending interaction, and bending force constants, respectively, and r (equilibrium distance) is mulfiplied to make f_{13} and f_{33} dimensionally similar to the others. Using matrix expression, Eq. (1-86) is written as

$$2V = \begin{bmatrix} \Delta r_1 & \Delta r_2 & \Delta \alpha \end{bmatrix} \begin{bmatrix} f_{11} & f_{12} & rf_{13} \\ f_{12} & f_{11} & rf_{13} \\ rf_{13} & rf_{13} & r^2 f_{33} \end{bmatrix} \begin{bmatrix} \Delta r_1 \\ \Delta r_2 \\ \Delta \alpha \end{bmatrix}. \tag{1-87}$$

Using matrix notation, the general form of Eq. (1-87) is written as

$$2V = \tilde{\mathbf{R}}\mathbf{F}\mathbf{R}, \tag{1-88}[18]$$

where $\mathbf{F}$ is the force constant matrix ($\mathbf{F}$-matrix), and $\mathbf{R}$ and its transpose $\tilde{\mathbf{R}}$ are column and row matrices, respectively, which consist of internal

[18] Hereafter, the bold-face letters indicate matrices.

coordinates. To take advantage of symmetry properties of the molecule, one must transform the $\mathbf{F}$-matrix into $\mathbf{F}_s$ via

$$\mathbf{F}_s = \mathbf{U}\mathbf{F}\tilde{\mathbf{U}}. \tag{1-89}$$

In the case of H_2O, $\mathbf{F}_s$ becomes:

$$\mathbf{F}_s = \begin{bmatrix} f_{11} + f_{12} & r\sqrt{2}f_{13} & 0 \\ r\sqrt{2}f_{13} & r^2 f_{33} & 0 \\ 0 & 0 & f_{11} - f_{12} \end{bmatrix} \tag{1-90}$$

Thus, the original 3×3 matrix is resolved into one 2×2 matrix (A_1 species) and one 1×1 matrix (B_2 species). In large molecules, such coordinate transformation greatly simplifies the calculation.

In the preceding, the potential energy was expressed in terms of the four force constants (stretching, stretching–stretching interaction, stretching–bending interaction, and bending). This type of potential field is called the *generalized valence force* (GVF) *field* and is most commonly used for normal coordinate calculations. In large molecules, however, the GVF field requires too many force constants, which are difficult to determine with limited experimental data. To overcome this difficulty, several other force fields have been developed, and some of these are given in reference books cited at the end of this chapter.

1.20.4 KINETIC ENERGY—G-MATRIX

The kinetic energy is not easily expressed in terms of internal (symmetry) coordinates. Wilson (55) has shown that

$$2T = \tilde{\dot{\mathbf{R}}}\mathbf{G}^{-1}\dot{\mathbf{R}}, \tag{1-91}$$

where $\dot{\mathbf{R}}$ is the time derivative of $\mathbf{R}$, $\tilde{\dot{\mathbf{R}}}$ is its transpose, and $\mathbf{G}^{-1}$ is the reciprocal of the G-matrix. G-matrix elements can be calculated by using Decius table (58). In the case of H_2O, it becomes

$$\mathbf{G} = \begin{bmatrix} \mu_3 + \mu_1 & \mu_3 \cos\alpha & -\dfrac{\mu_3}{r}\sin\alpha \\[2mm] \mu_3 \cos\alpha & \mu_3 + \mu_1 & -\dfrac{\mu_3}{r}\sin\alpha \\[2mm] -\dfrac{\mu_3}{r}\sin\alpha & -\dfrac{\mu_3}{r}\sin\alpha & \dfrac{2\mu_1}{r^2} + \dfrac{2\mu_3}{r^2}(1 - \cos\alpha) \end{bmatrix}. \tag{1-92}$$

Here, μ_1 and μ_3 are the reciprocal masses of the H and O atoms, respectively, and α is the bond angle. Again, it is possible to diagonize the **G**-matrix via coordinate transformation:

$$\mathbf{G_s} = \mathbf{U G \tilde{U}}$$

where $\mathbf{G_s}$ is the **G**-matrix that is expressed in terms of symmetry coordinates. In the case of H_2O, it becomes

$$\mathbf{G_s} = \begin{bmatrix} \mu_3(1 + \cos\alpha) + \mu_1 & -\dfrac{\sqrt{2}}{r}\mu_3\sin\alpha & 0 \\[2mm] -\dfrac{\sqrt{2}}{r}\mu_3\sin\alpha & \dfrac{2\mu_1}{r^2} + \dfrac{2\mu_3}{r^2}(1 - \cos\alpha) & 0 \\[2mm] 0 & 0 & \mu_3(1 - \cos\alpha) + \mu_1 \end{bmatrix} .\text{(1-93)}$$

1.20.5   SOLUTION OF SECULAR EQUATION

As stated in Section 1.6, normal vibrations are completely independent of each other. This means that the potential and kinetic energies in terms of normal coordinates (Q) must be written without cross terms. Namely,

$$2T = \mathbf{\tilde{Q}\dot{Q}},$$

$$2V = \mathbf{\tilde{Q}\Lambda Q},$$

where $\mathbf{\Lambda}$ is a diagonal matrix containing λ ($= 4\pi^2 c^2 \tilde{\nu}^2$) terms as diagonal elements.[19] On the other hand, the energy expressions in terms of internal (symmetry) coordinates contain cross terms such as $(\Delta r)(\Delta\alpha)$:

$$2T = \mathbf{\tilde{\dot{R}}G^{-1}\dot{R}}, \tag{1-91}$$

$$2V = \mathbf{\tilde{R}FR}. \tag{1-88}$$

To eliminate these cross terms, it is necessary to solve the secular equation of the form $|\mathbf{GF} - \mathbf{E}\lambda| = 0$ (55) where $\mathbf{E}$ is a unit matrix containing ones as the diagonal elements.[19] In the case of H_2O, this equation for the A_1 block becomes

$$|\mathbf{GF} - \mathbf{E}\lambda| = \begin{vmatrix} G_{11}F_{11} + G_{12}F_{21} - \lambda & G_{11}F_{12} + G_{12}F_{22} \\ G_{21}F_{11} + G_{22}F_{21} & G_{21}F_{12} + G_{22}F_{22} - \lambda \end{vmatrix} = 0, \text{(1-94)}$$

or

$$\lambda^2 - (G_{11}F_{11} + G_{22}F_{22} + 2G_{12}F_{12})\lambda + (G_{11}G_{22} - G_{12}^2)(F_{11}F_{22} - F_{12}^2) = 0. \tag{1-95}$$

[19] All the off-diagonal elements are zero in a diagonal as well as in a unit matrix.

For the B_2 vibration,

$$\lambda = F_{33}G_{33}.$$

1.20.6 CALCULATION OF FORCE CONSTANTS

In general, normal coordinate analysis is carried out on a molecule for which the atomic masses, bond distances and overall structures are known. Thus, the G-matrix can be readily calculated using known molecular parameters. Since the force constants are not known *a priori*, it is customary to assume a set of force constants that have been obtained for similar molecules, and to calculate vibrational frequencies by solving the secular equation $|\mathbf{GF} - \mathbf{E}\lambda| = 0$. Then, these force constants are refined until a set of calculated frequencies gives reasonably good agreement with those observed.

For the A_1 vibrations of H_2O, the G-matrix elements are calculated by using the following parameters:

$$\mu_1 = \mu_H = \frac{1}{1.008} = 0.99206,$$

$$\mu_3 = \mu_O = \frac{1}{15.995} = 0.06252,$$

$$r = 0.96(\text{Å}), \qquad \alpha = 105°,$$

$$\sin \alpha = \sin 105° = 0.96593,$$

$$\cos \alpha = \cos 105° = -0.25882.$$

If we assume a set of force constants,

$$f_{11} = 8.4280, \qquad f_{12} = -0.1050,$$

$$f_{13} = 0.2625, \qquad f_{33} = 0.7680,$$

we obtain a secular equation,

$$\lambda^2 - 10.22389\lambda + 13.86234 = 0.$$

The solution of this equation gives

$$\lambda_1 = 8.61475, \qquad \lambda_2 = 1.60914.$$

These values are converted into $\tilde{\nu}$ through the $\lambda = 4\pi^2 c^2 \tilde{\nu}^2$ relationship. The results are

$$\tilde{\nu}_1 = 3{,}824 \text{ cm}^{-1}, \qquad \tilde{\nu}_2 = 1{,}653 \text{ cm}^{-1}.$$

For the B_2 vibration, we obtain

$$\lambda_3 = G_{33}F_{33} = [\mu_1 + \mu_3(1 - \cos \alpha)](f_{11} - f_{12})$$

$$= 9.13681,$$

$$\tilde{\nu}_3 = 3,938 \text{ cm}^{-1}.$$

The calculated frequencies just obtained are in good agreement with the observed values, $\tilde{\nu}_1 = 3,825$ cm^{-1}, $\tilde{\nu}_2 = 1,654$ cm^{-1} and $\tilde{\nu}_3 = 3,936$ cm^{-1}, all corrected for anharmonicity (59). Thus, the set of force constants assumed initially is a good representation of the potential energy of the H_2O molecule. For large molecules, use of computer programs such as those developed by Schachtschneider (60) greatly facilitate the calculation.

1.21 Band Assignments and Isotope Shifts

Inspection of IR and Raman spectra of a large number of compounds shows that a common group exhibits its group vibrations in the same region regardless of the rest of the molecule. These "group frequencies" are known for a number of inorganic (29) and organic compounds (61,62). Figures 1-44 and 1-45 show group frequencies and corresponding normal modes for the

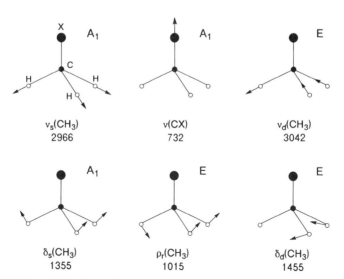

Figure 1-44   Normal vibrations of CH_3X-type molecule (frequencies are given for $X = Cl$): ν_s, symmetric stretching; ν_d, degenerate stretching; δ_s, symmetric bending; δ_d, degenerate bending; ρ_r, rocking.

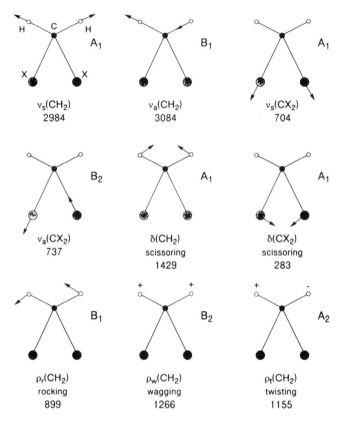

Figure 1-45   Normal vibrations of CH_2X_2-type molecule (frequencies are given for X = Cl): v_s, symmetric stretching; δ, symmetric bending (scissoring); ρ_r, rocking; ρ_w, wagging; ρ_t, twisting.

CH_3 and CH_2 groups, respectively. Notations such as v (stretching) and δ (bending) will be used throughout this book. Band assignments of other small molecules are tabulated by Shimanouchi (63). Using group frequency tables, it is possible to assign the observed spectra and to identify the atomic groups responsible for each group vibrations.

However, the concept of group frequencies is applicable only when the vibrations of a particular group are isolated from those of the rest of the molecule. If atoms of similar masses are connected by bonds of similar strength, the amplitudes of oscillation are similar for all atoms. For example, this situation occurs in a system like

In such a case, it is not possible to describe normal modes in terms of one local mode such as $v(C—C)$, $v(C—N)$ or $v(C—O)$. Instead, they are described as a mixing of these local modes ("vibrational coupling"). As will be shown in Chapter 4, Section 4.1.2, examples of such vibrational couplings are seen in metalloporphyrins and peptides.

In Section 1.6, we have shown the relationship between Cartesian and normal coordinates (Eq. (1-44)). Similar relationships exist between internal (symmetry) and normal coordinates:

$$R_1 = l_{11}Q_1 + l_{12}Q_2 + \cdots + l_{1N}Q_N,$$
$$R_2 = l_{21}Q_1 + l_{22}Q_2 + \cdots + l_{2N}Q_N,$$
$$\vdots \qquad \vdots$$
$$R_i = l_{i1}Q_1 + l_{i2}Q_2 + \cdots + l_{iN}Q_N.$$

Thus, the mixing ratio of individual coordinates in a given normal vibration (e.g., Q_1) is determined by the ratio

$$l_{11} : l_{21} : \cdots : l_{i1}.$$

If one of these values (e.g., l_{11}) is large relative to others, this normal vibration is assigned to the local mode, R_1. If both l_{11} and l_{21} are large relative to others, it is assigned to a coupled vibration between R_1 and R_2. The l_{iN} values are obtained once vibrational frequencies are calculated by the procedures described in the preceding section (29). Such calculations show that the $v_1(Q_1)$ at 3,825 cm^{-1} and $v_2(Q_2)$ at 1,653 cm^{-1} are almost pure $v(O—H)$ and $\delta(HOH)$, respectively.

Experimentally, band assignments are facilitated by the observation of isotope shifts. As shown in Table 1-3, the vibrational frequency of H_2 (4,160 cm^{-1}) is markedly downshifted (2,994 cm^{-1}) when H is replaced by D. The magnitude of this isotope shift is predicted by Eq. (1-40):

$$\frac{\tilde{v}_{H2}}{\tilde{v}_{D2}} = \sqrt{\frac{\mu_{D2}}{\mu_{H2}}} \cong \sqrt{2} = 1.414.$$

Obviously, this large shift originates in the mass effect; the mass of D is twice that of H. Such isotope shifts are seen in many isotopic pairs such as $^6Li/^7Li$, $^{10}B/^{11}B$, $^{12}C/^{13}C$, $^{14}N/^{15}N$, $^{16}O/^{18}O$, $^{32}S/^{34}S$ and $^{35}Cl/^{37}Cl$. As will be shown in Section 4.1.3, heavy metal isotopes ($^{58}Ni/^{62}Ni$ and $^{54}Fe/^{56}Fe$, etc.) are indispensable in assigning metal–ligand vibrations of coordination compounds (64).

References

1. F. P. Kerschbaum, *Z. Instrumentenk* **34**, 43 (1914).
2. B. Veskatesachar and L. Sibaiya, *Indian J. Phys.* **5**, 747 (1930).
3. J. H. Hibben, "The Raman Effect and Its Chemical Applicationm" Reinhold Publishing Corp., New York, 1939.
4 .F. H. Spedding and R. F. Stamm, *J. Chem. Phys.* **10**, 176 (1942).
5. D. H. Rank and J. S. McCartney, *J. Opt. Soc. Am.* **38**, 279 (1948).
6. H. L. Welsh, M. F. Crawford, T. R. Thomas, and G. R. Love, *Can. J. Phys.* **30**, 577 (1952).
7. N. S. Ham and A. Walsh, *Spectrochim. Acta* **12**, 88 (1958).
8. H. Stammreich, *Spectrochim. Acta* **8**, 41 (1956).
9. H. Stammreich, *Phys. Rev.* **78**, 79 (1950).
10. H. Stammreich, R. Forneris, and K. Sone, *J. Chem. Phys.* **23**, 1972 (1955).
11. H. Stammreich, R. Forneris, and Y. Tavares, *J. Chem. Phys.* **25**, 580, 1277 and 1278 (1956).
12. H. Stammreich, *Experientia* **6**, 224 (1950).
13. T. R. Gilson and P. J. Hendra, "Laser Raman Spectroscopy," pp. 1–266. Wiley-Interscience, London, 1970.
14. D. H. Rank and R. V. Wiegand, *J. Opt. Soc. Am.* **32**, 190 (1942).
15. R. F. Stamm and C. F. Salzman, *J. Opt. Soc. Am.* **43**, 126 (1953).
16. J. R. Ferraro, *in* "Raman Spectroscopy, Theory and Practice" (H. A. Szymanski, ed.), pp. 44–81. Plenum Press, New York, 1967.
17. J. R. Ferraro, R. Jarnutowski, and D. C. Lankin, *Spectroscopy* **7**, 30 (1992).
18. G. W. King, "Spectroscopy and Molecular Structure." Holt, Rinehart and Winston, New York, 1964.
19. H. Eyring, J. Walter, and G. E. Kimball, "Quantum Chemistry." John Wiley, 1944.
20. C. F. Shaw, III, *J. Chem. Educ.* **58**, 343 (1981).
21. T. G. Spiro and T. C. Strekas, *Proc. Nat. Acad. Sci.* **69**, 2622 (1972).
22. M. Zeldin, *J. Chem. Educ.* **43**, 17 (1956).
23. G. Herzberg, "Molecular Spectra and Molecular Structure. II. Infrared and Raman Spectra of Polyatomic Molecules." D. Van Nostrand, New York, 1945.
24. M. Hammermesh, "Group Theory and Its Applications to Physical Problems," p. 51. Addison Wesley, Reading, Massachusetts, 1972.
25. E. L. Muetterties, R. E. Merrifield, H. C. Miller, W. H. Knoth, and J. R. Downing, *J. Am. Chem. Soc.* **84**, 2506 (1962).
26. L. A. Paquette, R. L. Taransky, D. W. Balough, and G. Kentgen, *J. Am. Chem. Soc.* **105**, 5446 (1983).
27. H. W. Kroto, A. W. Allaf, and S. P. Balm, *Chem. Rev.* **91**, 1213 (1991).
28. F. A. Cotton, "Chemical Applications of Group Theory." John Wiley, New York, 1971.
29. K. Nakamoto, "Infrared and Raman Spectra of Inorganic and Coordination Compounds" (4th ed.). John Wiley, New York, 1986.
30. J. Tang and A. C. Albrecht, "Developments in the theories of vibrational Raman intensities," *in* "Raman Spectroscopy" (H. A. Szymanski, Ed.), Vol. 2. Plenum Press, New York, 1970.
31. A. C. Albrecht, *J. Chem. Phys.* **34**, 1476 (1961).
32. R. J. H. Clark and P. D. Mitchell, *J. Am. Chem. Soc.* **95**, 8300 (1973).
33. L. A. Nafie, P. Stein, and W. L. Peticolas, *Chem. Phys. Lett.* **12**, 131 (1971).
34. . Y. Hirakawa and M. Tsuboi, *Science* **188**, 359 (1975).
35. T. G. Spiro and X.-Y. Li, "Resonance Raman spectroscopy of metalloporphyrins," *in* "Biological Applications of Raman Spectroscopy" (T. G. Spiro, Ed.), Vol. 3. John Wiley, New York, 1988.
36. X.-Y. Li, R. S. Czernuszewicz, J. R. Kincaid, P. Stein, and T. G. Spiro, *J. Phys. Chem.* **94**, 47 (1990).

37. P. S. Wheatley, "The Determination of Molecular Structure" (2nd Ed.). Clarendon Press, Oxford, 1968.
38. G. Burns and A. M. Glazer, "Space Groups for Solid State Scientists," p. 81. Academic Press, 1978.
39. S. S. Mitra, "Optical Properties of Solids" (S. Nadelman and S. S. Mitra, Eds.), p. 333. Plenum Press, New York, 1969.
40. G. Turrell, "Infrared and Raman Spectra of Crystals." Academic Press, New York, 1972.
41. P. M. A. Sherwood, "Vibrational Spectroscopy of Solids," p. 15. Cambridge University Press, 1972.
42. S. Bhagavantum and T. Venkatarayudu, *Proc. Ind. Acad. Sci.* **A9**, 224 (1939); *A*13, 543 (1941).
43. R. S. Halford, *J. Chem. Phys.* **14**, 8 (1946).
44. D. F. Hornig, *J. Chem. Phys.* **16**, 1063 (1948).
45. H. Winston and R. S. Halford, *J. Chem. Phys.* **17**, 607 (1949).
46. R. L. C. Wyckoff, 'Crystal Structures," Vol. 2. John Wiley-Interscience, New York, 1964.
47. W. G. Fateley, F. R. Dollish, N. T. McDevitt, and F. F. Bentley, "Infrared and Raman Selection Rules for Molecular and Lattice Vibrations: The Correlation Method." John Wiley-Interscience, New York, 1972.
48. D. M. Adams and D. C. Newton, "Tables for factor group and point group analysis," Beckman-RHC Ltd., Surley House, U.K.; *J. Chem. Soc.*, 2822 (1970).
49. D. M. Adams, *Coord. Chem. Rev.* **10**, 183 (1973).
50. W. G. Fateley, N. T. McDevitt, and F. F. Bentley, *Applied Spectrosc.* **25**, 155 (1971).
51. R. Kopelman, *J. Chem. Phys.* **47**, 2631 (1967).
52. J. E. Bertie and J. W. Bell, *J. Chem. Phys.* **54**, 160 (1971).
53. S. P. Porto, J. A. Giordmaine, and T. C. Damen, *Phys. Rev.* **147**, 608 (1966).
54. I. Nakagawa and J. L. Walter, *J. Chem. Phys.* **51**, 1389 (1969).
55. E. B. Wilson, *J. Chem. Phys.* **7**, 1047 (1939); **9**, 76 (1941).
56. E. B. Wilson, J. C. Decius, and P. C. Cross, "Molecular Vibrations." McGraw-Hill, New York, 1955.
57. J. R. Ferraro and J. S. Ziomek, "Introductory Group Theory and Its Application to Molecular Structure" (2nd Ed.). Plenum Press, New York, 1975.
58. J. C. Decius, *J. Chem. Phys.* **16**, 1025 (1948).
59. J. R. Nielsen and L. H. Berryman, *J. Chem. Phys.* **17**, 659 (1949).
60. J. H. Schachtschneider, "Vibrational analysis of polyatomic molecules," Parts V and VI. Technical Reports 231-64 and 53-65. Shell Development Co., Emeryville, California, 1964 and 1965.
61. F. R. Dollish, W. G. Fateley, and F. F. Bentley, "Characteristic Raman Frequencies of Organic Compounds." John Wiley, New York, 1974.
62. D. Lin-Vien, N. B. Colthup, W. G. Fateley, and J. G. Grasselli, "The Handbook of Infrared and Raman Characteristic Frequencies of Organic Molecules." Academic Press, Boston, 1991.
63. T. Shimanouchi, "Tables of Molecular Vibrational Frequencies, Consolidated Volume," Nat. Stand. Ref. Data Ser., Nat. Bur. Stand., Vol. 39, June (1972). Also see *J. Phys. Chem. Ref. Data* **1**, 189 (1972); **2**, 121 and 225 (1973); and **3**, 269 (1974).
64. K. Nakamoto, *Angew. Chem. Int. Ed.* **11**, 666 (1972).

General References on Vibrational Spectroscopy

BASIC THEORY

1. G. Herzberg, "Molecular Spectra and Molecular Structure. Vol. I. Spectra of Diatomic Molecules." Van Nostrand, New York, 1951.

2. G. Herzberg, "Molecular Spectra and Molecular Structure. Vol. II. Infrared and Raman Spectra of Polyatomic Molecules." Van Nostrand, New York, 1945.
3. "Raman Spectroscopy: Theory and Practice" (H. A. Szymanski, Ed.), Vol. 1 (1967) and Vol. 2 (1970). Plenum Press, New York.
4. T. R. Gilson and P. J. Hendra, "Laser Raman Spectroscopy." John Wiley, New York, 1970.
5. D. Steele, "Theory of Vibrational Spectroscopy." Saunders, London, 1971.
6. J. A. Koningstein, "Introduction to the Theory of the Raman Effect." D. Reidel, Dortrecht, The Netherlands, 1973.
7. D. A. Long, "Raman Spectroscopy." McGraw-Hill, New York, 1977.
8. A. J. Sonnessa, "Introduction to Molecular Spectroscopy." Reinhold, New York, 1966.
9. C. N. Banwell, "Fundamentals of Molecular Spectroscopy" (2nd Ed.), McGraw-Hill, London, 1972.

NORMAL COORDINATE ANALYSIS

10. E. B. Wilson, J. C. Decius, and P. C. Cross, "Molecular Vibrations." McGraw-Hill, New York, 1955.
11. L. A. Woodward, "Introduction to the Theory of Molecular Vibrations and Vibrational Spectroscopy." Oxford University Press, London, 1972.

SYMMETRY AND GROUP THEORY

12. F. A. Cotton, "Chemical Applications of Group Theory" (3rd Ed.), John Wiley, New York, 1990.
13. J. R. Ferraro and J. S. Ziomek, "Introductory Group Theory and Its Application to Molecular Structure" (2nd Ed.). Plenum Press, New York, 1975.
14. J. M. Hollis, "Symmetry in Molecules." Chapman and Hall, London, 1972.
15. D. C. Harris and M. D. Bertolucci, "Symmetry and Spectroscopy." Oxford University Press, New York, 1978.

GENERAL APPLICATION

16. N. B. Colthup, L. H. Daly, and S. E. Wiberley, "Introduction to Infrared and Raman Spectroscopy" (3rd Ed.). Academic Press, Boston, 1990.
17. J. G. Grasselli, M. K. Snavely, and B. J. Bulkin, "Chemical Applications of Raman Spectroscopy." John Wiley, 1981.
18. "Infrared and Raman Spectroscopy" (E. G. Brame, Jr., and J. G. Grasselli, eds.), Vol. 1, Pts. A and B. Marcel Dekker, New York, 1976–77.
19. "Analytical Raman Spectroscopy" (J. G. Grasselli and B. J. Bulkin, eds.). John Wiley, New York, 1991.

INORGANIC, ORGANIC, AND ORGANOMETALLIC COMPOUNDS

20. K. Nakamoto, "Infrared and Raman Spectra of Inorganic and Coordination Compounds" (4th Ed.). John Wiley, New York, 1986.

21. J. R. Ferraro, "Low-frequency Vibrations of Inorganic and Coordination Compounds." Plenum Press, New York, 1971.
22. L. H. Jones, "Inorganic Vibrational Spectroscopy." Marcel Dekker, New York, 1971.
23. E. Maslowsky, Jr., "Vibrational Spectra of Organometallic Compounds." John Wiley, New York, 1976.
24. F. R. Dollish, W. G. Fateley, and F. F. Bentley, "Characteristic Raman Frequencies of Organic Compounds." John Wiley, New York, 1974.
25. D. Lin-Vien, N. B. Colthup, W. G. Fateley, and J. G. Grasselli, 'The Handbook of Infrared and Raman Characteristic Frequencies of Organic Molecules." Academic Press, Boston, 1991.

BIOLOGICAL COMPOUNDS

26. F. S. Parker, "Application of Infrared, Raman and Resonance Raman Spectroscopy in Biochemistry." Plenum Press, New York, 1983.
27. P. R. Carey, "Biochemical Application of Raman and Resonance Raman Spectroscopies." Academic Press, New York, 1982.
28. A. T. Tu, "Raman Spectroscopy in Biology." John Wiley, New York, 1982.
29. "Biological Applications of Raman Spectroscopy" (T. G. Spiro, ed.), Vols. 1, 2 and 3. John Wiley, New York, 1987–88.

CORRELATION THEORY

30. W. G. Fateley, F. R. Dollish, N. T. McDevitt and F. F. Bentley, "Infrared and Raman Selection Rules for Molecular and Lattice Vibrations: The Correlation Method." John Wiley-Interscience, New York, 1972.

ADVANCES SERIES

31. R. J. H. Clark and R. E. Hester, eds., "Advances in Infrared and Raman Spectroscopy," Vols. 1 to present. John Wiley, New York.
32. J. R. Durig, ed., "Vibrational Spectra and Structure," Vols. 1 to present. Elsevier, Amsterdam.
33. C. B. Moore, ed., "Chemical and Biochemical Applications of Lasers," Vols. 1 to present. Academic Press, New York.
34. "Spectroscopic Properties of Inorganic and Organometallic Compounds," Vols. 1 to present. The Chemical Society, London.
35. "Molecular Spectroscopy," Vols. 1 to present. The Chemical Society, London.

SPECIAL APPLICATIONS

36. A. Weber, ed., "Raman Spectroscopy of Gases and Liquids." Springer-Verlag, New York, 1979.
37. G. Turrell, "Infrared and Raman Spectra of Crystals." Academic Press, New York, 1972.
38. G. H. Atkinson, "Time-Resolved Vibrational Spectroscopy." Academic Press, New York, 1983.

39. D. B. Chase and J. F. Rabolt, eds; "Fourier Transform Spectroscopy from Concept to Experiment," Academic Press, Inc. Cambridge, MA. 1994.
40. J. R. Ferraro, "Vibrational Spectroscopy at High External Pressures. The Diamond Anvil Cell." Academic Press, New York, 1984.
41. C. Karr, ed., "Infrared and Raman Spectroscopy of Lunar and Terrestrial Minerals." Academic Press, New York, 1975.
42. W. B. Person and G. Zerbi, eds., "Vibrational Intensities in Infrared and Raman Spectroscopy." Elsevier, Amsterdam, 1982.
43. J. L. Koenig, "Spectroscopy of Polymers." American Chemical Society, Washington, D.C., 1992.
44. M. W. Urban. "Vibrational Spectroscopy of Molecules and Macromolecules on Surfaces," John Wiley, New York, 1993.

Chapter 2

Instrumentation and Experimental Techniques

The instrumentation for conventional Raman spectroscopy will be discussed in this chapter. Special techniques of Raman spectroscopy, including FT-Raman instrumentation, will be described in Chapter 3.

2.1 Major Components

Five major components make up the commercially available Raman spectrometer. These consist of the following:

(1) Excitation source, which is generally a continuous wave (CW) gas laser
(2) Sample illumination and scattered light collection system
(3) Sample holder
(4) Monochromator or spectrograph
(5) Detection system, consisting of a detector, an amplifier and an output device

Figure 2-1 illustrates a typical arrangement of these components (1). At present a wide variety of the systems shown in Fig. 2-1 exist. Some of the major additions in recent years consist of computers to control the instrument and to collect the data and the introduction of lasers capable of excitation in the UV and near-infrared regions.

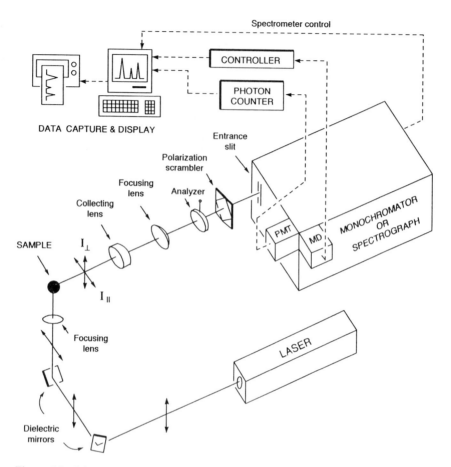

Figure 2-1 Schematic diagram of a typical dispersive Raman system. (Reproduced with permission from Ref. 1.)

2.2 Excitation Sources

Continuous wave (CW) lasers such as Ar^+ (351.1–514.5 nm), Kr^+ (337.4–676.4 nm) and He–Ne (632.8 nm) are commonly used for Raman spectroscopy. More recently, pulsed lasers such as Nd:YAG, diode, and excimer lasers have been used for time-resolved and UV resonance Raman spectroscopy (Chapter 3). Lasers are ideal excitation sources for Raman spectroscopy, mainly due to the following characteristics of the laser beam: (1) Single lines from large CW lasers can easily provide 1–2 W power, and pulsed lasers produce huge peak powers of the order of 10–100 MW (Tables 2-1 and 2-5). (2) Laser beams are highly monochromatic (band width, ~ 0.1 cm^{-1}, Ar$^+$ laser), and extraneous lines are much weaker. These can be eliminated easily by using filters or premonochromators. (3) Laser beams have small diameters (1–2 mm), which can be reduced to ~ 0.1 mm in diameter by using simple lens systems. Thus, all the radiant flux can be focused on a small sample, enabling fruitful studies of microliquids ($\sim \mu$L) and crystals (~ 1 mm^3). In the case of Raman microscopy (Section 3.7), sample areas as small as ~ 2 μm in diameter can be studied. (4) Laser beams are almost completely linearly polarized, and thus are ideal for measurements of depolarization ratios (Section 2.8). (5) It is possible to produce laser beams in a wide wavelength range by using dye lasers and other devices.

Reference 2 and Appendix 7 describe the principles and mechanisms involved in laser action.

2.2.1 CW GAS LASERS

The gas lasers operate mainly in the visible region of the electromagnetic spectrum. Figure 2-2 illustrates the basic components of a noble gas ion laser. A very high current discharge passes through Ar or Kr gas contained in the plasma tube. The outside of the tube is water jacketed to cool the tube. The discharge ionizes the gas and populates the excited state involved in the lasing. The ends of the tube are enclosed with Brewster windows that have an angle defined by tan $\theta = n$, where n is the index of refraction of the window material. For fused silica (quartz) in the visible region, θ is 55.6°. At Brewster's angle, the output beam is almost completely polarized in a fixed direction. The resonant cavity, which is defined by the semi-transparent output mirror and the high reflectance mirror, provides a mechanism for amplification of photons that are emitted parallel to the cavity axis; they are

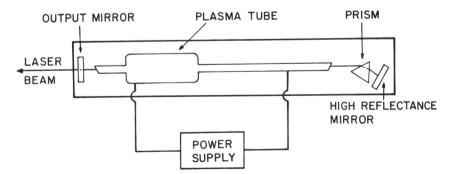

Figure 2-2 Schematic of a typical gas laser.

reflected by the two mirrors and interact with other excited ions. Stimulated emission (Appendix 7) produces photons of equal energy, phase and direction, and this process continues until an equilibrium between excitation and emission is reached. Both mirrors are coated to reflect the light of wavelength(s) of interest while transmitting all other light. The output mirror transmits a fraction of the energy that is stored in the cavity, and the transmitted radiation becomes the output beam of the laser. The prism is inserted between the two mirrors to force the laser to lase at a specific wavelength (single-line operation). The prism can be removed from the cavity, and the mirrors alone allow the laser to resonate simultaneously on a number of laser transitions (multiline operation). This provides the highest output powers and is used for exciting dye lasers.

Table 2-1 compiles the major wavelengths at which each of these lasers can be operated. Since the intensity of scattered light is governed by the v^4 rule (Section 1.15), the 488.0 nm line of an Ar^+ laser produces 2.8 times stronger scattering than the 632.8 nm line of a He–Ne laser if other conditions are equal. Tables 2-2 to 2-4 list the plasma lines of the gas lasers, which can be used for frequency calibration (Section 2.6). Figure 2-3 illustrates the spectrum of plasma lines from a Spectra-Physics Model 164-08 Ar^+ laser. Plasma lines may cause problems unless proper filtering is employed. Interference filters are commonly used to remove unwanted lines. However, they are somewhat inconvenient in that each exciting line requires its own filter and the filtering is usually not complete. Premonochromators (prism or grating type) are superior to interference filters, because they are applicable over a wide frequency range and can isolate the desired line from the rest with high efficiency. Plasma lines can be readily identified since they are quite sharp in contrast to Raman bands and disappear or shift upon changing the frequency of the exciting line.

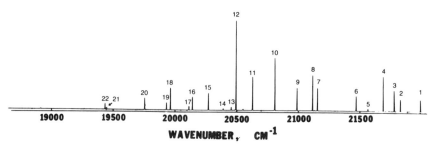

Figure 2-3 Spectrum of plasma lines emitted by a Spectra-Physics Model 164-09 argon-ion laser (band numbers refer to Table 2-2).

Table 2-1 Some Lasing Lines of Typical Gas Lasers in the Visible Region

Laser	Wavelength in Air (nm)	Wavenumber in Air (cm^{-1})	Typical Power (mW)
Ar-ion[a]	351.1–363.8 (UV)	28,481.9–27,487.6	100–400
	454.4 (blue–violet)	22,002.1	120
	457.9 (blue–violet)	21,838.8	350
	465.8 (blue)	21,468.4	200
	472.7 (blue)	21,155.1	300
	476.5 (blue)	20.986.4	750
	488.0 (blue)	20,491.8	1,500
	496.5 (blue–green)	20,141.0	700
	501.7 (green)	19,932.2	400
	514.5 (green)	19,436.3	2,000
	337.4 (UV)	29,638.4	200
	350.7 (UV)	28,514.4	1,200
	356.4 (UV)	28,058.4	600
Kr-ion[b]	406.7 (violet)	24,588.1	900
	413.1 (violet)	24,207.2	1,800
	415.4 (violet)	24,073.2	275
	468.0 (blue)	21,367.5	500
	476.2 (blue)	20.999.6	400
	482.5 (blue)	20,725.4	400
	520.8 (green–yellow)	19,201.2	700
	530.9 (green–yellow)	18,835.9	1,500
	568.2 (yellow)	17,599.4	1,100
	647.1 (red)	15,453.6	3,500
	676.4 (red)	14,784.2	900
	752.5 (near-IR)	13,289.0	1,200
	799.3 (near-IR)	12,510.9	300
He–Ne[c]	632.8 (red)	15,802.8	50
He–Cd[d]	441.6 (blue–violet)	22,644.9	40
	325.0 (UV)	30,769.2	10

[a] Power value for Spectra-Physics Model 2025.
[b] Power value for Coherent Innova 100-K3.
[c] Power value for Spectra-Physics Model 125A.
[d] Power value for Liconix Model 4240NB.
Note: The values in Tables 2-1 and 2-2 are expressed as wavelengths and wavenumbers in air. The difference between Δv (air) and Δv (vacuum) is usually less than 1 cm^{-1} and can be ignored in Raman spectroscopy. When molecular constants are calculated from absolute Raman frequencies, Δv (air) must be converted to Δv (vacuum).

2.2.2 DYE LASERS

Dye lasers are used to extend the wavelength range for Raman excitation. Basically, three types of dye lasers exist: those that are pumped by a CW gas laser, those pumped by a pulsed laser, and those pumped by a flash lamp. Relatively large volumes of organic dye solutions are required. Figure 2-4

Table 2-2 Some Plasma Lines from a Detuned Argon-Ion Laser

Line[d]	Wavelength in Air (Å)	Wavenumber in Air (cm^{-1})	Reference
1	4,545.05	22,001.96	a
2	4,579.35	21,837.16	a
3	4,589.93	21,786.82	b
4	4,609.56	21,694.04	a
5	?	?	—
6	4,657.89	21,468.95	a
7	4,726.86	21,155.69	a
8	4,735.93	21,115.18	b
9	4,764.89	20,986.84	b
10	4,806.07	20,807.02	b
11	4,847.90	20,627.49	b
12	4,879.86	20,492.39	c
13	4,889.03	20,453.96	c
14	4,904.75	20,388.40	c
15	4,933.21	20,270.78	c
16	4,965.07	20,140.70	c
17	4,972.16	20,111.98	c
18	5,009.33	19,962.75	c
19	5,017.16	19,931.59	c
20	5,062.04	19,754.88	c
21	5,141.79	19,448.48	c
22	5,145.32	19,435.14	c

[a] R. Beck, W. Englisch, and K. Gurs, "Tables of Laser Lines in Gases and Vapors, 2nd Ed., pp. 4–5. Springer-Verlag, New York, 1978.

[b] A. N. Zaidel, V. K. Prokofev, and S. M. Raiskii, "Tables of Spectrum Lines," pp. 299–301. Pergamon Press, New York, 1961.

[c] N. C. Craig and I. W. Levin, *Appl. Spectrosc.* **33**, 475 (1979).

[d] The numbers refer to those given in Fig. 2-3.

shows the wavelength range obtained by the Spectra-Physics Model 375 dye laser, pumped by argon-ion and krypton-ion lasers. Similar ranges may be obtainable from pulsed lasers and flash lamp pumped instruments. Band widths are typically 1 cm^{-1}, but can be reduced to 0.1 cm^{-1} by the use of fine tuning etalons. A recently developed solid-state laser utilizes a titanium–sapphire crystal: it is tunable in the 700 to 1,030 nm range and can provide 3 W output power when pumped by a 20 W Ar$^+$ laser.

2.2.3 OTHER LASERS

In FT-Raman spectroscopy (Section 3.8), the Nd:YAG laser line at 1,064 nm (CW operation) is used as the source. In the pulsed operation, its second (532 nm), third (355 nm) and fourth (266 nm) harmonics are available for

Table 2-3 Some Plasma Lines from a Krypton-Ion Laser[a]

Wavelength in Air (nm)	Relative Intensity	Wavelength in Air (nm)	Relative Intensity
522.95	600	587.09	750
530.87	2,300	599.22	1,000
533.24	2,000	624.02	700
544.63	900	657.01	1,000
546.82	1,100	721.31	600
552.29	1,050	728.98	900
556.86	1,000	740.70	800
557.03	550	752.45	600
563.50	1,400	758.74	550
567.28	570	760.15	600
568.19	3,500	784.07	520
569.03	2,000	785.48	500
575.30	1,000	799.32	700
577.14	1,700	805.95	600

[a] These lines (relative intensity > 500) were chosen from the table of C. Julien and C. Hirlimann, *J. Raman Spectrosc.* **9**, 62 (1980).

Table 2-4 Principal Plasma Lines from a He–Ne Laser[a]

Wavelength in Air (nm)	Relative Intensity	Wavelength in Air (nm)	Relative Intensity
638.299	53	706.519	31
640.108	> 100	717.394	4
640.975	31	724.517	5
644.472	30	728.135	11
650.653	50	748.887	0.5
659.895	41	753.577	0.4
667.815 ⎫	91	754.405	0.3
667.828 ⎭		777.730	5
671.704	36	794.318	0.1
692.947	19	813.641	0.2

[a] Reproduced with permission of Spex Industries.

time-resolved and UV resonance Raman spectroscopy. By combining these harmonics with dye lasers and hydrogen (H_2, D_2) Raman shifters,[1] it is now possible to cover the whole range between 185 and 880 nm without a gap (3, 4). Table 2-5 lists the characteristics of these and other lasers that are currently available.

[1] A long gas cell containing high-pressure H_2 gas (e.g., 0.75 m, 6.8 atm).

ARGON-ION & KRYPTON-ION PUMPED DYES (Spectra-Physics)

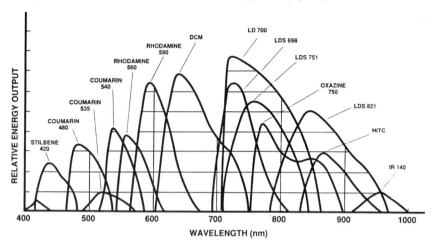

Figure 2-4 Output powers and wavelengths obtainable from a Spectra-Physics Model 375 dye laser pumped by an argon-ion and krypton-ion laser. (Reproduced with permission.)

Table 2-5 Other Laser lines

Laser	Wavelength	Type and Typical Power
Solid State Laser		
Ruby	694.3 nm	Pulsed, up to 100 MW
Nd:YAG[a]	1,064 nm (near IR)	CW/pulsed, up to 100 MW
Diode	3,500 to 380 cm^{-1} (IR)	CW/pulsed, up to 25 W
Gas Laser		
Nitrogen	337.1 nm (UV)	Pulsed, 100 to 1,000 kW
Carbon dioxide	9 to 11 μm (IR)	CW/pulsed, up to 10 MW
Excimer (XeCl)[a]	308 nm (UV)	Pulsed, up to 40 MW

[a]For more information, see Table 3-2.

2.3 Sample Illumination

Since the Raman scattering is inherently weak, the laser beam must be properly focused onto the sample, and the scattered radiation efficiently collected. The focusing of the laser onto the sample can be readily achieved because of the small diameter of the laser beam (~ 1 mm). Excitation and collection from the sample can be accomplished by using several optical configurations, such as the 90° and 180° scattering geometries illustrated in Figures 2-5a and 2-5b, respectively. As will be shown in Section 2.7, an optical configuration with an oblique angle (backscattering) is also common. Figure 2-6 shows a configuration that avoids the use of a collecting lens. It has advantages when the measurement is made in the UV region.

Collection optics consist of an achromatic lens system with a collecting lens and a focusing lens (Fig. 2-1). The light-gathering power of a lens is expressed in terms of F number, defined by

$$F = \frac{f}{D}, \tag{2-1}$$

where f is the focal length of the lens and D is its diameter. The smaller the F, the larger the light-gathering power. It is important to match this F value

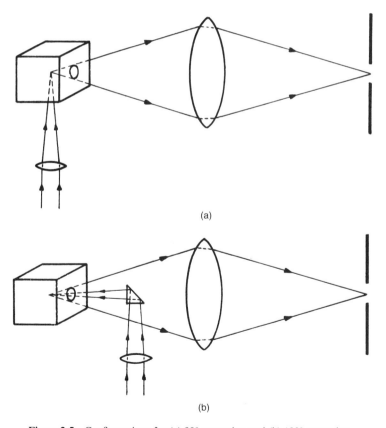

(a)

(b)

Figure 2-5 Configurations for (a) 90° scattering and (b) 180° scattering.

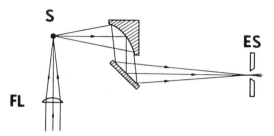

Figure 2-6 Collection optics with an elliptical collection mirror. FL, focusing lens; S, sample; ES, entrance slit.

to that of the monochromator (see Section 2.4). If not, full utilization of the collected light and the grating surface cannot be achieved.

Preliminary focusing of the incoming laser beam may be achieved by removing the sample holder and observing the focus position by interposing a thin layer of lens paper in the beam. The focal point may be readily observed by moving the paper along the beam direction. The position of the focus is adjusted by changing the position of the lens. The final adjustment of this lens can be made by observing the effect of its position on a Raman signal. Considerable time can be saved if the laser-focusing lens is achromatic, since the foregoing procedure does not need to be repeated when changing the laser wavelength. During this procedure, the laser power should always be kept at a minimum and *care should be taken to protect eyes during the procedure.* An excellent review article concerning the radiation hazards of lasers is available (5).

Image positioning of the irradiated sample on the entrance slit of the monochromator is most important and difficult. This image can be seen by holding a small section of a 3 × 5 file card near the slit. Again, caution should be exercised since *reflections at this point may seriously damage your eyes.* For 90° scattering, a bright image may sometimes be observed. However, this is not due to Raman scattering but to fluorescence or reflections of the laser beam from glass, quartz, or the surface of the sample. Since the image is rather weak, the room must be kept as dark as possible during this procedure.

2.4 The Monochromator

In a single monochromator, extraneous light that bounces around the spectrometer overlaps the weak Raman scattered light. This is caused mainly by undiffracted light scattered from the face of the grating. Such stray light could be reduced considerably by arranging two spectrometers in tandem so that the output of one was purified by the second. Thus, the construction of double monochromators began. Figure 2-7 is a schematic of a Spex 1403/4 double monochromator. A triple monochromator has even greater stray light rejection than a double monochromator and allows observation of Raman bands located very close to the Rayleigh line. Figure 2-8 depicts schematics of the Spex Triplemate, in which a double monochromator is coupled with a spectrograph. This arrangement is used for multichannel detection of Raman signals. Recently, high-performance Raman systems have been built by combining a single-stage spectrograph, charge-coupled device (CCD) detectors (see Section 2.5), and several kinds of filters that reject Rayleigh scattering efficiently (6).

The F value of a Raman spectrometer is also determined by Eq. (2-1). In this case, F is the focal length of the collimator mirror, and D is usually calculated by

$$\tfrac{1}{4}\pi D^2 = L^2, \tag{2-2}$$

where L is the side length of a square grating. To make F smaller, f should be smaller and D should be larger. However, the resolution decreases as f becomes small. To maintain a good resolution, then, D must be large, and this requires a large and expensive grating. For these reasons, most Raman

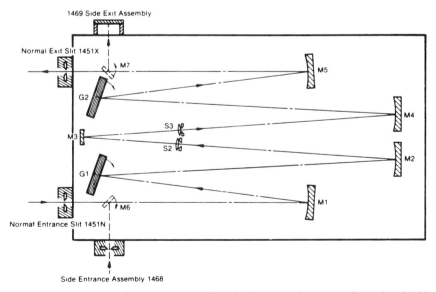

Figure 2-7 Schematic of Spex Model 1403/4 double monochromator. (Reproduced with permission.)

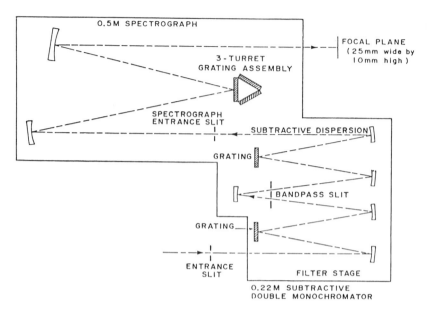

Figure 2-8   Schematic of Spex Model 1877 triple monochromator. The exit slit is removed for multichannel detection (see Section 2.5). (Reproduced with permission.)

spectrometers have F values between 5 and 10. For example, the Spex Model 1403/4 double monochromator ($f = 0.85$ m) has an F value of 7.8 with gratings 110×110 mm in size. Gratings determine the resolution of a spectrometer to a large extent. The more grooves per millimeter, the better the dispersion and the greater the resolution. The signal loss caused by improved resolution can be compensated by widening the slit width. Using an 1,800 grooves/mm grating, the Spex Model 1403 double monochromator can cover the range from 31,000 to 11,000 cm^{-1} with a cosecant drive system. However, this range would decrease with gratings of higher-density grooves (2,400 and 3,600 grooves/mm).

The slit width and monochromator advancing speed (increments between data points) are important in obtaining maximum resolution. The effect of changing the slit width (SW) is clearly seen in Fig. 2-9a, where the 459 cm^{-1} band of CCl_4 has been scanned at 1, 2, 3, and 4 cm^{-1} bandpasses (BP).[2] The signals were made comparable by lowering the laser power along the series, since the signal is proportional to $(P_0) \times (SW)^2$, where P_0 is the incident power. Rapid scanning can also cause distortion of the true spectral pattern, as is shown in Fig. 2-9b. A simple way to detect this problem is to rerecord

[2] $BP = D_L^{-1} \times (SW)$, where D_L^{-1} is the inverse linear dispersion of the grating and SW is the mechanical slit width.

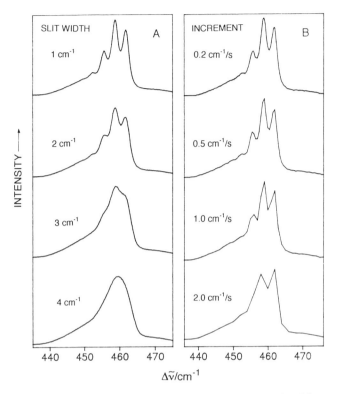

Figure 2-9 Raman spectra of CCl_4 (488.0 nm excitation) obtained under different conditions using a Spex Model 1403 double monochromator equipped with 1,800 grooves/mm gratings and a Hamamatsu R928 photomultiplier. (a) The effect of spectral band pass (0.2 cm^{-1} increments per data point). (b) The effect of size of increments between the data points (1 cm^{-1} slitwidth; accumulation time for all spectra is 1 second per data point.)

the spectrum at a slower scan speed with smaller increments and look for significant changes in the band patterns.

Mechanical backlash of the wavenumber reading is another problem that may be encountered when recording a spectrum on some instruments. Finally, the temperature of the monochromator should be kept constant since band position may vary by as much as 3 cm^{-1}. Although the monochromator is normally thermostated above room temperature, it is recommended that a laboratory thermometer be kept in the monochromator for easy monitoring.

2.5 Detection

Since Raman signals are inherently weak, the problems involved with detection and amplification are severe. Most of the very early work was done with photographic detection using long exposure times. Furthermore, the time to develop plates and examine them with a microphotometer rendered Raman spectroscopy unfit as a routine technique. This situation has changed considerably since the development of strong laser sources and sensitive detection techniques. Several detection techniques that are commonly used are described next.

2.5.1 Photon Counting

The Raman scattered light coming out from the exit slit of the monochromator is collected and focused on a photomultiplier (PM) tube, which converts photons into an electrical signal. The PM tube consists of a photocathode that emits electrons when photons strike it; a series of dynodes, each of which emits a number of secondary electrons when struck by an electron; and an anode that collects these electrons as an output signal. Figure 2-10 shows a schematic of a typical PM tube where the dynodes and the photocathode are interconnected by registers (not shown) that distribute a high voltage among them. This high voltage determines the total gain of the tube. The quantum efficiency of a primary electron being emitted from the photocathode is wavelength-dependent. Figure 2-11 shows the quantum efficiency and photocathode sensitivity of typical PM tubes as a function of wavelength.

For good sensitivity, proper care of the PM tube, its optics, and its housing are essential. The background noise is the primary limiting factor in PM tube performance. This is called the "dark current" and is primarily caused by the spurious emission of electrons from the photocathode and dynodes and by dielectric leakage across the PM tube base pins and resistor chain.

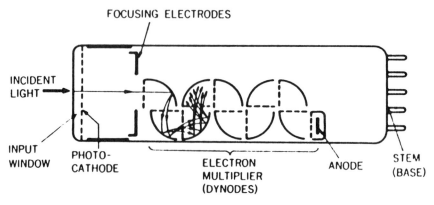

Figure 2-10 Schematic of a head-on type PM tube. (Reproduced with permission from Hamamatsu.)

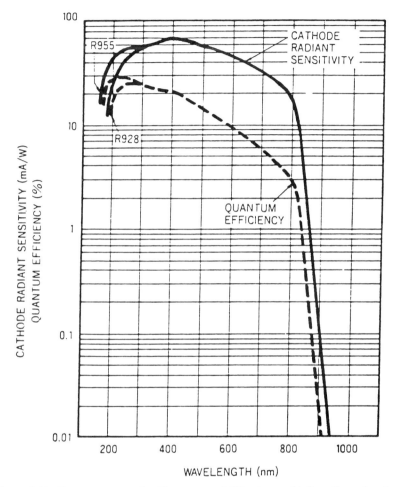

Figure 2-11 Response curve for Hamamatsu R928 photomultiplier. (Reproduced with permission.)

These spurious emissions can be reduced (a large portion is of thermonic origin) if the tube is thermoelectrically cooled via the Peltier effect so the weak Raman signal may be easily measured. Because of cooling, however, condensation of water in the lens system just in front of the tube and in the base may become a problem. Then, the detector system should be disassembled and dried under vacuum in a dark room.

The electron pulses (Raman signal) from the PM tube may be processed by the direct-current (dc) amplification or photon-counting (PC) method. In the former, the electron pulses from the PM tube are averaged over time, and the resulting dc current is directly amplified and measured by a picoammeter or electrometer. However, this method is no longer used since the PC method gives much better sensitivity.

In the PC method, the electron pulses caused by individual photons reaching the photocathode are measured. Photon counting has advantages in that a substantial portion of the dark signal is electronically discriminated from phonon pulses, allowing the ultimate sensitivity of the detector to be increased, typically, by a factor of 10 over a dc system. These systems have a disadvantage in that the maximum signal is limited to a photon count rate at which phonon events do not overlap. In practice, this "pulse pileup" limit is around 150×10^6 photon s^{-1}, which corresponds to about 800 nanoamps. Thus, this system is applicable to all but the strongest of Raman signals.

For the photon-counting mode an optimum resolution can be maintained by coordinating integration time with monochromator scan rate or interval. For each Raman apparatus, careful optimization of the PM tube's high voltage and pulse discriminator levels based on signal-to-noise ratios must be performed.

2.5.2 PHOTODIODE ARRAY DETECTION

In normal Raman measurements, the detection of Raman signals is made for each frequency, and the Raman spectrum is obtained through scanning the entire frequency range. This "single-channel" technique is time-consuming, and it is not suitable when the compound is unstable or short-lived. Simultaneous detection of Raman signals in the entire frequency range can be made by using multichannel detection. (See Refs. 7–8.)

Multichannel photon detectors consist of an array of small photosensitive devices that can convert an optical image into a charge pattern that can be read as a Raman spectrum. Several types of multichannel detectors are commercially available. The silicon vidicon and silicon diode array are based on the silicon diode detector. Vidicon is an image-sensing vacuum tube (TV tube), and its round target is 16 mm in diameter and contains more than 15,000 photodiodes per square millimeter. On the other hand, the silicon diode array is a linear array of diodes, each of which is 25 μm wide and 2.5 mm high. A typical photodiode array detector shown in Figure 2-12a consists of 1,024 such diodes in an array ~2.5 cm long. When the dispersed radiation strikes such a multichannel detector, a charge pattern develops that is related to the intensity of radiation along the focal plane, and this charge pattern is detected and ultimately converted into a spectrum. Use of an intensifier (multichannel plate) in front of a photodiode array detector increases the sensitivity of the order of $10^3 \sim 10^4$ (intensified silicon photodiode, ISPD). The sensitivity of a vidon can be increased by adding an intensifier that

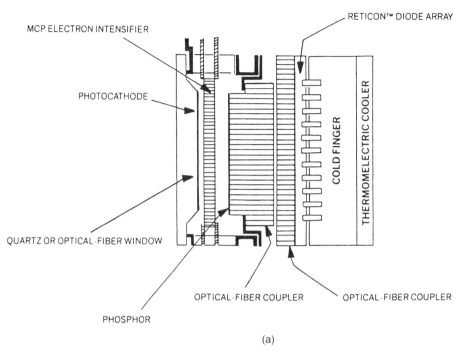

(a)

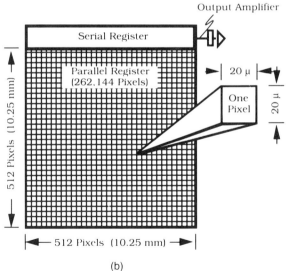

(b)

Figure 2-12 (a) Schematic of a photodiode array detector. (Reproduced with permission from Princeton Instruments.) (b) Schematic of a CCD detector (Photometrices PM512). (Reproduced with permission from Ref. 10.)

converts the photon image into a corresponding photoelectron image (intensified silicon intensified target, ISIT). ISPD is preferred to ISIT because of its longer exposure time (~ 100 vs. ~ 1 s).

2.5.3 CHARGE-COUPLED DEVICE DETECTION

In recent years the charge-coupled device (CCD) has been increasingly used in Raman spectroscopy (9, 10). The CCD is an optical-array detector based on silicon-metal-oxide semiconductor research. As shown in Figure 2-12b the CCD consists of two-dimensional arrays of $> 10^6$ pixels (silicon diode), each pixel ranging in size from 6 to 30 μm. The major advantages of the CCD relative to other multichannel detectors are the low readout noise, which makes optical intensification unnecessary, and high quantum efficiency and sensitivity in a wide wavelength range (120–1,000 nm). Thus, the CCD coupled with near IR laser excitation (dye laser, diode laser) can be used to measure Raman spectra of fluorescent compounds (see FT-Raman, Section 3.8). Complete utilization of a large number of detector elements in the CCD for spectroscopic work is under way. For example, time-resolved spectra can be obtained by utilizing a horizontal strip of the CCD and shifting its position up vertically as a function of time (11). However, disadvantages of using a CCD for Raman spectroscopy should also be noted (10).

2.5.4 OTHER DETECTION DEVICES

In FT-Raman spectroscopy, IR laser lines such as 1,064 nm line of the Nd:YAG laser are used for excitation. To detect such IR radiation, several detectors with the required sensitivity are available. Most of the FT-Raman instruments use indium gallium arsenide (IGA) detectors (12). For a detector 1 mm in size, a noise equivalent power (NEP) of 10^{-14} or less is usually attained when the detector is cooled. However, this detector suffers from a reduced long wavelength cutoff at 77 K. Thus, one can obtain spectra only up to 3,000 cm^{-1}. It does have an advantage in that the dark current is negligible. This allows direct coupling of dc to the first stage amplifier, and allows for both dc and ac components of the interferogram to be monitored.

Very high purity germanium detectors can be used in the Raman spectral range. The NEP of this type of detector approaches 10^{-15}. This detector system can operate to 3,500 cm^{-1}. Unfortunately, it is susceptible to cosmic radiation, and care must be taken to filter such radiation (13). An InSb detector has also been tested (14). The InSb element responds to 5 μm with a NEP of 10^{-12}, if used unfiltered. The major use of this detector will be with lasers that operate further into the infrared (12). Other detectors have been proposed, such as the silicon IGA with 256 elements, and platinum silicide with 1,024 elements (12). The quantum efficiency of these detectors drops off to 10%, but the spectral range extends to 5 μm. It is obvious that the search for improved and more efficient detectors continues. For further discussion on detectors for Raman use, see Ref. 14.

2.6 Instrument Calibration

2.6.1 FREQUENCY CALIBRATION[3]

Raman spectrometers are similar to any spectroscopic device. The wavenumber or wavelength readings on the instrument are not to be taken at face value. It is recommended that the instrument be calibrated each time prior to measurements. The time involved in calibration depends on the accuracy desired for a particular experiment. Scanning Raman spectrometers are typically calibrated for frequency by one of the following methods. For calibration of multichannel Raman spectrometers, see Ref. 15.

(a) Internal Standards

When accuracy of ~ 1 cm^{-1} is required, internal standards may be employed. These can be frequencies of solvent bands or the bands of added noninteracting solutes. Bands due to the compounds being measured are compared with the frequencies of the internal standard. However, care must be taken so that significant band shifts do not occur because of chemical interaction between the substance under study and the reference itself. In addition to its simplicity this method has a distinct advantage over the other methods in that the frequencies determined from the position of a band relative to the internal standard are essentially temperature-independent. It should be noted that the absolute readings from the monochromator may change from day to day as much as 2 to 3 cm^{-1} if the temperature control inside the monochromator is malfunctioning.

(b) Indene

If additional accuracy is desired (on the order of 0.5 cm^{-1}), then indene may be used (16). Indene has also been used as a frequency calibrant for IR spectrophotometers. Before use, it should be purified by vacuum distillation and stored in sealed capillary or a NMR tube. Figure 2-13 shows the Raman spectrum of indene, and Table 2-6 lists the frequencies that are recommended for use in calibration.

---

[3] As stated in Section 1.1, the term "frequency" is used interchangeably with "wavenumber."

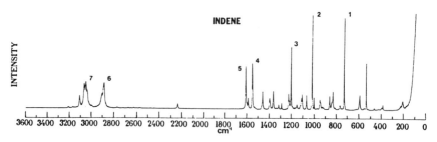

Figure 2-13 Raman spectrum of indene. The frequencies of the numbered bands are listed in Table 2-6.

Table 2-6 Recommended Frequencies for Calibration from the Spectrum of Indene

Band[a]	Wavenumber (cm^{-1})
1	730.4 ± 0.5
2	$1,018.3 \pm 0.5$
3	$1,205.6 \pm 0.5$
4	$1,552.7 \pm 0.5$
5	$1,610.2 \pm 0.5$
6	$2,892.2 \pm 1$
7	$3,054.7 \pm 1$

[a]The numbers refer to Fig. 2-13.

(c) Laser Plasma Lines

Table 2-2 lists the principal plasma lines of the argon-ion laser, some of which may be used for calibration (16–18). To observe these lines, the laser beam should be detuned and the scattered radiation be collected from a Kimax melting point tube (19). This method gives a calibration better than 1 cm^{-1} accuracy. Table 2-3 lists plasma lines of the krypton-ion laser (20).

(d) Neon Emission Lines

If a neon lamp is available, the Ne emission lines may be used to obtain high frequency calibration in a wide frequency range. Figure 2-14 shows the spectra taken with a Ne lamp. Tables 2-7 and 2-8 list the Ne frequencies that are useful for calibrating the Raman spectra obtained by excitation with a He–Ne and a Kr-ion laser, respectively. For complete listing of neon emission lines, see Strommen and Nakamoto (General References).

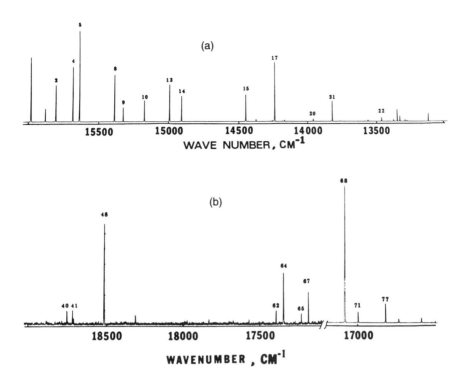

Figure 2-14   Neon lamp emission spectrum. Band numbers refer to Table 2-7 in (a) and to Table 2-8 in (b).

2.6.2   INTENSITY CALIBRATION

The intensity of a Raman signal is governed by a number of factors including incident laser power, frequency of the scattered radiation, efficiency of the grating and detector, absorptivity of the materials involved in the scattering, molar scattering power of the normal mode and the concentration of the sample. This situation is further complicated by the fact that many of these parameters are frequency-dependent as indicated in the following equation:

$$I = K(v) \times A(v) \times v^4 \times I_0 \times J(v) \times C, \qquad (2\text{-}3)$$

where I is the intensity of a Raman line, $K(v)$ describes the overall spectrometer response, $A(v)$ is the self-absorption of the medium, v the frequency of the scattered radiation, I_0 the intensity of the incident radiation, $J(v)$ a molar scattering parameter, and C the concentration of the sample. The v^4 term dominates the intensity if the remaining terms do not differ appreciably. Thus, a laser beam of a higher frequency is preferred to obtain a stronger

Table 2-7   Calibration Lines of a Ne Lamp[a,c]

Line[b]	Wavelength in Air (Å)	Wavenumber in Air (cm^{-1})
1	6,328.1646	15,802.3702
2	6,334.4279	15,786.7453
3	6,351.8618	15,743.4156
4	6,382.9914	15,666.6356
5	6,402.2460	15,619.5185
6	6,421.7108	15,572.1743
7	6,444.7118	15,516.5977
8	6,506.5279	15,369.1802
9	6,532.8824	15,307.179
10	6,598.9529	15,153.9193
11	6,652.0925	15,032.8637
12	6,666.8967	14,999.4824
13	6,678.2764	14,973.9235
14	6,717.0428	14,887.5038
15	6,929.4672	14,431.1239
16	7,024.0500	14,236.8007
17	7,032.4128	14,219.8706
18	7,051.2937	14,181.7948
19	7,059.1079	14,166.096
20	7,173.9380	13,939.3455
21	7,245.1665	13,802.3053
22	7,438.8981	13,442.8512

[a] Useful for Raman shift spectra (0 to 2,400 cm^{-1} region) obtained by 632.8 nm excitation of a He–Ne laser.
[b] Refer to Fig. 2-14a.
[c] Source: K. Burns, K. B. Adams, and J. Longwell, *J. Opt. Soc. Am.* **40**, 339 (1950)

Raman signal. Since most of these factors are not known, it is extremely difficult to determine the absolute intensity defined by Eq. (2-3). However, the relative amount of a sample in solution can be determined easily by measuring the relative intensity of a Raman band.

First, a working curve is prepared from the spectra of a series of solutions that contain varying amounts of the sample under consideration and a constant amount of a noninteracting standard. For aqueous solutions, ClO_4^- is ideal since it is chemically inert and has a strong Raman band at 928 cm^{-1}. Then a relative intensity is calculated for each solution by dividing the intensity of the strongest band of the sample by that of the internal standard. Then the sample relative intensity is given by

$$I_{rel} = \frac{K(v) \times A(v) \times v^4 \times J(v) \times C}{K(v') \times A(v') \times (v')^4 \times J(v') \times C'}, \tag{2-4}$$

Table 2-8 Calibration Lines of a Ne Lamp[a,c]

Line[b]	Wavelength in Air (Å)	Wavenumber in Air (cm^{-1})
40	5,330.7775	18,758.9897
41	5,341.0938	18,722.7567
42	5,343.2834	18,715.0844
43	5,349.2038	18,694.3709
44	5,360.0121	18,656.6743
45	5,372.3110	18,613.9633
46	5,374.9774	18,604.7294
47	5,383.2503	18,576.1379
48	5,400.5616	18,516.5928
49	5,412.6490	18,475.2420
50	5,418.5584	18,455.0931
51	5,433.6513	18,403.831
52	5,448.5091	18,353.6447
53	5,494.4158	18,200.2971
54	5,533.6788	18,071.1609
55	5,538.6510	18,054.9379
56	5,562.7662	17,976.6678
57	5,652.5664	17,691.0792
58	5,656.6588	17,678.2803
59	5,662.5489	17,659.8916
60	5,689.8163	17,575.2599
61	5,719.2248	17,484.8871
62	5,748.2985	17,396.4522
63	5,760.5885	17,359.3375
64	5,764.4188	17,347.8027
65	5,804.4496	17,228.1623
66	5,811.4066	17,207.538
67	5,820.1558	17,181.6706
68	5,852.4878	17,086.7507
69	5,868.4183	17,040.3667
70	5,872.8275	17,027.5732
71	5,881.8950	17,001.3235
72	5,902.9623	16,942.082
73	5,902.7835	16,941.16
74	5,906.4294	16,930.7027
75	5,913.6327	16,910.0797
76	5,918.9068	16,895.0118
77	5,944.8342	16,821.327

[a] Useful for Raman shift spectra (0 to 2,000 cm^{-1} region) obtained by 530.9 nm excitation of a Kr-ion laser.

[b] Refer to Fig. 2-14b.

[c] Source: K. Burns, K. B. Adams, and J. Longwell, *J. Opt. Soc. Am.* **40**, 339 (1950).

where all the terms involving v' indicate those of the internal standard, and C and C' denote the concentrations of the sample and the internal standard, respectively. Since the lead terms remain constant, the resulting equation is

$$I_{rel} = \text{constant} \times C. \tag{2-5}$$

Thus, a standard working curve can be obtained as in any other quantitative technique.[4]

The Raman intensity plotted against the exciting laser wavelength is called an *excitation profile* (EP). Excitation profiles such as that shown in Fig. 3-27 (Chapter 3) provide important information about electronic excited states as well as symmetry of molecular vibrations. The intensity of a Raman line is maximized if strict resonance conditions are met (Section 1.15). When constructing EPs, the frequency dependence of $J(v)$ is of interest. It is difficult, however, to determine the J dependence on v from intensity changes since K and A also vary with v.

The self-absorption term in Eq. (2-4) is expressed as (21)

$$\frac{A(v)}{A(v')} = \frac{\varepsilon(v')}{\varepsilon(v)} \cdot \frac{1 - \exp[-\varepsilon(v)C \times l]}{1 - \exp[-\varepsilon(v')C \times l]}, \tag{2-6}$$

where $\varepsilon(v)$ is the molar absorptivity at v, C the concentration of the absorber and l the path length. However, the correction of the spectrometer response is more involved. For detailed procedure, see Strommen and Nakamoto (General References).

2.7 Sampling Techniques

Marked differences are seen between infrared and Raman spectroscopy in sampling techniques. In infrared spectroscopy, sampling techniques for routine measurements are relatively simple. In contrast, Raman sampling techniques are intricate and versatile, and individual workers employ a variety of sampling techniques developed for their needs. Some of these techniques are described below.

2.7.1 COLORLESS COMPOUNDS

If the sample is colorless, the Raman spectrum can be easily obtained by sample irradiation with a laser beam whose wavelength is in the visible region (normal Raman scattering). The major advantages of Raman over infrared

---

[4]For more information on quantitative analysis, see Ref. 20 (General Reference section of Chapter 1).

spectroscopy are twofold: (1) The sample is contained or sealed in a glass (Pyrex) tube because Raman-scattered light in the visible region is not absorbed by glass. Thus, the Raman spectra of hygroscopic, corrosive, or oxygen-sensitive compounds can easily be measured by sealing them in glass tubes. However, some glass tubing gives rise to fluorescence or spikes if it is contaminated with rare earth salts. Use of a container or pelletization (see Section (c), Solids) can be avoided by using a Raman microprobe (Section 3.7) if the sample is a stable solid. (2) Raman spectra of aqueous solutions can be easily measured, since water is a very weak Raman scatterer, as contrasted to infrared spectroscopy where water is very strongly absorbing.

(a)   Gases

The sample gas is normally contained in glass tubing of diameter 1–2 cm and thickness ~ 1 mm. The gas can be sealed in a small capillary tube whose diameter is slightly larger than that of the laser beam (~ 1 mm). For weak Raman scatterers, an external resonating setup is used to increase their Raman intensity by multiple passing of the laser beam through the sample (Fig. 2-15a).

(b)   Liquids

Liquid samples may be sealed in ampoules, tubing, or capillaries, depending on the amount of the sample available (Fig. 2-15b). For micro quantities ($\sim 10^{-9}$ liter), capillaries as small as 0.5–0.1 mm bore and ~ 1 mm in length have been used. Use of a large cylindrical cell, such as that shown in Fig. 2-15c, reduces local heating and allow more accurate determination of depolarization ratios (Section 2.8). Strong spike noise may appear if the solution contains solid particles.

(c)   Solids

Depending upon the amount of the sample, powdered samples may be packed in ampoules or capillaries, and their Raman spectra measured in the same manner as has been described for liquid samples. The KBr pellet technique similar to that used in infrared spectroscopy is also useful. First, 200 mg of KBr powder is compressed to provide a support. Then, the ground sample diluted with KBr is spread evenly over the support and the die, and pressed to obtain a double-layer KBr pellet (Fig. 2-15d). If necessary, an internal standard (Section 2.6) may be mixed with the sample at this stage. This technique requires a small amount of the sample, and it reduces the decomposition of the sample by local heating. If large single crystals are obtained, it is possible to carry out detailed analysis of polarizability tensors via measurements of polarized spectra along the three principal axes of the crystal fixed on a goniometer head (Section 1-19).

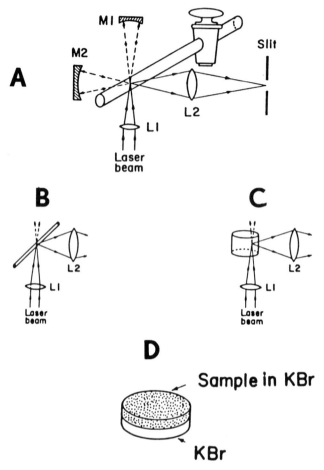

Figure 2-15 (a)Gas cell with external resonating mirrors. (b) Capillary cell for liquids. (c) Cylindrical cell for liquids. (d) KBr pellet for solid samples.

2.7.2 COLORED COMPOUNDS

For colored samples that absorb the energy of the laser beam, decomposition by local heating may occur. In this case, several procedures are available in addition to the simple reduction of laser power. These include: (1) changing the laser wavelength, (2) defocusing the laser beam on the sample, (3) diluting the sample concentration in a pellet or in solution to avoid absorption by the sample, (4) cooling the sample, (5) rotating the sample, and (6) rotating or oscillating the laser beam on a fixed sample. These techniques are extremely

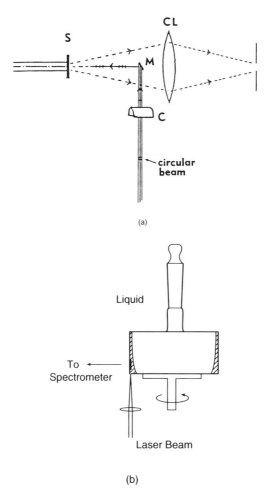

(a)

(b)

Figure 2-16 (a) Use of a cylindrical lens for line focusing. *C*, cylindrical lens; *M*, mirror; *S*, sample; *CL*, collection lens. (b) Rotating cylindrical cell.

important in recording resonance Raman spectra that are obtained by deliberately tuning the laser excitation into strong absorption bands of samples (Section 1.15). One may also insert a cylindrical lens between the laser and the sample (Fig. 2-16a). The beam is then focused on the sample over a length of 10 to 25 mm instead of a few microns. This "line focus" method can reduce power density per unit area by a factor of as much as 1/1000 (22).

(a)   *Rotating Sample Techniques* (23)

Liquids. Figure 2-16b illustrates a rotating cylindrical cell used for colored solution (24). It is symmetrically glued onto a circular piece of brass, which has a central rod that fits into a chuck connected to a motor rotating at 0 to 3,000 rpm. Although the cell has a volume of ~65 mL, only ~15 mL of liquid are necessary as the centrifugal force during rotation drives the liquid to the outer part of the cell. It is necessary that the laser beam be focused to this area (near the wall) to minimize the absorption of Raman-scattered light by the liquid itself (self-absorption). The laser beam must not fall on the glass wall because that will cause spurious lines originating within the glass to be observed. Since the laser beam must be aimed at the bottom of the cell, it is necessary to use a cell with minimum distortion at the corners. The following equation has been proposed to estimate the optimum concentration that is required to minimize self-absorption (25):

$$A_{opt} = 1/(2kr). \tag{2-7}$$

Here, A_{opt} is the optimum absorbance of the solution, k is $\log(e) = 2.303$, and r is the path length (cm) of the scattered radiation inside the cell. The equation was derived on the assumption that r is equal to the path length to the point of scattering within the cell. If $r = 0.5$ cm, $A_{opt} = 0.434$.

In another technique, the solution is circulated through a capillary cell by a peristaltic pump (26). If part of the circulating loop is immersed in a constant-temperature bath, it is possible to measure the spectrum over a wide temperature range. A more sophisticated technique (27) allows the measurements of redox potentials and electronic spectra as well as Raman spectra using a circulating cell.

Solids. (Resonance) Raman spectra of solid samples can be measured using pellets such as those previously described, and rotating them using the technique depicted in Fig. 2-17a. This rotating device is commercially available from Raman instrument manufacturers. Figure 2-17b illustrates a special die with a ring-shaped, grooved disk directly connected to the rotating shaft. The sample is packed in the groove and thus requires much smaller amounts (28). If the width of the groove is reduced to 1 mm and the sample is placed on top of a powdered KBr layer, it is possible to obtain the Raman spectra of samples as small as 1 mg (29).

Gases. Although there has not been much work done in the area of strongly absorbing vapors, the description of a rotating cell for absorbing vapors at high temperatures appeared in the literature (30).

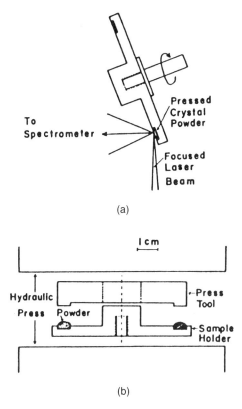

(a)

(b)

Figure 2-17 (a) Rotating device for solid samples, and (b) apparatus for making ring-shaped powder pellets. (Reproduced with permission from Ref. 24.)

(b)   Surface Scanning Techniques

In some cases, it is desirable to rotate the sample and keep it cool at the same time. A cell that allows the measurement of Raman spectra of rotating samples at liquid nitrogen temperatures has been designed (31). A rotating surface-scanning cell that can be used for obtaining Raman spectra of anaerobic solid samples cooled by a stream of cold nitrogen is available (32). A rotating vacuum cell for spectroscopic studies of surface phenomena has been designed (33). A universal rotating system for recording Raman spectra of rotating liquid or solid samples and difference spectra as well as for automatic scanning of the depolarization ratio has been constructed (34). Sometimes it is more convenient to oscillate the laser beam along one direction (35) or rotate it on the sample rather than to rotate the sample. Thus, Raman spectra have been measured by flicking the laser beam rapidly on the sample

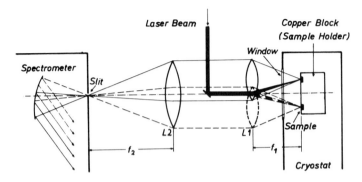

Figure 2-18 Schematic diagram of the rotating surface scanning technique. L_1, rotating lens that focuses the laser beam on the surface of the sample and that simultaneously collects the backscattered Raman light; L_2, focusing lens that focuses the Raman light on the spectrometer slit; f, the focal length of the lens. (Reproduced with permission from Ref. 37.)

at liquid nitrogen temperature (36). Using a setup such as that shown in Fig. 2-18. Raman spectra of samples cooled by a cryostat have been measured by using the rotating surface scanning technique (37).

2.7.3 SPECIAL CELLS

In addition to the cells already described, there are many other cells that are suited to special applications.

(a) Thermostated Cells

Biological molecules such as proteins and nucleic acids undergo conformational changes if the temperature is changed during the measurement. Several cells have been designed to maintain the desired temperature: 5 to 95°C (38) and room temperature to 100°C (39, 40).

(b) High-Temperature Cells

Raman spectroscopy has been utilized to study the structures of glasses, ceramic materials and molten salts at high temperatures. Figure 2-19 illustrates a high-temperature cell that was employed for Raman studies in the 295–483 K range (41). Two types of cells were designed for metal salts that melt at higher temperatures (up to 1,000°C); one is a windowless cell (42, 43), and the other is a graphite cell with diamond windows (44). A rotating cell for gaseous compounds ($\sim 300°C$) is also available (30).

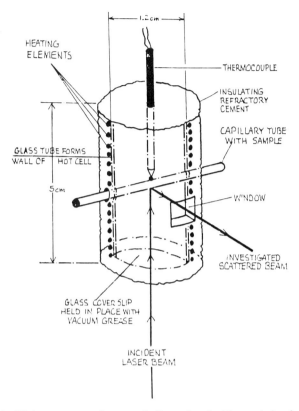

Figure 2-19   High-temperature Raman cell. (Reproduced with permission from Ref. 41.)

(c)   Low-Temperature Cells

A very simple device has been designed to obtain Raman spectra in a wide range of temperature (-150 to $\sim 200°C$) (45). A Dewar cell for low-temperature liquids (77 to 300 K) is available (46). Figure 2-20a shows a simple Dewar cell. Liquid nitrogen and organic slushes are used as coolants. In some cases. Raman scattering from the glass or quartz appears between 500 and 200 cm^{-1}. This problem can be circumvented by using a cell that allows the observation of Raman scattering directly from the surface of a frozen solution kept at 77 K (47). Figure 2-20b shows the configuration used for obtaining Raman spectra of liquids cooled by a cryo-cooler (48). The sample solution is contained in a mini-bulb (~ 0.4 mL), and any temperature between $-80°C$ and room temperature can be obtained by controlling the temperature of a cold tip.

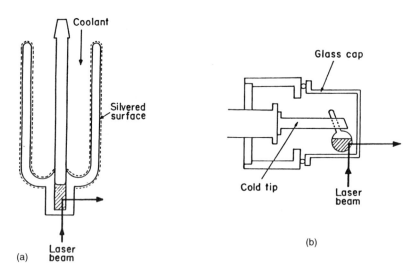

Figure 2-20   (a) Simple low-temperature Dewar cell. (b) Mini-bulb configuration.

(d)   UV Resonance Cell

In UV-resonance Raman studies, UV lines such as the fourth harmonic (266 nm) of the Nd:YAG laser are used for excitation. Under prolonged illumination by focused UV radiation, quartz and other UV-transparent materials tend to become fluorescent. To avoid the use of window materials and to minimize sample damage by strong UV light, several sampling techniques, such as the fluid jet stream technique (49) and the thin-film technique (50), have been developed.

(e)   Fiber Optics

Recently, McCreery and co-workers (51–53) developed a fiber optic Raman probe in which the laser beam is introduced and the Raman scattered light is collected via fiber optics (typically 200 μm in diameter). This technique requires no alignment of sample with input beam and no collection optics, and the sample can be placed a great distance from the spectrometer. For example, fiber optics have been employed to measure Raman scattering at low temperatures and high magnetic fields (54). For industrial applications of fiber optics in Raman spectroscopy, see Section 4.4.7.

The temperature of the sample may be estimated from the intensity ratios of the Stokes and anti-Stokes Raman lines via

$$\frac{I(\text{Stokes})}{I(\text{anti-Stokes})} = \frac{(\tilde{v}_0 - \tilde{v}_m)^4}{(\tilde{v}_0 + \tilde{v}_m)^4} \exp\{hc\tilde{v}_m/kT\}. \tag{2-8}$$

Here, $\tilde{v}_0$ is the wavenumber of the laser line, $\tilde{v}_m$ is the wavenumber of a band of the solvent or sample, h is Planck's constant, c is the velocity of light, k is Boltzmann's constant, and T is absolute temperature. A more convenient equation may be written as

$$T = \frac{-\tilde{v}_m \times 1.43879}{\left[\ln \dfrac{I(\text{anti-Stokes})}{I(\text{Stokes})} + 4\ln \dfrac{\tilde{v}_0 - \tilde{v}_m}{\tilde{v}_0 + \tilde{v}_m}\right]} \tag{2-9}$$

However, this equation is not applicable when anti-Stokes lines are very weak. It should be noted that these equations are applicable only for the spectra obtained under off-resonance conditions. For the spectra obtained under resonance conditions, see Ref. 55.

2.7.4 BACKSCATTERING GEOMETRY

The backscattering geometry (135° or 180°) has several advantages over the commonly used 90° scattering geometry. Figures 2-21a and b show a simple and versatile design that allows the rotation and cooling of the sample

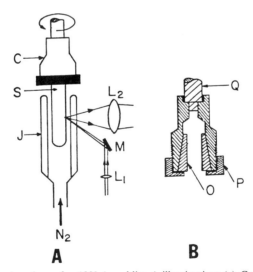

Figure 2-21 Sample spinner for 180° (or oblique) illumination: (a) Complete system with evacuated Pyrex jacket (J) surrounding the sample tube. S, cold (or warm) gas such as N_2 is passed through J to control the sample temperature. L_1, L_2, and M are lenses and a mirror; C is the sample chuck. (b) Details of the sample chuck. O, split nylon cone; P, knurled aluminum nut attached to aluminum body of chuck; Q, spinner shaft. (Reproduced with permission from Ref. 56.)

simultaneously (56). Also, sample replacement and laser beam focusing on the sample can be done easily and quickly. Certain advantages of backscattering include (1) ease of correction for self-absorption in highly colored solutions, (2) the ability to measure Raman scattering and UV-visible absorption simultaneously (57), (3) the ability to obtain single-crystal Raman spectra on small crystals with only one good face for each orientation, whereas two are required for 90° scattering, and (4) the ability to obtain low-temperature spectra of very small samples (47, 58). There are some disadvantages, such as the appearance of the Raman scattering background caused by various glasses and the danger of specular (mirror like) reflection from the glass container into the monochromator (180° backscattering). The intensity of the former may be minimized by using lenses with short focal depths, while both can be circumvented by scattering directly from the surface of a sample (solids, frozen solutions) (32).

2.7.5 FLUORESCENCE PROBLEMS

The sample (or impurities) may absorb the laser beam and emit it as fluorescence (Fig. 1.8). If this occurs, Raman spectra are obscured by a broad and strong fluorescence band. The intensity of the latter could be as much as 10^4 greater than the Raman signal. There are several methods of minimizing this problem. If impurities in the sample are causing fluorescence, the sample should be purified or irradiated by high-power laser beams for a prolonged time so that fluorescent impurities are bleached out.

If the sample itself is fluorescent, the first thing to do is to change the exciting wavelength. By shifting it to longer wavelength, fluorescence may be reduced greatly. FT-Raman (micro) spectroscopy (Section 3.8) is ideal since it employs an exciting line in the IR region (1,064 nm of Nd:YAG laser), where electronic transitions are rare. Addition of quenching agents such as potassium iodide (solution, solids) (59) or mercury halides (vapor) (60) is also effective in some cases. Repetitive scanning coupled with background subtraction is also found to be effective (61). It is also possible to discriminate fluorescence by using pulsed lasers since the lifetime of Raman scattering (10^{-12}–10^{-13} s) is much shorter than that of fluorescence (10^{-7}–10^{-9} s). Thus, an electronic gate may be employed to record the former preferentially (62).

A CCD Raman spectrometer (Section 2.5) coupled with a 10 mW He–Ne laser has been used to eliminate fluorescence because the long wavelength excitation by the He–Ne laser is not likely to cause fluorescent transitions (63). Because of its directional property, CARS (Section 3.1.4) is also effective in avoiding fluorescence interference.

2.8   Determination of Depolarization Ratios

As discussed in Section 1.9, depolarization ratios provide valuable information concerning the symmetry properties of Raman-active vibrations and are, therefore, indispensable for making band assignments. Figure 1-18 (and also Fig. 2-41) shows an experimental configuration for depolarization measurements in 90° scattering geometry. In this case, the polarizer is not used because the incident laser beam is almost completely polarized in the z direction. If a premonochromator is placed in front of the laser, a polarizer must be inserted to ensure complete polarization. The scrambler (crystal quartz wedge) must always be placed after the analyzer since the monochromator gratings show different efficiencies for $\perp$ and $\parallel$ polarized light. For information on precise measurements of depolarization ratios, see Refs. 64–67.

The depolarization ratio (ρ_p) is the ratio of the intensity obtained with perpendicular ($\perp$) and parallel ($\parallel$) polarization:

$$\rho_p = \frac{I_\perp}{I_\parallel}. \tag{2-10}$$

In the case of normal Raman scattering, its value ranges $0 < \rho_p < \frac{3}{4}$ for totally symmetric vibrations (*polarized*) and $\frac{3}{4}$ for non-totally symmetric vibrations (*depolarized*). However, some bands exhibit *inverse polarization* (ip) ($\frac{3}{4} < \rho_p < \infty$) under resonance conditions (Section 1.15). Figure 2-22 shows the Raman spectra of CCl_4 obtained with 90° scattering geometry. In this case, the ρ_p values obtained were 0.02 for the totally symmetric (459 cm^{-1}) and 0.75 for the non-totally symmetric modes (314 and 218 cm^{-1}). For ρ_p values in other scattering geometry, see Ref. 68.

Although polarization data are normally obtained for liquids and single crystals, it is possible to measure depolarization ratios of Raman lines from solids by suspending them in a material with similar index of refraction (69). The use of suspensions can be circumvented by adding carbon black or CuO (70). The function of dark (black) additives appears to be related to a reduction of the penetration depth of the laser beam, causing an attenuation of reflected or refractive radiation, which is scrambled relative to polarization.

2.9   Raman Difference Spectroscopy

Raman difference spectroscopy (71, 72) is a valuable technique for subtracting solvent bands from solution spectra and for determining small shifts of solute bands due to isotopic substitution or interaction with other molecules. Figure 2-23 illustrates a cylindrical rotating cell that is divided into two equal parts:

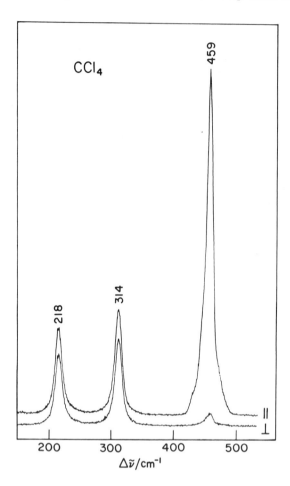

Figure 2-22 Raman spectrum of CCl_4 (500–200 cm^{-1}) in parallel and perpendicular polarization (488 nm excitation).

one containing the solution and the other containing only the solvent (73, 74). By rotation of the cell, the laser beam irradiates the solution and the solvent alternately. To record the difference spectrum, an electronic system containing a gated differential amplifier must be constructed.

If the frequency shift has occurred only via the solute–solvent interaction, the frequency shift is calculated by using the equation (75)

$$\Delta\tilde{\nu} = 0.385\Gamma(I_d/I_0). \qquad (2\text{-}11)$$

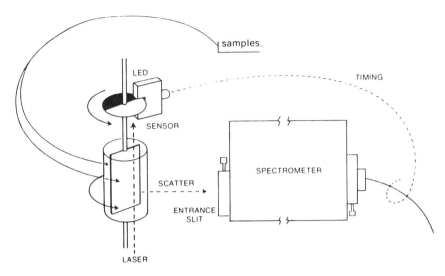

Figure 2-23 Divided rotating cell connected to a Spex 1475 difference/ratio generator. (Reproduced with permission.)

Here, Γ is the observed line width ($\Delta \tilde{\nu} \ll \Gamma$), I_d is the peak-to-valley intensity in the difference spectrum, and I_0 is the peak height of the Raman line. Although this equation is for Lorentzian-shaped bands, the results are approximately the same for Gaussian-shaped bands (the constant 0.385 becomes 0.350). In the case of carbon disulfide–benzene mixtures, the smallest shift observed was -0.06 cm^{-1}, and the associated error was ± 0.02 cm^{-1} (76). A convenient rotating system that can be used for (1) difference spectroscopy, (2) normal rotating sample techniques (solid and solution), and (3) automatic scanning of the depolarization ratios as a function of the wave number has been designed (34).

Two elaborate devices for difference spectroscopy have been constructed (77); in one, an NMR tube is divided into two equal sections (each containing solutions of 0.5 mL or less) and rotated by an NMR spinner. Then, difference spectra are measured using backscattering geometry. In the other, a tuning fork equipped with a pair of sample cups is attached to a cold finger of a cryostat, and the difference spectrum is obtained by oscillating two small frozen samples horizontally via magnetic devices. Figure 2-24 shows the Raman spectra of SO_4^{2-} ion in K_2SO_4 and Na_2SO_4 frozen solutions obtained by the former method. Using Eq. (2-11), $\Delta \tilde{\nu}$ is calculated to be -0.4 cm^{-1}. This shift is clearly due to the difference in interionic interactions between the two solutions.

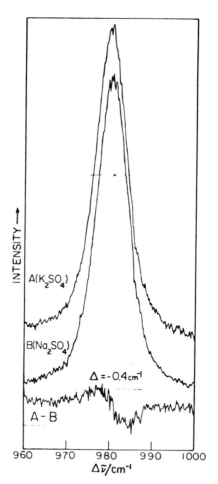

Figure 2-24 Top: The v_1 (A_1) band of SO_2^{-4} in (a) K_2SO_4 and in (b) Na_2SO_4 frozen solutions. Both spectra were obtained with 488 nm excitation (Ar-ion laser) and 5 cm^{-1} slit width while the spectrometer was advanced in 0.2 cm^{-1} per second increments. Bottom: Raman difference spectrum of K_2SO_4 vs. Na_2SO_4 ($A - B$). (Reproduced with permission from Ref. 77. Copyright © 1985 John Wiley & Sons, Ltd.)

General References

D. P. Strommen and K. Nakamoto, "Laboratory Raman Spectroscopy." John Wiley, New York, 1984.

D. J. Gardiner and P. R. Graves (eds.), "Practical Raman Spectroscopy." Springer-Verlag, Berlin, 1989.

References

1. R. S. Czernuszewicz, "Resonance Raman spectroscopy of metalloproteins with CW laser excitation," *in* "Methods in Molecular Biology: Physical Methods of Analysis" (C. Jones, B. Mulloy, and A. H. Thomas, eds.), Vol. 17. Humana Press, New Jersey, 1993.
2. J. Hecht, "The Laser Guidebook." McGraw-Hill, New York, 1986.
3. V. Wilke and W. Schmidt, *Appl. Phys.* **16**, 151 (1978).
4. V. Wilke and W. Schmidt, *Appl. Phys.* **18**, 177 (1979).
5. J. Marshall, "The radiation hazards of lasers and a guide to their control in the laboratory," *in* "Advances in Infrared and Raman spectroscopy" (R. J. H. Clark and R. E. Hester, eds.), Vol. 10, p. 145. John Wiley, New York, 1983.
6. M. Kim, H. Owen, and P. R. Carey, *Appl. Spectrosc.* **47**, 1780 (1993) and references therein.
7. Y. Talmi, *Appl. Spectrosc.* **36**, 1 (1982); *American Lab.*, March, 1978.
8. D. G. Jones, *Anal. Chem.* **57**, 1057A (1985); 1207A (1985).
9. P. M. Epperson, J. V. Sweedler, R. B. Bilhon, G. R. Sims, and M. B. Denton, *Anal. Chem.* **60**, 327A (1988).
10. J. E. Pemberton, R. L. Sobocinski, M. A. Bryant, and D. A. Carter, *Spectroscopy* **5**, 26 (1990).
11. R. B. Bilhorn, P. M. Epperson, J. V. Sweedler, and M. B. Denton, *Appl. Spectrosc.* **41**, 1125 (1987).
12. B. Chase, "Raman spectroscopy: From the visible to the infrared," *in* "Proceedings of XII International Conference on Raman Spectroscopy" (J. R. Durig and J. F. Sullivan, eds.), pp. 11–14. J. Wiley & Sons, New York, 1990.
13. B. Schrader, *in* "Practical Fourier Transform Infrared Spectroscopy—Industrial and Laboratory Chemical Analysis" (J. R. Ferraro and K. Krishnan, eds.), pp. 168–202. Academic Press, San Diego, 1990.
14. R. K. Chang, *in* "Spectroscopic and Diffraction Techniques in Interfacial Electrochemistry" (C. Gutierrez and C. A. Melendres, eds.), Kluwer Academic Publ., Dordrecht, The Netherlands, 1990.
15. H. Hamaguchi, *Appl. Spectrosc. Rev.* **24**, 137 (1988).
16. P. J. Hendra and E. J. Loader, *Chem. Ind.* 718 (1968).
17. R. Beck, W. Englisch, and K. Gurs, "Tables of Laser Lines in Gases and Vapors," 2nd Ed., pp. 4–5. Springer-Verlag, New York, 1978.
18. A. N. Zaidel, V. K. Prokofev, and S. M. Raiskii, "Tables of Spectrum Lines," pp. 299–301. Pergamon Press, New York, 1961.
19. M. C. Craig and I. W. Levin, *Appl. Spectrosc.* **33**, 475 (1979).
20. C. Julien and C. Hirlimann, *J. Raman Spectrosc.* **9**, 62 (1980).
21. F. Inagaki, M. Tasumi, and T. Miyazawa, *J. Mol. Spectrosc.* **50**, 286 (1974).
22. H. H. Eysel and S. Sunder, *Appl. Spectrosc.* **34**, 89 (1980).
23. W. Kiefer, "Recent Techniques in Raman Spectroscopy," *in* "Advances in Infrared and Raman Spectroscopy" (R. J. H. Clark and R. E. Hester, eds.), Vol. 3. John Wiley, New York, 1977.
24. W. Kiefer and H. J. Bernstein, *Appl. Spectrosc.* **25**, 500 (1971).
25. T. C. Strekas, D. H. Adams, A. Packer, and T. G. Spiro, *Appl. Spectrosc.* **28**, 324 (1974).
26. W. H. Woodruff and T. G. Spiro, *Appl. Spectrosc.* **28**, 74 (1974).
27. J. L. Anderson and J. R. Kincaid, *Appl. Spectrosc.* **32**, 356 (1978).
28. W. Kiefer and H. J. Bernstein, *Appl. Spectrosc.* **25**, 609 (1971).
29. G. J. Long, L. J. Basile, and J. R. Ferraro, *Appl. Spectrosc.* **28**, 73 (1974).
30. R. J. H. Clark, O. H. Ellestad, and P. D. Mitchell, *Appl. Spectrosc.* **28**, 575 (1974).
31. H. Homborg and W. Preetz, *Spectrochim. Acta* **32A**, 709 (1976).
32. R. S. Czernuszewicz, *Appl. Spectrosc.* **40**, 571 (1986).

33. F. R. Brown, L. E. Makovsky, and K. H. Rhee, *Appl. Spectrosc.* **31**, 563 (1977).
34. W. Kiefer, W. J. Schmid, and J. A. Topp, *Appl. Spectrosc.* **29**, 434 (1975).
35. J. A. Koningstein and B. F. Gachter, *J. Opt. Soc. Am.* **63**, 892 (1973).
36. R. J. H. Clark and P. C. Turtle, *Inorg. Chem.* **17**, 2526 (1978).
37. N. Zimmerer and W. Kiefer, *Appl. Spectrosc.* **28**, 279 (1974).
38. G. J. Thomas, Jr., and J. R. Barylski, *Appl. Spectrosc.* **24**, 463 (1970).
39. D. E. Irish, T. Jarv, and O. Puzic, *Appl. Spectrosc.* **31**, 47 (1977).
40. J. W. Fox and A. T. Tu, *Appl. Spectrosc.* **33**, 647 (1979).
41. D. Duval and R. A. Condrate, *Appl. Spectrosc.* **42**, 701 (1988).
42. A. S. Quist, *Appl. Spectrosc.* **25**, 80 (1971).
43. B. Gilbert, G. Momantov, and G. M. Begun, *Appl. Spectrosc.* **29**, 276 (1975).
44. S. K. Ratkje and E. Rytter, *J. Phys. Chem.* **78**, 1449 (1974).
45. F. A. Miller and B. M. Harney, *Appl. Spectrosc.* **24**, 291 (1970).
46. C. W. Brown, A. E. Hopkins, and F. P. Daly, *Appl. Spectrosc.* **28**, 194 (1974).
47. R. S. Czernuszewicz and M. K. Johnson, *Appl. Spectrosc.* **37**, 297 (1983).
48. K. Nakamoto, Y. Nonaka, T. Ishiguro, M. W. Urban, M. Suzuki, M. Kozuka, Y. Nishida, and S. Kida, *J. Am. Chem. Soc.* **104**, 3386 (1982).
49. G. A. Reider, K. P. Traar, and A. J. Schmidt, *Appl. Optics* **23**, 2856 (1984).
50. L. D. Ziegler and B. Hudson, *J. Chem. Phys.* **74** 982 (1981).
51. R. L. McCreery, P. J. Hendra, and M. Fleischmann, *Anal. Chem.* **55**, 146 (1983).
52. S. D. Schwab and R. L. McCreery, *Anal. Chem.* **56**, 2199 (1984).
53. S. D. Schwab and R. L. McCreery, *Appl. Spectrosc.* **41**, 126 (1987).
54. E. D. Isaacs and D. Heiman, *Rev. Sci. Instruments* **58**, 1672 (1987).
55. K. T. Schomaker, O. Bangcharoenpaurpong, and P. M. Champion, *J. Chem. Phys.* **80**, 4701 (1984).
56. D. F. Shriver and J. B. R. Dunn, *Appl. Spectrosc.* **28**, 319 (1974).
57. E. G. Rodgers and D. P. Strommen, *Appl. Spectrosc.* **35**, 215 (1981).
58. B. M. Sjöberg, T. M. Loehr, and J. Sanders-Loehr, *Biochem.* **21**, 96 (1982).
59. J. M. Friedman and R. M. Hochstrasser, *Chem. Phys. Lett.* **33**, 225 (1975).
60. V. A. Maroni and P. T. Cunningham, *Appl. Spectrosc.* **27**, 428 (1973).
61. T. M. Loehr, W. E. Keyes, and P. A. Pincus, *Anal. Biochem.* **96**, 456 (1979).
62. R. P. Van Duyne, D. L. Jeanmaire, and D. F. Shriver, *Anal. Chem.* **46**, 213 (1974).
63. J. E. Pemberton and R. L. Sobocinski, *J. Am. Chem. Soc.* **111**, 432 (1989).
64. C. D. Allemand, *Appl. Spectrosc.* **24**, 348 (1970).
65. F. G. Dijkman and J. H. Van der Maas, *Appl. Spectrosc.* **30**, 545 (1976).
66. G. Fini and P. Mirone, *Appl. Spectrosc.* **29**, 230 (1975).
67. J. R. Scherer and G. F. Bailey, *Appl. Spectrosc.* **24**, 259 (1970).
68. D. A. Long, "Raman Spectroscopy." McGraw-Hill, New York, 1977.
69. P. J. Hendra, "Raman instrumentation and sampling," *in* "Laboratory Methods in Infrared Spectroscopy" (R. G. Miller, ed.), p. 249. Heyden, London, 1971.
70. D. P. Strommen and K. Nakamoto, *Appl. Spectrosc.* **37**, 436 (1983).
71. L. Laane, "Raman difference spectroscopy," *in* "Vibrational Spectra and Structure" (J. R. Durig, ed.), Vol. 12, Elsevier, Amsterdam, 1983.
72. B. P. Asthana and W. Kiefer, "Vibrational line profiles and frequency shift studies by Raman spectroscopy," *in* "Vibrational Spectra and Structure" (J. R. Durig, ed.), Vol. 20. Elsevier, Amsterdam, 1992.
73. W. Kiefer, *Appl. Spectrosc.* **27**, 253 (1973).
74. W. Kiefer, "Recent techniques in Raman spectroscopy," *in* "Advances in Infrared and Raman Spectroscopy" (R. J. H. Clark and R. E. Hester, eds.), Vol. 3, p. 1. Heyden, London, 1977.
75. J. Laane, *J. Chem. Phys.* **75**, 2539 (1981); *Appl. Spectrosc.* **37**, 474 (1983).
76. J. Laane and W. Kiefer, *Appl. Spectrosc.* **35**, 267 (1981).
77. J. F. Eng, R. S. Czernuszewicz, and T. G. Spiro, *J. Raman Spectrosc.* **16**, 432 (1985).

Chapter 3

Special Techniques

Thus far, we have reviewed basic theories and experimental techniques of Raman spectroscopy. In this chapter we shall discuss the principles, experimental design and typical applications of Raman spectroscopy that require special treatments. These include nonlinear Raman spectroscopy, time-resolved Raman spectroscopy, matrix-isolation Raman spectroscopy, Raman microscopy, high-pressure Raman spectroscopy, FT-Raman spectroscopy, surface-enhanced Raman spectroscopy and Raman spectro-electrochemistry. The applications of Raman spectroscopy discussed in this chapter are brief in nature and are shown to illustrate the various techniques. Chapter 4 is devoted to a more extensive discussion of Raman applications to indicate the breadth and usefulness of the Raman technique.

3.1 Nonlinear Raman Spectroscopy

As stated in Chapter 1, the induced dipole moment (P) is expressed as

$$P = \alpha E + \tfrac{1}{2}\beta E^2 + \tfrac{1}{6}\gamma E^3 + \dots . \tag{3-1}$$

Here, E is the strength of the applied electric field (laser beam), α the polarizability and β and γ the first and second hyper-polarizabilities, respectively. In the case of conventional Raman spectroscopy with CW lasers (E, 10^4 V cm^{-1}), the contributions of the β and γ terms to P are insignificant

135

since $\alpha \gg \beta \gg \gamma$. Their contributions become significant, however, when the sample is irradiated with extremely strong laser pulses ($\sim 10^9$ V cm^{-1}) created by Q-switched ruby or Nd-YAG lasers (10–100 MW peak power). These giant pulses lead to novel spectroscopic phenomena such as the hyper-Raman effect, stimulated Raman effect, inverse Raman effect, coherent anti-Stokes Raman scattering (CARS), and photoacoustic Raman spectroscopy (PARS). Figure 3-1 shows transition schemes involved in each type of nonlinear Raman spectroscopy. (See Refs. 1–7.)

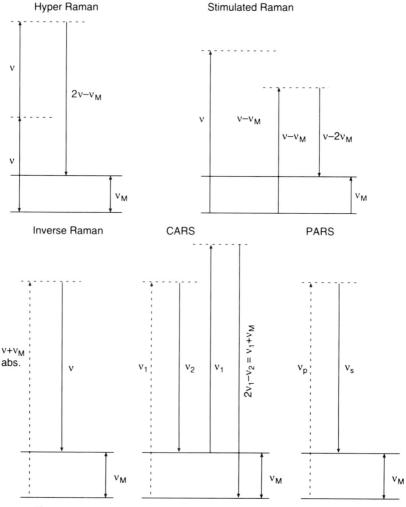

Figure 3-1   Transition schemes involved in nonlinear Raman spectroscopy.

3.1.1 HYPER-RAMAN EFFECT

When the sample is illuminated by a giant pulse of frequency v, the scattered radiation contains frequencies of $2v$ (hyper-Rayleigh scattering) and $2v \pm v_M$ (Stokes and anti-Stokes hyper-Raman scattering), where v_M is a frequency of a normal vibration of the molecule. Clearly, this is Raman scattering caused by two incident photons ($2v$) of the laser. Experimentally, this phenomenon is rather difficult to observe since only $\sim 10^{-12}$ of the radiation (v) is converted to $2v \pm v_M$ and since the intensity of the incident radiation can be increased only to a certain limit beyond which the stimulated Raman scattering (see below) becomes dominant. Hyper-Raman spectroscopy has several advantages over normal Raman spectroscopy because of a difference in selection rules. As already discussed in Section 1.7, a vibration is Raman-active if at least one of the components of the polarizability tensor changes during the vibration. Similarly, a vibration is hyper-Raman active if one of the components of the hyper-polarizability tensor changes during the vibration. Table 3-1 compares symmetry properties of these two components for the point group D_{6h} (benzene). It is seen that some vibrations that are not IR or Raman-active become hyper-Raman-active (B_{1u}, B_{2u}, and E_{2u}). It is also seen that some Raman-active vibrations are not hyper-Raman-active (E_{1g}, E_{2g}), while all IR-active vibrations are hyper-Raman-active (A_{2u}, E_{1u}). Similar effects are noted for other point groups. Thus, the hyper-Raman spectrum contains all the frequency information obtained from an IR spectrum.

Table 3-1   Selection Rules for IR, Raman and Hyper-Raman Spectra of Benzene (D_{6h}).

Symmetry Species	μ	α	β	Number of Normal Modes
A_{1g}		$\alpha_{xx} + \alpha_{yy}, \alpha_{zz}$		2
A_{2g}				1
B_{1g}				0
B_{2g}				2
E_{1g}		$(\alpha_{yz}, \alpha_{zx})$		1
E_{2g}		$(\alpha_{xx} - \alpha_{yy}, \alpha_{xy})$		4
A_{1u}				0
A_{2u}	z		$\beta_{yyz} + \beta_{zxx}, \beta_{zzz}$	1
B_{1u}			$\beta_{xxx} - 3\beta_{xyy}$	2
B_{2u}			$\beta_{yyy} - 3\beta_{xxy}$	2
E_{1u}	(x, y)		$(\beta_{xxx} + \beta_{xyy},$ $\beta_{yyy} + \beta_{xxy})$ $(\beta_{zzx}, \beta_{yzz})$	3
E_{2u}			$(\beta_{yyz} - \beta_{zxx}, \beta_{xyz}$	2

3.1.2   Stimulated Raman Effect

In normal Raman scattering, laser (v) irradiation on the sample results in "spontaneous" Raman scattering ($v - v_M$), which is very weak. If the electric field of the laser exceeds $\sim 10^9$ V cm^{-1}, the hyper-Raman scattering mentioned earlier is superseded by "stimulated" Raman scattering, which generates a strong coherent beam at Stokes frequency, ($v - v_M$) (2). Figure 3-2 shows a typical arrangement used for the observation of the stimulated Raman effect. Here, the giant laser radiation (v) is focused on the sample (benzene), and the scattered light is observed along the direction of the incident beam. If a color-sensitive film is placed in the direction perpendicular to the incident beam, one observes the concentric colored rings shown in Fig. 3-2. Interestingly, only one normal mode (v_M), which is the strongest in a normal Raman spectrum, is extremely strongly enhanced in the stimulated Raman effect. In benzene, it is the 992 cm^{-1} band (E_g). In fact, $\sim 50\%$ of the incident beam is converted into the first Stokes line, $v - v_M$ of this mode. Since this line is so intense, it acts as a source to excite the second Stokes line, $(v - v_M) - v_M = v - 2v_M$, and this line again acts as the source for the third Stokes line, and so forth. Thus, the concentric colored rings observed correspond to frequencies v, $v - v_M$, $v - 2v_M$, $v - 3v_M$, and so forth. It should be noted that the $2v_M$ thus observed is exactly two times v_M and not the first overtone of v_M (no anharmonicity correction). The high conversion efficiency of the stimulated Raman effect can be used to generate many laser lines of a variety of frequencies (for example, the H_2 Raman shifter; see Section 2.2).

3.1.3   Inverse Raman Effect

Suppose that a compound has a Raman-active vibration at v_M. If it is illuminated by a probe laser (v) simultaneously with a pump continuum covering the frequency range from v to $v + 3{,}500$ cm^{-1}, one observes an absorption at $v + v_M$ in the continuum together with emission at v. Clearly, the absorbed energy, $h(v + v_M)$, has been used for excitation (hv_M) and emission of the extra energy (hv). This upward transition is called the *inverse Raman effect* since the normal anti-Stokes transition occurs downward. Because the inverse Raman spectrum can be obtained in the lifetime of the pulse, it may be used for studies of shortlived species (Section 3.2). It should be noted, however, that the continuum pulse must also have the same lifetime as the giant pulse itself. Thus far, the inverse Raman effect has been observed only in a few compounds, because it is difficult to produce a continuum pulse at the desired frequency range.

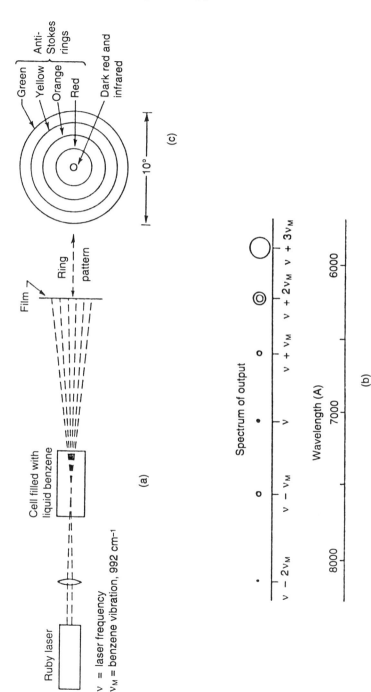

Figure 3-2 (a) Experimental setup for stimulated Raman spectroscopy, (b) a diagram showing the stimulated Raman spectrum of benzene, and (c) anti-Stokes rings of stimulated Raman spectrum of benzene. (Reproduced with permission from Ref. 1.)

3.1.4 COHERENT ANTI-STOKES RAMAN SPECTROSCOPY (CARS)

When the sample is irradiated by two high-energy laser beams with frequencies v_1 and v_2 ($v_1 > v_2$) in a collinear direction (Fig. 3-3), these two beams interact coherently to produce the strong scattered light of frequency $2v_1 - v_2$. If v_2 is tuned to a resonance condition such that $v_2 = v_1 - v_M$ where v_M is a frequency of a Raman-active mode of the sample, then a strong light of frequency $2v_1 - v_2 = 2v_1 - (v_1 - v_M) = v_1 + v_M$ is emitted (Fig. 3-1). This multi-photon process is called *coherent anti-Stokes Raman spectroscopy (CARS)* (3).

The advantages of CARS include the following: (1) Since the CARS light ($v_1 + v_M$) is coherent and emitted in one direction with a small solid angle, it can be detected easily and efficiently without a monochromator. Furthermore, fluorescence interference can be avoided because of this directional property. (2) The CARS frequency ($v_1 + v_M$) is higher than v_1 or v_2. Thus, it is on the anti-Stokes side of the pump frequency (v_1), whereas the fluorescence is on the Stokes side; hence, this condition also discriminates fluorescence. (3) Since CARS signals are very strong, gaseous compounds in very low concentrations can be detected. (4) Selection rules different from those of normal Raman spectroscopy are applicable to CARS. All Raman-active modes are CARS-active. In addition, many vibrations that are Raman-inactive and, in some cases, IR-inactive, become active in CARS. The main disadvantage of CARS is its high cost.

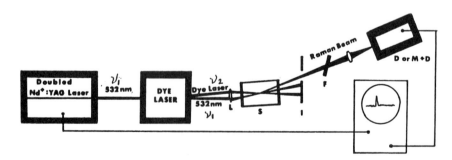

Figure 3-3 Initial apparatus for measuring anti-Stokes emission using a frequency-doubled Nd:YAG pumped dye laser. L is a short focal lens (3–4 cm); S is the sample; I is an iris for spatially filtering the two exciting beams; F is a wideband interference filter; D is the detector (usually a PIN diode); M is a monochromator (not usually necessary). Not shown are the PAR-160 box car integrator, chart recorder, and dye laser scan drive used to record spectra. (Reproduced with permission from Ref. 1.)

3.1.5 Photoacoustic Raman Spectroscopy (PARS)

The principle of photoacoustic Raman spectroscopy (4) is similar to that of CARS. When two laser beams, v_p (pump beam) and v_s (Stokes beam), impinge on a gaseous sample contained in a cell (Fig. 3-4), these two beams interact when the resonance condition, $v_p - v_s = v_M$, is met, where v_M is a frequency of a Raman-active mode. This results in the amplification of the Stokes beam and the attenuation of the pump beam. Each Stokes photon thus generated brings the molecule up to the excited state, and collisional deactivation of these excited state molecules increases their translational energy. This change in the translational energy results in a change in the pressure of the sample

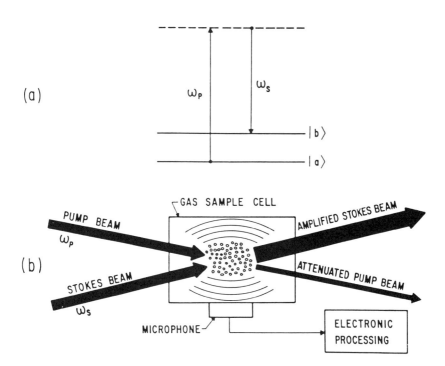

Figure 3-4 Schematic representation of the photoacoustic Raman scattering (PARS) process. (a) A simple energy level diagram illustrating the Raman interaction that occurs in the PARS process. (b) Basic elements of the PARS experimental arrangement. The pump beam is attenuated and the Stokes beam is amplified by the stimulated Raman process that takes place where the beams overlap in the gas sample cell. For each Stokes photon created by the Raman process, one molecule is transferred from the lower state to the upper state of the transition. Collisional relaxation of these excited molecules produces a pressure change that is detected by a microphone. (Reproduced with permission from Ref. 4.)

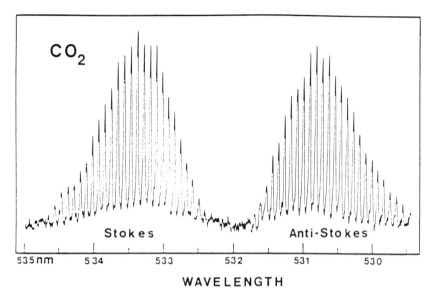

Figure 3-5 Photoacoustic rotational Raman spectrum of CO_2 at a pressure of 80 kPa (600 torr). The rotational line spacing is about 3.1 cm^{-1}. Laser powers of the pump and Stokes beams were 3.3 MW and 120 kW, respectively. (Reproduced with permission from Ref. 4.)

in the cell that can be detected by a microphone. Use of such an acoustic detective device is unique among spectroscopic techniques. By scanning v_s (using a dye laser), the pressure change is measured and converted into a spectrum. As an example, Fig. 3-5 shows the rotational Raman spectrum of CO_2 obtained by the PARS. The absence of the strong Rayleigh band makes it particularly useful for the study of low-energy rotational transitions of gaseous compounds.

3.2 Time-Resolved Raman (TR²) Spectroscopy

Recent technical developments in laser Raman spectroscopy have made it possible to measure the Raman spectra of short-lived transient species, such as electronically excited molecules, radicals and exciplexes, which have lifetimes on the order of nano- (10^{-9}) and pico- (10^{-12}) seconds. These short-lived species may be generated by electron pulse radiolysis, photo-excitation and rapid mixing. However, the application of electron pulse radiolysis is limited in its adaptability and selectivity, while rapid mixing is limited by mixing rates, normally to a resolution on the order of milliseconds. Thus, photoexcitation is most widely used.

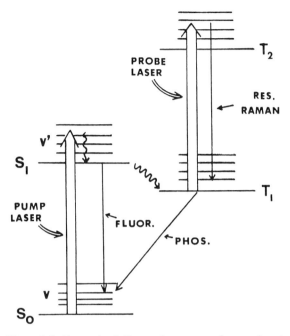

Figure 3-6   Energy level diagram in pump–probe experiment.

Figure 3-6 illustrates the basic scheme involved in a typical pump–probe experiment. First, molecules are excited from S_0 (singlet ground state) to S_1 (singlet excited state) by a pump laser of frequency ν_0. Molecules excited to S_1 undergo nonradiative decay (intersystem crossing) to T_1 (triplet state). Since the pump pulse width is much narrower than the lifetime of the T_1 state (milli $\sim$ microseconds), excitation to the S_1 state by a pump laser increases the population of molecules on T_1, which may become sufficient to observe the Raman spectrum of the T_1 state molecule with a probe laser (ν_1). If ν_1 is chosen to meet the resonance condition as shown in Fig. 3-6, we can take advantage of the resonance Raman scattering discussed in Section 1.15 (extraordinary strong enhancement, selectivity and low sample concentration). Thus, time-resolved resonance Raman (TR^3) spectroscopy is ideal for obtaining Raman spectra of excited state molecules. Some compounds may have ν_1 that is close to ν_0. In such a case, it is possible to obtain TR^3 spectra using a single laser. (See Refs. 8–11.)

Pulsed lasers such as Nd:YAG and excimer lasers are commonly used for the pump–probe experiment just mentioned. Some characteristics of these lasers are listed in Table 3-2. Although the fundamental of the Nd:YAG laser is at 1,064 nm, this frequency can be multiplied by using nonlinear crystals,

Table 3-2 Some Characteristics[a] of Nd:YAG and Excimer Lasers

Nd:YAG: Rep. rate 2–40 Hz, linewidth < 1 cm^{-1} (<0.2 cm^{-1} with intra-cavity etalon). Beam divergence <0.5 mrad.

Wavelength, nm	1,064	532	355	266
Energy/pulse, mJ	280	110	50	20
Pulse width, ns	9	7	6	5
Peak power, MW	30	15	8	4
Average power, W				
(at 10 Hz)	2.8	1.1	0.5	0.2

Excimer: Rep. rate 0.1–100 Hz, beam divergence 5–10 mrad, beam size 10 × 25 mm, linewidth 10–30 cm^{-1}.

Laser medium	XeF	N$_2$	XeCl	KrF	KrCl	ArF	F$_2$
Wavelength, nm	351	337	308	249	222	193	157
Energy/pulse, mJ	400	16	500	1,000	100	500	15
Pulse width, ns	14	6	10	16	9	14	6
Peak power, MW	28	3	50	60	11	35	2.5
Average power, W							
(at 10 Hz)	4	0.1	5	10	1	5	0.1

[a] Representative figures for good commercially available lasers.

such as KDP (potassium dideuterium phosphate), to obtain the second (532 nm), third (355 nm) and fourth (266 nm) harmonics. Furthermore, a wide range of UV-visible pulsed radiation can be generated from these harmonics by pumping a dye laser or using a Raman shifter (Section 2.2).

Figure 3-7 shows the arrangement for obtaining TR^3 spectra of carotenoid excited states used by Dallinger *et al.* (12). In one of their experiments, Laser II (355 nm) was used to produce T_1 state molecules via excitation of a sensitizer (anthracene) and subsequent energy transfer to carotenoids. Laser I (532 nm or 555–610 nm dye laser) was used to obtain RR spectra of carotenoid triplet states. The proper time delay between the pump and probe pulses was determined by the time required to accumulate enough population in the T_1 state. In this experiment, a long delay (~microseconds) was necessary so that two separate lasers were employed for excitation and probing. Such a delay can be accomplished electronically by triggering two lasers in sequence (timing circuitry, Fig. 3-7). If a relatively short time delay is required, an optical delay line, such as that shown in Fig. 3-7, may be employed (for example, a three-meter difference in path length causes a time delay of 10 ns). In some cases, such a delay is not necessary since the excited state of interest can be created within a single pulse width (~10 ns). Then, the leading edge of the pulse is used for pumping and the rest for probing.

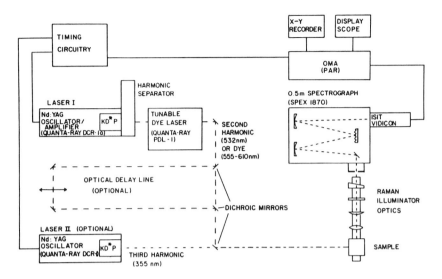

Figure 3-7 Simplified diagram of the experimental apparatus for the optically sensitized triplet generation/TR³ studies. Requisite time delays were obtained either by optical delay or by the two-laser experimental configuration. The optical delay line is not to scale; its actual length was approximately 120 ft. (Reproduced with permission from Ref. 12. Copyright 1981 American Chemical Society.)

Although the peak power of the pump laser must be high, the power of the probe laser should be kept low to avoid nonlinear effects (multiphoton absorption, stimulated Raman scattering; see Section 3.1) and dielectric breakdown (ionization of molecules) that damage the sample. Thus, signal averaging of many pulses (high repetition rate) is made to obtain acceptable S/N ratios. Multichannel detectors such as an intensified silicon photoiode array (ISPD) and an intensified silicon intensified target (ISIT) are used because of their efficiency in data acquisition and handling (Section 2.5). In conjunction with such a multichannel detector, a polychromator such as a Spex Model 1870 spectrograph is used. A triple polychromator (Spex Model 1877, Section 2.4) gives a better stray light rejection.

Figure 3-8 shows the RR spectra of canthaxanthin in the S_0 and T_1 state obtained by Dallinger *et al.* (12). The ground state spectrum (trace 'a') exhibits two intense bands at 1,519 and 1,155 cm^{-1}, which are assigned to the in-phase $v(C=C)$ and $v(C-C)$, respectively, of the conjugated chain. The excited state (TR³) spectrum (trace 'c'), shows that these bands are shifted by ~ 27 cm^{-1} to lower frequencies (1,491 and 1,129 cm^{-1}, respectively) relative to the ground-state spectrum. M.O. calculations predict that, in the T_1 state, the C—C bonds are shortened while the C=C bonds are lengthened. Then, the

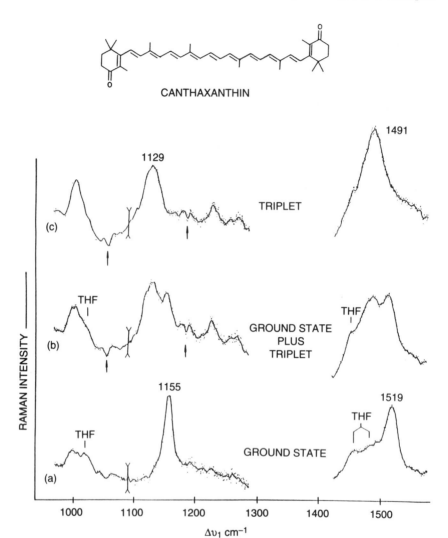

CANTHAXANTHIN

Figure 3-8 Resonance Raman spextra of canthaxanthin in the S_0 and T_1 states. Spectra were obtained with the vidicon spectrograph by using 555 nm probe wavelength and optical excitation of anthracene (at 355 nm) for triplet production. Trace 'a', groundstate spectrum, no 355 nm excitation pulse; trace 'b', superposition of S_0 and T_1 spectra; trace 'c', T_1 spectrum, obtained by approximately normalized subtraction of trace 'a' from trace 'b'. Three different vidicon frames (along the frequency axis) are shown; intensities are not to scale. The solid line is obtained by an 11-point quartic running smooth of the observed points. The negative base line excursions denoted by the vertical arrows are OMA artifacts. (Reproduced with permission from Ref. 12. Copyright 1981 American Chemical Society.)

$1,519 \text{ cm}^{-1}$ band should be downshifted, whereas the $1,155 \text{ cm}^{-1}$ band should be upshifted, in going from the S_0 to the T_1 state. However, both bands were found to be downshifted. This apparent contradiction has led Dallinger *et al.* (12) to conclude that the interaction force constant between the (C=C) and ν(C—C) vibration (Section 1.20) changes sign in going from the S_0 to the T_1 state.

Recently, the $[\text{Ru(bpy)}_3]^{2+}$ ion and related complexes have attracted much attention because of their excited-state redox properties, which may be utilized as solar-energy conversion devices. The TR^3 spectrum of this ion can be obtained by using a single laser line (Nd:YAG third harmonic at 355 nm) since the S_0–S_1 transition absorbs this line considerably and since the absorption maximum of the $T_1 \rightarrow T_n$ transition is near 360 nm. The T_1 state population can be built up via efficient and rapid intersystem crossing from the S_1 to the T_1 state because of the short lifetime of the S_1 state (~ 10 ps) and the long lifetime of the T_1 state (~ 600 ns). The TR^3 spectrum of the $[\text{Ru(bpy)}_3]^{2+}$ ion was originally obtained by Bradley *et al.* (13). These workers found that the TR^3 spectrum consists of two series of bpy vibrations; one series of bands is the same as that observed in the S_0 state and the other is close to those of Li$^+$(bpy$^-$). These experiments were repeated by Mallick *et al.* (14) and are shown in Fig. 3-9. Thus, the T_1 state may be formulated as $[\text{Ru(III)(bpy)}_2(\text{bpy}^-)]^{2+}$. Namely, the electron is localized on one bpy ligand rather than delocalized over all three bpy ligands. TR^3 spectroscopy has also been applied to biological compounds (Section 4.2.1).

3.3 Matrix-Isolation Raman Spectroscopy

The matrix-isolation (MI) technique was largely developed by Pimentel and coworkers (15) to study the IR spectra of unstable (free radicals and reaction intermediates) as well as stable species isolated in inert gas matrices. In this method, gaseous samples and an inert matrix gas, such as Ar or Kr, are mixed and deposited on an IR transparent window (e.g., alkali halide crystal) cooled to 10–20 K by a cryostat. Since the mixing ratio (sample/gas) is 1:500 or higher, the sample (solute) molecules are completely isolated from each other in the frozen gas matrix. Thus, MI spectra are similar to those of the gaseous phase; no intermolecular interaction is present, and no lattice modes are observed (although weak interaction between the solute and the inert gas is noted). Furthermore, MI spectra are simpler than gas-phase spectra because only a few or no rotational transitions are observed, a result of steric restriction of molecular rotation in the matrix. The resulting sharpness of observed bands tends to resolve closely located bands. The MI technique is also applicable to solid samples as long as they can be vaporized without

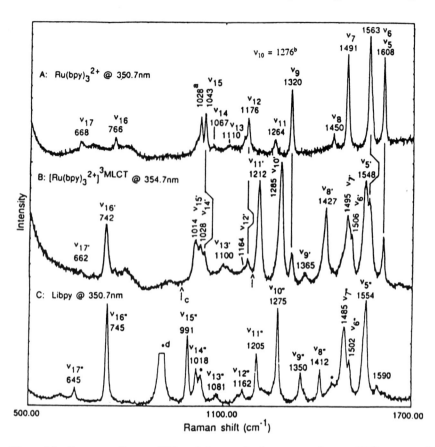

Figure 3-9 Resonance Raman (RR) and time-resolved resonance Raman (TR^3) spectra of [Ru(bpy)$_3$]$^{2+}$. (a) RR spectrum using 350.7 nm excitation (Ar$^+$ laser); (b) TR^3 spectrum using 354.7 nm excitation; (c) RR spectrum of Li(bpy) using 350.7 nm excitation (Ar$^+$ laser). (Reproduced with permission from Ref. 14. Copyright 1990 American Chemical Society.)

decomposition. Extensive research has been carried out in the field of matrix isolation IR spectroscopy, and the results are thoroughly reviewed in several monographs (16–19).

Technically, Raman spectroscopy is more difficult to apply to low-temperature matrices than IR spectroscopy for the following reasons (20, 21): (1) Since Raman signals are inherently weak, relatively high concentrations of the sample or relatively wide slit widths are required. The former may cause the formation of dimeric and polymeric species, while the latter leads to the loss of resolving power of the monochromator. (2) If one increases the laser power to obtain stronger Raman signals, the matrix temperature will rise because of local heating by the laser beam, and this will accelerate the diffusion of solute molecules in the matrix. (1) and (2) can be circumvented if Raman spectra are obtained under resonance conditions (Section 1.15 and the following examples). (3) The quality of the Raman spectra obtained depends on the quality of the matrix prepared; "clear matrices" give better results than "frosty matrices." However, preparation of the former is time-consuming. (4) The matrix itself or oil contamination from the diffusion pump may cause fluorescence. In spite of these problems, matrix isolation Raman spectroscopy has advantages over its IR counterpart, as listed in Section 1.8. The first matrix isolation laser-Raman experiment was carried out by Shirk and Claassen (22) in 1971.

The experimental setup for matrix Raman spectroscopy is essentially the same as that for matrix IR spectroscopy. The major difference lies in optical geometry. Namely, backscattering geometry must be employed in Raman spectroscopy since the matrix gas and sample vapor are deposited on a cold metal (Cu, Al) surface. Figure 3-10 shows the optical arrangement used by Proniewicz et al. (23). Here, the miniature oven technique (24) was employed to vaporize the solid samples, and a cylindrical lens was used to produce a line-focused image on the matrix so that the local heating effect due to the laser beam is minimized (Section 2.7). As stated in Section 1.9, depolarization ratios are highly important in making band assignments in Raman spectra. These values can be obtained in inert gas matrices, although their values are somewhat higher than true values. For example, the ρ value for the totally symmetric mode of CCl_4 at 459 cm^{-1} in an Ar matrix is ~ 0.14, although it should be close to zero (25). This is largely caused by scrambling of the polarized scattered light by the matrix and the lack of complete randomization of molecular orientation in a frozen matrix.

Since the results obtained by matrix isolation Raman spectroscopy have been reviewed extensively (18–20), only two typical examples are discussed here to show the utility of this technique. It should be noted that both works took advantage of resonance Raman spectroscopy to detect Raman signals using low laser power.

Andrews and co-workers (26) studied IR and resonance Raman spectra of alkali metal atom–halogen reaction products in inert gas matrices. For example, these workers (26) reacted Cs atom vapor with Cl_2/Ar (1/100) to produce $Cs^+ (Cl_2)^-$. This $(Cl_2)^-$ ion has an absorption maximum near 365 nm that tails out to 500 nm. Thus, the 457.9 nm line of an Ar-ion laser (75 mW) was used to resonance-enhance the $(Cl_2)^-$ vibration. As seen in Fig. 3-11, this anion, $[Cl_2^{35}]^-$, exhibits its fundamental at 259.0 cm^{-1} followed by a long series of overtones up to the eighth (1,984.5 cm^{-1}). Each band is accompanied by a side band due to its isotopic counterpart, $(^{35}Cl^{37}Cl)^-$ on the low-frequency side, and the intensities of both overtone series decrease

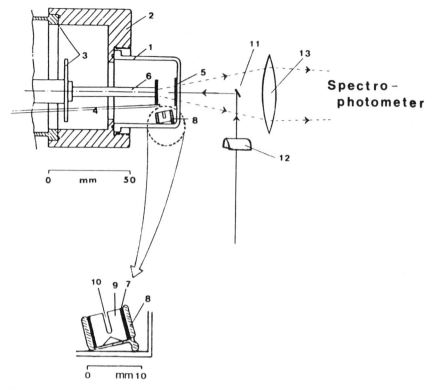

Figure 3-10 Schematic drawing of matrix-isolation apparatus for resonance Raman measurements. 1, glass envelope; 2, aluminum sleeve; 3, refrigerator; 4, gas line; 5, steel screen; 6, cold tip; 7, aluminum radiation shield; 8, Pyrex cup; 9, spectroscopic-grade spark graphite rod; 10, Pyrex capillary tube with sample; 11, small mirror; 12, cylindrical lens; 13, collecting lens. 5 is placed to prevent sample deposition in the optical path. It must be removed from the path by using an external magnet once sample deposition on 6 is completed.

progressively in going to the higher overtone. As discussed in Section 1.15, this behavior is typical of the A-term resonance. The frequencies corrected for anharmonicity, anharmonicity constant and dissociation energy have been calculated based on the observed frequencies of the overtone series.

In some cases, matrix-isolation laser-Raman spectroscopy can be utilized to produce unstable species via laser photolysis and to measure their resonance Raman spectra simultaneously in the same matrix using the same laser. For example, Proniewicz et al. (23) measured the resonance Raman spectra of $Fe(TPP-d_8)$ ($TPP-d_8:d_8$ analogue of tetraphenylporphyrinato anion) in O_2 matrices at ~ 30 K using the 406.7 nm line of a Kr-ion laser. As the first step, two types of O_2 adducts are formed via co-condensation of $Fe(TPP-d_8)$ vapor with O_2.

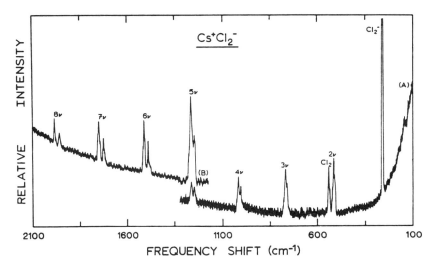

Figure 3-11 Resonance Raman spectrum of matrix-trapped $Cs^+Cl_2^-$ (75 mW, 457.9 nm excitation). (a) 0.3×10^{-9} A range, 3-s rise time; (b) 0.1×10^{-9} A range, 10-s rise time (reproduced with permission from Ref. 26. Copyright 1975 American Chemical Society.)

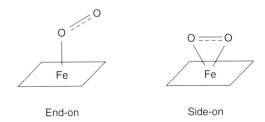

The end-on type exhibits the $\nu(O_2)$, $\nu(Fe-O_2)$ and $\delta(Fe-O-O)$ at 1,195, 508 and 345 cm^{-1}, respectively, whereas the side-on type exhibits the $\nu(O_2)$ and $\nu_a(Fe-O)$ at 1,105 and 407 cm^{-1}, respectively. These bands are seen in Fig. 3–12A, which was obtained by using only 0.2 mW laser power.

As the laser power is increased, however, all these bands become weaker and disappear almost completely at 8 mW. In contrast, the bands at 853 and 815 cm^{-1}, which were originally weak, become stronger with increasing laser power. The latter two bands can be assigned to the $\nu(Fe=O)$ of oxyferrylporphyrin and its π-cation radical, respectively, which are produced by the O—O bond cleavage of the dioxygen adducts mentioned earlier. The 815 cm^{-1} band becomes weaker in going from trace C to D because the π-cation radical is converted to the non–π-cation radical at higher laser power.

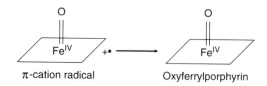

π-cation radical   Oxyferrylporphyrin

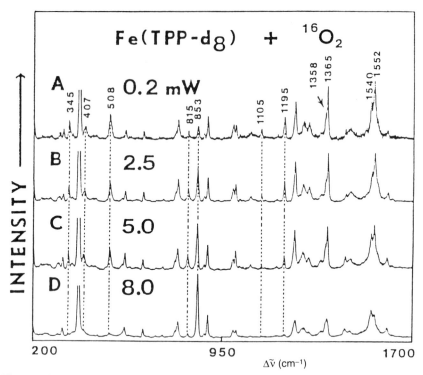

Figure 3-12   Resonance Raman spectra of Fe(TPP-d_8) co-condensed with $^{16}O_2$ at 30K (406.7 nm excitation). (a) 0.2; (b) 2.5; (c) 5.0; (d) 8.0 mW. These spectra are composites of four sections measured separately. (Reproduced with permission from Ref. 23. Copyright 1991 American Chemical Society.)

All of the preceding band assignments have been confirmed by $^{16}O_2/^{18}O_2$ and $^{54}Fe/^{56}Fe$ substitutions, and oxygen isotope scrambling techniques.[1] As will be shown in Section 4.2.2, oxyferrylporphyrin and its π-cation radical serve as model compounds of horseradish peroxidase Compounds II and I, respectively.

---

[1] In oxygen isotope scrambling experiments, a mixture of $^{16}O_2/^{16}O^{18}O/^{18}O_2$ ($\sim 1/2/1$ ratio) is reacted with iron porphyrin. Such a mixture can be obtained by electrical discharge of an equimolar mixture of $^{16}O_2$ and $^{18}O_2$.

3.4 High-Pressure Raman Spectroscopy

Subjecting various stages of matter to high external pressure, and utilizing a spectroscopic technique as the diagnostic tool to determine the changes that have occurred, has proven very successful. Infrared and Raman spectroscopies have been the most useful of the diagnostic tools utilized. Valuable information about intermolecular interactions, phase transitions, structural changes, vibrational assignments and conversions of insulators (semiconductors) to metals is obtainable when matter is subjected to pressure. Pressure-induced frequency shifts are often accompanied by intensity changes and can be used to identify the nature of vibrations and provide correct vibrational assignments (27). A diamond anvil cell (DAC) capable of reaching 5.7 megabar pressures and $\sim 4,000°$K temperature, has been developed, which allows studies to be made concerning the behavior of various minerals occurring in the depths of the earth (28). Since the core–mantle interface has a pressure of ~ 1.6 Mbar and 3,000–4,000 K, simulation of core–mantle reactions is possible.

3.4.1 PRINCIPLES

The pressure technique involves a pressure device (DAC) that can transmit the pressure to the sample under study. If spectroscopic methods are chosen for diagnostic purposes, it is a requirement to use windows on the pressure device that are hard and transmit the irradiating light in the particular wavelength of the elctromagnetic spectrum being studied. The window of choice for IR and Raman studies is Type IIa diamond.[2] It is the hardest material known and is transmissive for laser Raman studies. Additionally, it is an excellent thermal conductor as well. The pressure device must be compact and fit into the sample compartment of the spectrometer. The DAC fulfills all of these criteria and has been extensively used since its discovery by Weir and Van Valkenburg in 1959 (29). The interface to the Raman spectrometer is readily accomplished (unlike the IR experiment where beam condensers are required in the dispersive instrument, although presently this requirement is unnecessary in the Raman/DAC experiment), and normal commercial Raman instrumentation can be used.

Pressure calibration is necessary in pressure work, and this is accomplished by incorporating a small ruby crystal with the sample under study. The Ruby scale (30) was developed by the National Bureau of Standards (now the NIST) in 1972, and the sharp Ruby R_1 fluorescent line has been calibrated vs. pressure by NIST, and is suitable even up to megabar pressures (31).

---

[2] Type IIa diamonds are better suited for optical studies because fewer impurities are present. They show absorptions at ~ 3 and 4–5.5 μm and are transparent into the FIR to ~ 10 cm^{-1}.

Further details on the Ruby scale are provided in the following section on instrumentation.

3.4.2 INSTRUMENTATION

A conventional Raman instrument is suitable for high-pressure Raman measurements. The pressure device can be a piston-cylinder cell or a DAC. The first Raman studies with a piston-cylinder cell were made in 1957 (32). The DAC was first used in Raman studies in 1968 (33, 34). Since that time, the DAC has undergone several modifications. Figure 3-13 shows the details of the first DAC, which consists of pistons A and B, as well as C, the hardened steel insert; D, the presser plate; E, the lever arm; G, the screw; and H, the calibrated spring. In the DAC (Fig. 3-13), pressure is applied by turning the knobs to compress the springs. The pressure is transmitted along the lever arms, and this compresses the bottom plate pushing the two pistons holding the diamonds together and affecting the pressure. Figure 3-14 shows the present ultrahigh-pressure cell capable of reaching megabar pressures. Similar transmission of pressure applies to the ultrahigh-pressure cell. Figure 3-15 shows a typical forward scattering geometry in a laser DAC experiment.

Laser radiation at 632.8 and 476.5 nm is recommended, as radiation at 514.5 and 488.0 nm can cause excessive fluorescence problems.

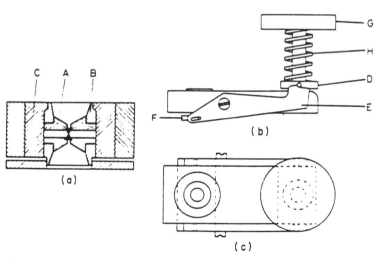

Figure 3-13 Diamond anvil high-pressure cell. (a) Detail of diamond cell. (b) Side view. (c) Front view. A and B, parts of piston; C, hardened steel insert; D, presser plate; E, lever; G, screw; H, calibrated spring. (Reproduced with permission from Ref. 27.)

Figure 3-14   Superhigh-pressure cell for megabar use.

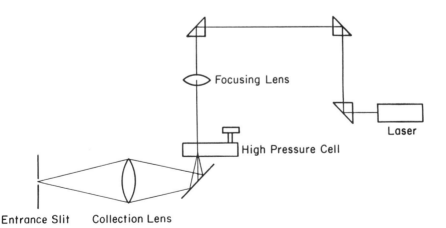

Figure 3-15   Detail of the scattering geometry in laser Raman DAC experiment. Figure not drawn to scale. (Reproduced with permission from Ref. 33.)

Calibration of the pressure is best accomplished using the Ruby scale. Generally this is done using a metal gasket between the two diamond windows surrounding the sample in which a liquid (such as Nujol or Teflon oil) is added to produce hydrostatic pressure. The technique measures the pressure dependence of the sharp Ruby R_1 fluorescence transitio at 692.8 nm, although the R_2 band at 694.2 nm can also be used. The Ruby fluorescence is induced by the blue excitation of the Ar^+ (488.0 nm) or the He–Cd (441.6 nm) lasers.

3.4.3 APPLICATIONS

A few applications illustrating the pressure effects on materials using Raman spectroscopy as the diagnostic tool are presented.

(a) Solid State Phase Transitions with Pressure

Phase Transitions in Solids. Solid state phase transitions with pressure are quite common (27). Solid H_2S was subjected to pressure up to 20 GPa at 300 K, and measured by Raman spectroscopy (35). Figure 3-16a shows the Raman spectra in the stretching vibration region under pressure. The symmetric stretching mode v_1 shows a red-shift and broadens with an increase in pressure. At ~ 11 GPa, the broad band narrows and v_3 (antisymmetric stretch) appears on the high-frequency side of v_1, indicative of a phase transition occurring. The v_2 bending vibration at $\sim 1{,}160$ cm^{-1} is only slightly affected by an increase in pressure. At the second phase transition, another v_2 vibration appears at $\sim 1{,}250$ cm^{-1} that shifts toward higher frequency with pressure. All five lattice modes blue-shift with pressure, typical of a molecular solid. Figure 3-16b records the pressure dependence of the intramolecular and lattice vibrations.

Pressure Changes in Solid Coordination Compounds. Adams *et al.* (36) examined decacarbonyl dimanganese, $Mn_2(CO)_{10}$, and decacarbonyl dirhenium, $Re_2(CO)_{10}$, with pressure in a DAC, and followed the change in the carbonyl and low-frequency region with Raman spectroscopy. In the pressure conversion from the D_{4d} to the D_{4h} isomer, a carbonyl vibration (E_2 doublet) centered at 2,020 cm^{-1} becomes a single band. Three other bands in this region in the high-pressure phase appear below 2,000 cm^{-1}. The high-pressure phase can be ascribed to a phase transition to the staggered isomer in $Mn_2(CO)_{10}$ and $Re(CO)_{10}$ (36). Figure 3-17 shows the spectral data obtained with pressure. Transformation occurs for the $Mn_2(CO)_{10}$ at 8 kbar and for $Re_2(CO)_{10}$ at 5 kbar. Two further phase transformations occur below 140 kbar for both carbonyls.

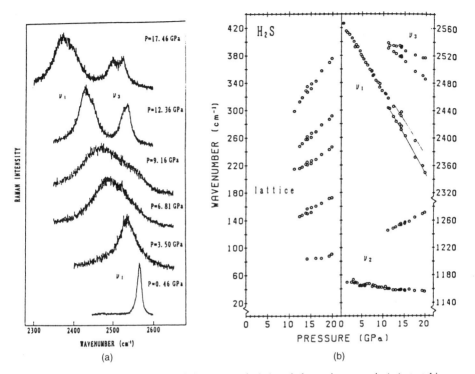

Figure 3-16 (a) Raman spectra of the symmetric (v_1) and the anti-symmetric (v_3) stretching modes in solid H_2S at various pressures. The phase transition occurs at about 11 GPa. (b) Pressure dependence of the intramolecular and the lattice vibrational frequencies in solid H_2S at 300 K. (Reproduced with permission from Ref. 35.)

Evidence for Metallization of hydrogen at Megabar Pressures. The behavior of condensed hydrogen has been a subject of considerable interest since Wigner and Huntington (37) in 1935 discussed the possibility of a high-pressure metallic phase of hydrogen existing. Theoreticians have derived equations of state for both molecular and metallic phases to predict transition pressures. In recent years the effort to search for metallic hydrogen has been stimulated from attempts to understand the planetary surfaces and interiors of Jupiter, Saturn, and other outer planets.

Metallic hydrogen was first reported by the Russians in 1972 (38). However, this was not verified. Recently, optical and Raman observations by Mao and Hemley (39) of solid hydrogen, at pressures of 2.5 megabar at 77 K, have provided evidence that the metallic phase of hydrogen had been achieved. Electronic excitations in the visible region are observed at 2 megabar. Raman scattering data illustrated that the hydrogen solid was stable to ~2 kbar at

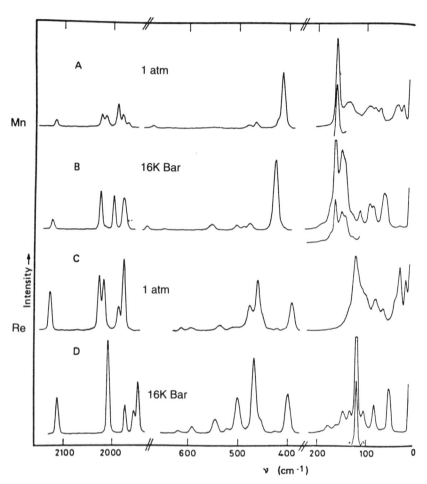

Figure 3-17 Raman spectra of $Mn_2(CO)_{10}$ at (a) ambient pressure and (b) 16 kbar, and of $Re_2(CO)_{10}$ at (c) ambient pressure and (d) 16 kbar in diamond anvil cell. (Reproduced with permission from Ref. 36.)

77 K, as evidenced by the presence of the H—H stretching mode (39). The upper trace of Fig. 3-18 shows the Raman spectrum of the H—H stretching vibration in solid hydrogen at 158 GPa and 295 K. The insulator-to-metal transition occurs at 77 K and is readily observed as two bands appear, arising from two phases (the metallic and the insulator phase). At 158 GPa, the H_2 vibron[3] in solid hydrogen is observed at $\sim 4,020$ cm^{-1}. Supporting evidence has come from optical transmission and reflectance measurements. Some

---

[3] Interaction of electrons with the H—H stretching vibration.

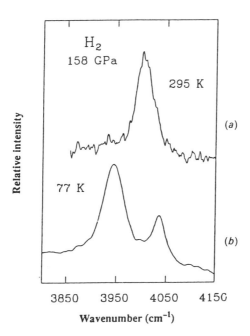

Figure 3-18 Raman spectra of the intramolecular H—H stretching vibration (vibron) in solid hydrogen at 158 GPa and 77 (B) and 295 K (A). A sloping background signal has been subtracted from the high-temperature spectrum. The estimated random errors in pressure and temperature are ±1 GPa and ±2 K (low temperature). (Reproduced with permission from Ref. 39.)

controversy has developed as Silvera *et al.* (40), repeating the work of Mao and Hemley, find the results suggestive, but do not establish the metallization of solid hydrogen. Ruoff and co-workers (41) also have raised objections and believe that Mao and Hemley have observed evidence for aluminum metal instead of metallic hydrogen—the aluminum coming from a chemical reaction of ruby powder (for pressure calibration) under the conditions of the experiment.

Experiments with Materials Important to Problems in Earch Science. Many pressure-induced phase transformations in minerals may be responsible for the structural features of the Earth's interior. Raman scattering measured at high pressures in the 100-GPa (10^3 kbar) range has been found to be a powerful tool (28).

Studies of magnesium silicate perovskite at high pressure are of considerable interest because of the abundance of perovskites and related materials in the Earth's lower mantle (42–44). Raman spectra of single crystals of $MgSiO_3$-perovskite were measured (45). Spectra from 650 to 150 cm^{-1}

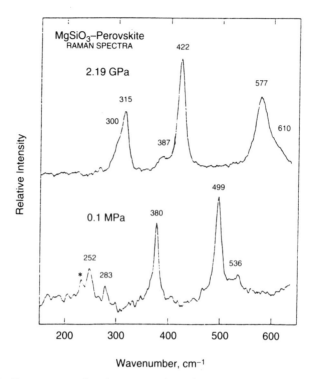

Figure 3-19 Raman spectra (low-frequency region) of MgSiO$_3$-perovskite at 0.1 MPa (zero pressure) and 21.9 GPa. The spectra were measured at 298 ($\pm$2) K. Plasma lines from the ion laser are indicated by an asterisk. (Reproduced with permission from Ref. 45.)

are shown in Fig. 3-19. Six bands were observed in the 600–200 cm^{-1} region at ambient pressure. Changes in the pressure spectrum were observed, and three main bands at 315, 422 and 577 cm^{-1} are seen in this region. All of the bands showed a blue shift with pressure (499 → 577 cm^{-1}), and no soft modes[4] were found. Vibrational assignments were made for the stronger Raman bands.

Further examples of useful applications to the high-pressure technique on matter, followed by Raman spectroscopy, may be found in Refs. 46–54.

3.5   Surface-Enhanced Raman Spectroscopy (SERS)

Considerable interest has developed in recent years concerning the nature of adsorbed molecules on metal surfaces (e.g., electrode surfaces).

---

[4] A soft mode is a temperature-dependent, low-frequency, unstable mode having a large oscillator strength.

Determinations of the nature of the adsorbate molecule on the surface and the frequencies of bands associated with adsorbed molecules are important data. Information concerning the strength of the adsorbate–surface interactions and the conformation of the adsorbed molecule becomes obtainable. Previously, much of this surface information has been provided by surface-sensitive infrared techniques, such as ATR and reflectance methods. In 1974, Fleischmann *et al.* (55) reported the Raman spectrum of a monolayer of pyridine adsorbed on a silver electrode surface. In 1977, Jeanmaire and Van Duyne (56) and Albrecht and Creighton (57) noted that the scattering intensity from the adsorbed species is 10^5–10^6 times stronger than that of nonadsorbed species. Thus, SERS research had begun.

It has now been demonstrated that many molecules adsorbed on appropriately prepared metal surfaces display Raman cross-sections several orders of magnitude greater than the corresponding quantity for an isolated molecule or from a solution. Together with other surface-sensitive techniques, SERS has catalyzed the study of condensed phases on surfaces. It has demonstrated promise as a vibrational probe of *in situ* gas–solid, liquid–solid, and solid–solid environments, as well as a high-resolution probe of vacuum–solid interfaces.

The surface enhancement (58) of an adsorbate molecule obtained by the SERS technique has been the reason for the increased interest. Because of the large enhancement, small sample volumes can be used, and current detection limits are in the picomole-to-femtomole range. The large amount of data obtainable and its surface selectivity and sensitivity make the SERS technique a welcome addition to the battery of tools available for surface studies.

3.5.1 PRINCIPLES

Normal Raman laser excitation in the visible and NIR region (59) can be used to obtain the SERS effect. The substrate surface is extremely important in providing the necessary enhancement to make the technique as valuable as it has become. A number of substrates have been used (60). These include evaporated silver films deposited on a cold surface at elevated temperature (~ 390 K) on a glass substrate, photochemically roughened surfaces (e.g., silver single crystals subjected to iodine vapor, which roughens the surface), grating surfaces, and mechanically abraded and ion-bombarded silver surfaces. A conventional Raman spectrometer is suitable for the measurement of the SERS effect.

The mechanisms involved whereby surface enhancement spectra are obtainable are still the subject of considerable controversy (60). At least two mechanisms have been proposed (60, 61). If one examines the relationship

$P = \alpha E$ (more thoroughly discussed in Chapter 1), it is clear that any enhancement must come from an enhancement of α (molecular polarizability) or from E (the electric field), since the intensity of Raman scattering is proportional to the square of the induced electric dipole moment P. The enhancement associated with E is termed the electromagnetic effect. This enhancement occurs because the local electromagnetic field at the surface of a metal is significantly changed from that in the incident field, because of the metal influence, and is pronounced when fine metal particles or rough metal surfaces are involved. In this case, the light excitation at the surface of the metal excites conduction electrons and generates a surface plasma resonance (sometimes called a plasmon resonance). This causes the roughness feature of the metal to be polarized and the electromagnetic field in the interior of the particle at the surface to increase significantly from the applied field.

The other enhancement is termed the chemical enhancement and can result from a charge-transfer (C—T) or bond formation of the metal and adsorbate, which can increase α, the molecular polarizability (62). For the pyridine/Ag system, the C—T band appeared on adsorption of pyridine on Ag and disappeared reversibly on desorption of pyridine (63). It was observed that the stronger the C—T band, the stronger the SERS spectrum.

3.5.2 INSTRUMENTATION

For SERS measurements on condensed-phase interfaces, a conventional Raman spectrophotometer can be used. The laser beam in the visible or NIR region may be directed to the surface via 90°, 180° or backscattering geometry. A double monochromator with a diode array detector can be used, or alternatively, a triple monochromator with the aforementioned detector may be utilized, to give better stray light performance. To maximize the enhancement effect, the incident wavelength appropriate for a particular substrate is necessary. Krypton-ion or dye lasers operating in the red region are used for copper and gold substrate surfaces. The argon-ion laser line at 514.5 nm is suited for a silver substrate surface. The maximum enhancement is said to be at ~ 750 nm (58). Thus, the ν^4 relationship (Chapter 1) is not applicable in SERS.

As was mentioned in the previous section, the nature of the substrate is extremely critical in obtaining the maximum enhancement. For visible Raman excitation, the noble metals, such as Ag and Au, and the alkali metals are the substrates of choice. For other regions, other substrates are more suitable (e.g., Ge or Pt in the IR region). Probably the most common substrates used for SERS are the colloidal suspensions of silver or gold particles (~ 5–20 nm in diameter), and electrochemically roughened silver electrodes.

Since stationary electrodes are employed in most SERS experiments, a relatively small number of adsorbed molecules are continuously irradiated by laser beams. When exciting lines are within strong absorption bands of the adsorbed species, surface-enhancement resonance Raman spectra (SERRS) are obtained. However, this may lead to decomposition of such species due to local heating. Use of a cylindrical rotating electrode can circumvent this problem (64).

3.5.3 APPLICATIONS

Only a few examples of applications of SERS technique will be presented.

(a) Trace Analyses

SERS has been found to be useful for trace analytical applications. Organophosphorus compounds used in insecticides have been detected at the nanogram level (65). The Raman spectra of these compounds are relatively specific, and they can be readily identified. Figure 3-20 shows SERS spectra of several organophosphorus compounds. The spectra were obtained using SERS-active substrates of silver-covered microspheres on cellulose and glass surfaces. The technique has also proven fruitful in identifying ground water contaminants of an organic nature (66). Figure 3-21 shows the SERS spectra of a mixture of ground water contaminants. The spectra were obtained using a SERS substrate in the form of a silver electrode.

(b) Biological Molecules

SERS spectra of several dipeptides adsorbed on silver colloidal particles have been obtained (67). Figure 3-22 illustrates the SERS spectra of $\sim 10^{-5}$ M dipeptides adsorbed on colloidal silver. Applications of SERS to other biological molecules are found in Section 4.1.7.

(c) Catalysis

SERS has been used in monitoring catalytic reactions (60). Figure 3-23 illustrates the measurement of catalytic formation of SO_3^{2-} and SO_4^{2-} on silver powder surfaces by exposure to SO_2 gas. It was concluded that part of the SO_3^{2-} was oxidized to SO_4^{2-}, giving a new absorption band at 962 cm^{-1} (attributed to SO_4^{2-}), and the SO_3^{2-} was thermally desorbed.

(d) Characterization of Modified Electrodes and Electrochemical Processes

SERS has also been found to be valuable in characterizing modified electrodes

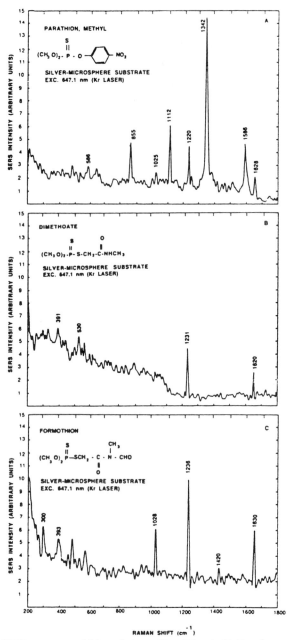

Figure 3-20 SERS spectrum of (a) methylparathion (26 ng), (b) dimethoate (23 ng), and (c) formotion (26 ng). (Reproduced with permission from Ref. 65. Copyright 1987 American Chemical Society.)

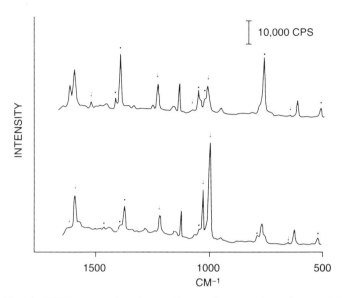

Figure 3-21 The SERS spectra of a mixture of contaminants at electrode potentials of -1.0 (top) and -0.6 (bottom) V and an excitation wavelength of 514.4 nm. The arrows and the dots indicate the major peaks due to pyridine and quinoline, respectively. (Reproduced with permission from Ref. 66. Copyright 1987 American Chemical Society.)

and electrochemical processes (68). SERS was used to observe the sequential electrochemical formation of adsorbed O^{2-}, OH^- and H_2O on a silver electrode in 0.1–0.001 M alkali chloride solution containing micromolar concentrations of $KMnO_4$ (69). A peak at 615 cm^{-1} has been assigned to a mixed oxide designated by (Ag, Mn)—O. The (Ag, Mn)—O species are successfully protonated to produce (Ag, Mn)—OH_2. A peak at ~ 590 cm^{-1} disappears, and the hydrate (Ag, Mn)OH_2 with a peak at about 475 cm^{-1} forms, as the V_{SCE} is swept cathodically. Oxidation of the silver surface by MnO_4^- produces (Ag, Mn)O, and the surface roughness that is needed for the SERS effect. Figure 3-24 illustrates the SERS spectra of this system.

For additional reading on the SERS technique, see Refs 70–77.

3.6 Raman Spectroelectrochemistry

Raman spectroelectrochemistry (78, 79) is a field in which one studies electrogenerated species on electrode surfaces, in electrode diffusion layers and bulk solution by Raman spectroscopy. Thus, the surface-enhanced Raman scattering (SERS) discussed in the preceding section is part of Raman

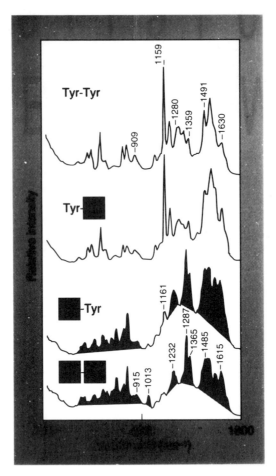

Figure 3-22 SER spectra of $\sim 10^{-5}$ M dipeptides adsorbed on colloidal silver. (Reproduced with permission from Ref. 67. Copyright 1989 American Chemical Society.)

spectroelectrochemistry. Here, we discuss Raman spectroscopic studies on electrogenerated species in bulk solution and in electrode diffusion layers. Since no enhancement from SERS is expected and since the concentrations of these electrogenerated species are rather low, it is imperative to take advantages of resonance Raman (RR) scattering (Section 1.15).

3.6.1 INSTRUMENTATION

Figure 3-25a illustrates a typical cell for obtaining RR spectra of electrogenerated species in bulk solution (80). The laser beam is focused on

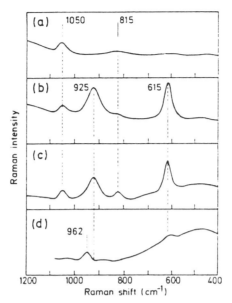

Figure 3-23 SER spectra from finely divided Ag powder. (a) Fresh powder in He atmosphere, $T = 295$ K; (b) like (a), but exposed to SO_2 gas for 1–2 min; (c) like (b), but kept in a SO_2/O_2 (ratio 1 to 5) containing atmosphere after initial exposure; (d) like (c), but slowly heated to 380 K. Spectra have been taken with 514.5 or 488 nm radiation and 4 cm^{-1} bandpass. (Reproduced with permission from Ref. 60.)

a position that is well removed from the working electrode surface. Sufficient concentration of the species under investigation is accumulated via controlled potential coulometric electrogeneration. The species generated must be stable over the time required to record a spectrum so that no interference from decomposition products can occur. If electrogenerated species are stable only at low temperatures, their RR spectra must be measured using a low-temperature bulk electrolysis cell (81).

Figure 3-25b shows a "sandwich" cell for recording RR spectra of electrogenerated species in the diffusion layer (80). The laser beam is reflected from a Pt working electrode and the scattered light is measured via backscattering geometry. Spectral interference from the bulk solution can be avoided by choosing a proper system for RR excitation (*vide infra*). To create a steady-state concentration profile in the diffusion layer, a square-wave voltage is applied to the working electrode immersed in an unstirred solution containing electroactive species, solvent and supporting electrolyte.

(a) *Bulk solution*

Using the cell shown in Fig. 3-25a, Jeanmaire and Van Duyne (82) measured the RR spectrum of the tetracyanodimethane anion radical (TCNQ$^-$).

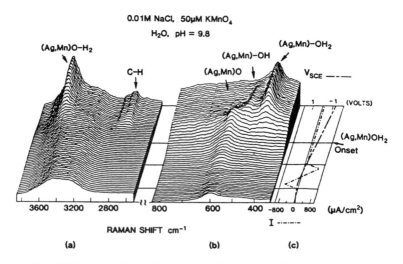

Figure 3-24 SERS spectra of (Agn, Mn)—O, (Ag, Mn)—OH, and (Ag, Mn)—OH$_2$ adsorbed on a Ag electrode as a function of V_{SCE}. The solution is 0.01 MNaCl for the third ORC and the voltage interval is 35 mV. (a) The O—H$_2$ stretching mode region. The enhanced spectrum due to adsorbed (Ag, Mn) O—H$_2$ is superimposed on the unenhanced Raman spectrum of bulk H$_2$O. (b) The (Ag, Mn)—O stretching mode region. (c) The "unfolded" voltammogram showing the voltage current characteristics of the cell corresponding to the evolution of the SERS spectra. The laser power at $\lambda = 514.5$ nm is 100 mW. (Reproduced with permission from Ref. 69. Copyright 1986 American Chemical Society.)

The radical was electrogenerated by the reaction

$$TCNQ + e^- \rightleftarrows TCNQ^-$$

The TCNQ$^-$ radical thus obtained is completely stable for at least 3 hr, and its electronic spectrum shows strong absorption bands between 950 and 550 nm. Thus, the RR spectrum of TCNQ$^-$ was obtained by using the 647.1 nm line of a Kr$^+$ laser (Fig. 3-26a). On the other hand, TCNQ has a strong absorption band near 400 nm. Therefore, its preresonance spectrum, shown in Fig. 3-26b, was obtained by the 457.9 nm line of an Ar$^+$ laser. In both spectra, all strong and medium intensity bands were found to be polarized (totally symmetric). The vibrational frequency shifts in going from TCNQ

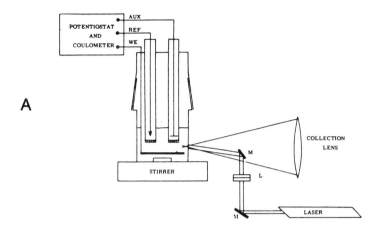

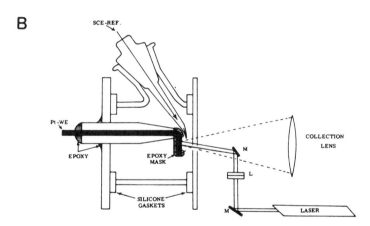

Figure 3-25 Resonance Raman spectroelectrochemistry cells and back scattering geometry. (a) Controlled potentional electrolysis cell; (b) "sandwich" cell for semi-infinite diffusion conditions. (Reproduced with permission from Ref. 80. Copyright 1975 American Chemical Society.)

to its anion radical can be explained in terms of π bond order changes, which can be calculated by MO methods. These workers (83) also carried out detailed excitation profile studies on several bands of $TCNQ^-$. Each profile contains 90 intensity points per 100 cm^{-1}, which is the most complete profile obtained for any compounds studied thus far. As shown in Fig. 3-27, the electronic absorption spectrum of $TCNQ^-$ has almost no structure, while its RR excitation profiles show many details in the same region.

Figure 3-28 shows the RR spectra of O=Fe(IV)(TMP) (TMP: tetra-mesitylporphyrinato anion) and its ^{18}O analogue generated at $-40°C$ by

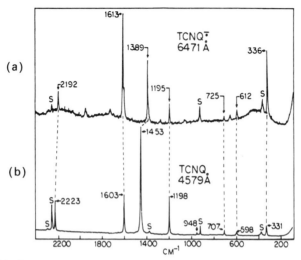

Figure 3-26 Resonance Raman spectra of TCNQ and electrogenerated TCNQ⁻. [TCNQ]
1.09 mM, laser power = 20 mW, bandpass = 1.2 cm⁻¹; [TCNQ⁻] = 2.24 mM, laser
power = 74 mW, bandpass = 2.2 cm⁻¹. TCNQ⁻ was electrogenerated by controlled potential
caulometry at −0.10 V vs. SCE in 0.1 M TBAP/CH_3CN. All spectra were scanned at 50 cm⁻¹
min⁻¹ using a 1.00 s counting interval. Plasma lines were removed at 4579 Å with an interference
filter and at 6471 Å with a Claassen filter. S denotes a normal Raman band of the solvent
(acetonitrile). No normal Raman bands are observed for the supporting electrolyte (TBAP,
tetrabutylammonium perchlorate). (Reproduced with permission from Ref. 82. Copyright
1976 American Chemical Society.)

electrooxidation of Fe(III)(TMP)(OH) in CH_2Cl_2. These spectra were
obtained by using a low-temperature cell mentioned earlier (81). The band
at 841 cm⁻¹ is due to the $\nu(Fe=O)$ of O=Fe(TMP) since it is absent in the
original solution and since it is shifted to 805 cm⁻¹ by ¹⁶O/¹⁸O substitution
(theoretical isotope shift, 38 cm⁻¹). Cooling was necessary because
O=Fe(TMP) is unstable and reacts readily with CH_2Cl_2 to form
Fe(III)(TMP)Cl at higher temperatures.

(b) Diffusion Layer

Using the experimental setup and conditions mentioned earlier, Jeanmaire
and Van Duyne (84) studied one-electron oxidation of N,N,N',N'-tetramethyl-
p-phenylenediamine (TMPD) to its radical cation (TMPD·⁺, Würster's Blue).

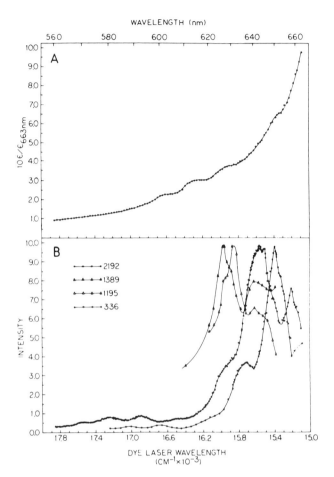

Figure 3-27 Comparison of TCNQ⁻ electronic absorption spectrum and resonance Raman excitation profiles. (a) Electronic absorption spectrum from 15,000 to 17,850 cm⁻¹. The extinction coefficient, ε, scale is normalized with respect to ε at 663.0 nm (15,083 cm⁻¹) = 3.0 × 10³ M⁻¹ cm⁻¹. (b) Superposition of the v_2 (2,192 cm⁻¹). v_4 (1,389 cm⁻¹), v_5 (1,195 cm⁻¹) and v_9 (336 cm⁻¹) excitation profiles. The relative intensity scale has been scaled to 0.0 to 10.0 for all four spectra. (Reproduced with permission from Ref. 83. Copyright 1976 American Chemical Society.)

The RR spectrum of TMPD·⁺, confined to the diffusion layer, was obtained by applying a square-wave voltage (repetition rate, 10 Hz). The excitation was made at 612 nm (CW dye laser) since TMPD·⁺ has its absorption maximum at this wavelength. TMPD in bulk solution does not interfere with the spectrum because it absorbs only in the UV region. Quantitative analysis of the time dependence of the forward and reverse portions of

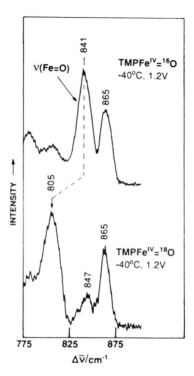

Figure 3-28   Resonance Raman spectra of O = Fe(IV)(TMP) (top trace) and its ^{18}O isotope analogue (bottom trace), generated at $-40°C$ by electrooxidation of Fe(III)(TMP)(OH) at 1.2 V in CH_2Cl_2 containing OH^- and $^{18}OH^-$, respectively. Both spectra were obtained *in situ* via backscattering from the low-temperature Raman spectroelectrochemical cell using 406.7 nm excitation (~ 50 mW) and 8 cm^{-1} slit widths. (Reproduced with permission from Ref. 81. Copyright 1988 John Wiley & Sons, Ltd.)

TMPD$^{\cdot+}$ shows excellent agreement with the expected linear relationship between the RR intensity of the strongest band at 1,628 cm^{-1} (C=C stretch of benzene ring) and the calculated time factor.

3.7   Raman Microscopy

Raman microscopy was developed in the 1970s. Delhaye (85) in 1975 made the first micro Raman measurement. Simultaneously, Rosasco (86, 87) designed a Raman microprobe instrument at the National Bureau of Standards (now the NIST). This early work established the utility of Raman spectroscopy for microanalysis. The technique provides the capability of obtaining analytical-quality Raman spectra with 1 μm spatial resolution using samples in the picogram range. Commercial instruments became available. Today, Instruments S. A. markets the MOLE (Molecular Optics Laser Examiner), and Spex Industries the Micramate.

3.7.1 PRINCIPLES

The major limitations in the design of a Raman microprobe are related to the feeble Raman effect and the minute sample size. It is necessary to optimize the Raman signal, and this is accomplished by taking care in the development of the fore optical configuration to provide a high numerical operative and detector system.

The fore optical configuration is extremely important in optimization of the Raman microprobe. A high numerical aperture (NA) is necessary to collect the light scattered over a large solid angle to assure that more Raman scattered light from the sample is detected. A large-aperture collector is used, which minimizes elastic and inelastic scattering from the substrate. The substrate (88) must have a weak Raman or fluorescence spectrum in the region of interest. Periclase (MgO) can be used as the substrate, although glass slides are routinely used at present. It has no Raman spectrum and has good thermal properties. A microscope glass slide can be used in some cases. The substrate must be optically polished to provide visible observation of the micro-sized crystal. Additionally, it is important to align the excitation beam onto the substrate to prevent specular reflectance[5] from entering the Raman scattering path. A spatial filter is used to minimize other sources of spectral interferences. The design of the microprobe is made to provide totally independent optical paths for the excitation radiation and the scattered radiation, and the two paths are then coupled by the scattering properties of the particle being studied.

For monochannel instruments an efficient micro analyzer is necessary to provide enough stability to allow the measurement to be made, especially for long periods of time.

3.7.2 EXPERIMENTAL

Figure 3-29 illustrates the optical schematic of the Spex Micramate Raman microprobe. The sample being studied is placed on the stage of the microscope and is illuminated by light from the transmission illuminator. The focus on the sample is adjusted by viewing from the optical viewpoint and adjusting the objective. The illuminator lamp is switched off and the laser beam is

---

[5] Specular reflectance is the reflectance spectrum obtained from a flat, clean surface (e.g., a mirror).

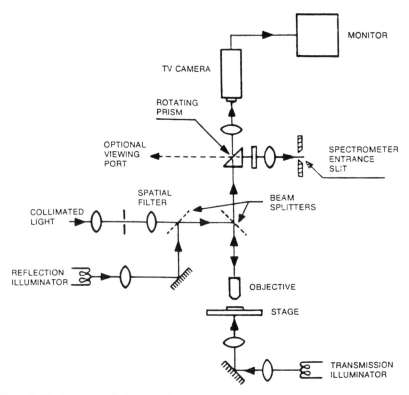

Figure 3-29 Optical path for laser input optics and transfer of the Raman signal to the monochromator for microprobe sampling (e.g., the Spex Micramate). (Reproduced with permission from Spex Industries.)

directed to the beam splitter. The optical viewpoint is turned off and the TV camera is switched on by rotating the prism. The scattered light from the sample is collected by the objective and sent into the spectrometer. The microscope is a modified Zeiss 20 three-turret with detached stage/base. A cooled photomultiplier detector and photon-counting processing is used to provide the necessary sensitivity and low noise.

3.7.3 APPLICATIONS

Only a few applications of the Raman microprobe will be presented. The interested reader can refer to the references cited at the end of this section for further applications.

The Raman microprobe has provided applications in a number of diverse areas of science. Generally, the areas of applications fall into two major categories:

(1)   finger-print identification of microscopic contaminants, and

(2)   characterization of new materials.

(a)   Surface Contaminant Identification

The presence of organic contaminants as small as 1 μm or films as thin as 1 μm on silicon wafers during the manufacturing process of integrated circuits can be readily identified (89). These contaminants can affect the performance of the device and must be identified. Figure 3-30 shows the identification of possible Teflon contaminants. Other techniques, such as IR, x-ray diffraction, Auger and electron microprobe, are insensitive in identifying the nature of the contaminant.

(b)   Biological Compounds

The Raman microprobe has been used to detect foreign bodies in various tissues (89). Figure 3-31 shows spectra of lymph node tissue of 5 μm size,

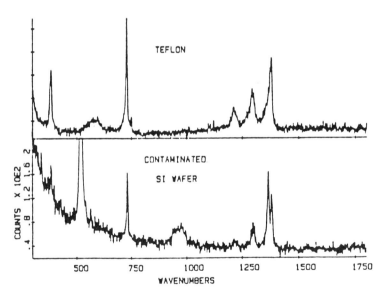

Figure 3-30   Raman microprobe spectrum of fluorinated hydrocarbon contaminant on silicon wafer that had been polished and plasma-etched (lower) and Raman spectrum of polytetra-fluoroethylene (upper). Laser, 135 mW at 514.5 nm. Slits, 300 μm. Time, 0.5 s per data point. (Reproduced with permission from Adar, F., *in* "Microelectronics Processing: Inorganic Materials Characterization" (L. A. Casper, ed.), ACS Symposium Series Vol. 295, pp. 230–239. American Chemical Society, Washington, D.C., 1986. Copyright 1986 American Chemical Society.)

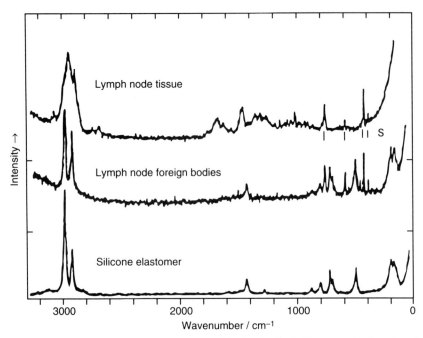

Figure 3-31   RMP spectra obtained in a study of foreign bodies in biopsy section lymph node tissue; S, features of sapphire substrate. (Reproduced with permission from Ref. 89.)

which was obtained by biopsy from a patient. The foreign body was identified as a particle of silicon rubber (dimethyl siloxane). For more biological and medical applications, see Section 4.2.7.

(c)   Inclusions in Solid Inorganic Materials

The nature of solid, liquid or gaseous inclusions that may be found within transparent inorganic glass or crystalline materials can be determined by Raman microprobe techniques without breaking up the sample (90). Other analytical techniques, such as mass spectroscopy or electron microscopy, that may be used to obtain such informqtion require destruction of the original sample. This capability of the microprobe is useful if one wants to analyze inclusions in a material before and after a sample treatment. The only limitation is that the position of the inclusion in the material must be located within the working distance of the objective lens in the microscope (90). Figure 3-32 illustrates the Raman microprobe spectra that were measured for a bubble inclusion that was formed as a defect in a $NaPO_3$ glass during nitriding of the glass with ammonia (91). The glass had been nitrided to increase its mechanical strength. The resulting rotational Raman spectra

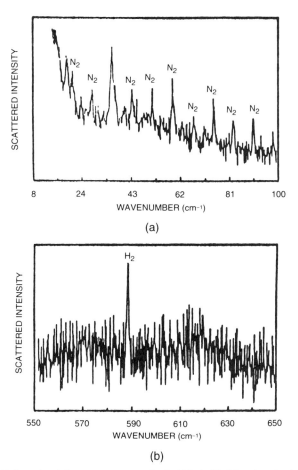

Figure 3-32 (a) Rotational Raman spectrum of a bubble in $NaPO_3$ glass in the 5–100 cm^{-1} region indicating N_2. (b) Rotational Raman spectrum of a bubble in $NaPO_3$ glass in the 550–650 cm^{-1} region indicating H_2. (Reproduced with permission from Ref. 91.)

clearly indicate that the ammonia remaining in the bubble had decomposed into molecular nitrogen and hydrogen during this treatment. If a new solid phase had been formed on the surface of the bubble during the treatment, its nature could also have been identified without destroying the sample by focusing the laser beam of the microprobe on it during spectral measurement.

(d) Surface Mapping

Advances in the Raman microprobe technique have afforded the opportunity to map molecular and crystalline phases on the surface of a sample (89). This is accomplished by integrating a computer-controlled microscope stage into the system software. The sample can be imaged through its Raman signal with 1 μm spatial resolution. This mapping capability can be achieved by combining 3-D display programs with signals acquired with a photomultiplier (PM) tube or with a linear image-intensified silicon photodiode (ISPD). Applications of Raman surface mapping are growing rapidly, and the technique is especially suited for studies of surfaces of semiconductors, diamond films, ceramics and mineralogical samples. Applications, including biological materials, appear to be practical as well.

A weakness of current microprobe systems is that the scattering power at laser wavelengths is low.

For other applications of Raman microscopy refer to Refs. 88, 89, and 95.

3.7.4   FT-RAMAN MICROSCOPY

Messerschmidt and Chase (92) recently demonstrated that it was feasible to use a microscope to obtain the FT-Raman effect. Transfer of laser energy occurs through collection optics in their instrument. Figure 3-33 illustrates the spectrum of a single strand (12 μm in diameter) of Kevlar polymer as

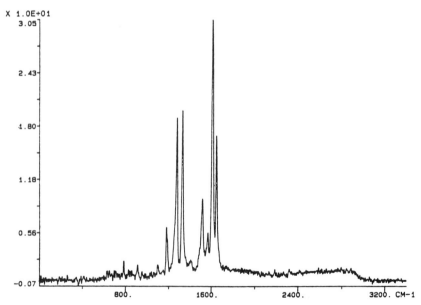

Figure 3-33   Micro FT-Raman spectrum of a single strand of Kevlar fiber with a diameter of 12 μm. (Reprinted with permission from R. G. Messerschmidt and D. C. Chase, *Appl. Spectrosc.* **43**, 11 (1989).)

accomplished by FT-Raman microscopy. Kevlar is difficult to measure by conventional Raman spectroscopy because of fluorescence effects.

Recently, Bruker has introduced an FT-Raman microscopy that is an accessory to the Bruker FRA 106 FT-Raman module, which is an accessory to an FT-IR spectrometer (93). The coupling between the microscope and the Raman module is made by NIR-fiber optics. In the wavelength range of the Raman experiment excited by a Nd:YAG laser, the fiber optics transmission is at a maximum, thus allowing the experiment to be successful (94). Spatial resolution down to 5 μm can be achieved. The technique appears to be capable adjunct to FT-IR microscopy.

For a general discussion on the Raman microprobe, see Ref. 95.

3.8   FT-Raman Spectroscopy[a]

When the conventional Raman effect was discovered by Sir C. N. Raman, the expectations were very high. It was believed that the technique would be adequately applied in the analytical field. Those expectations were not realized, primarily because of fluorescence problems, where fluorescence completely masks the Raman spectrum (see discussion in Chapter 1 on fluorescence). The advent of FT-Raman makes Raman spectroscopy more useful in chemical analyses.

A number of methods have surfaced to reduce or eliminate fluorescence. These are discussed in Section 2.7.5. A new technique can now be added to this list (96–98). This technique involves obtaining the Raman effect using a Nd:YAG laser, which lases at 1.064 μm (9,395 cm^{-1}), and interfacing the Raman module to the FT-IR instrument (97). The technique was first suggested in 1964 by Chantry and Gebbie (96). Because of a lack of technological developments at that time, it took more than 20 years before the technique became viable. At present all FT-IR instruments can be coupled with a Raman accessory to obtain Raman spectra and utilize the FT-IR capabilities with computer manipulations and software programs developed for infrared spectroscopy. These include Bomem, Bruker, Digilab, Nicolet, and Perkin-Elmer FT-IR instruments with interfacing Raman modules. Spex Industries, Inc. markets an FT-R/IR system as well as an FT-R version.

Besides aiding in the fluorescence problems in Raman scattering, FT-Raman can be of aid in other problems of conventional Raman spectroscopy. Raman spectroscopy suffers from a lack of frequency precision, and therefore good spectral subtractions are not possible. In addition, high-resolution experiments are difficult to achieve with conventional Raman spectroscopy. For all three problems, FT-Raman has been shown to be an improvement to conventional Raman spectroscopy.

[a] See General Reference 39 in Chapter 1.

In the conventional Raman experiment, the noise level of the photomultiplier detector is proportional to the square root of the light intensity striking it. Although the S/N ratio increases as the square root of the resolution elements (frequency range), the use of a multiplex spectrometer (interferometer) allows all the scattered energy to bear on the detector simultaneously. The noise increases at the detector by the same amount as the S/N increase for multiplexing, thus cancelling out the two effects (98). For this reason, it was originally believed that FT-Raman spectroscopy would not be too helpful (99). However, the feasibility of FT-Raman has now been demonstrated.

3.8.1 PRINCIPLE

Conventional Raman spectroscopy measures intensity versus frequency or wavenumber. FT instruments, on the other hand, measure the intensity of light of many wavelengths simultaneously. The latter is often referred to as a time-domain spectroscopy. This spectrum is then converted into a conventional spectrum by means of Fourier transformation using computer programs. The waveform in the FT-Raman experiment is illustrated in Fig. 3-34a. Figure 3-34b shows the Fourier transform integral, which produces a peak at frequency ω_0. The sum of the two waveforms of different frequencies in the FT experiment is represented in Fig. 3-34c. The result of the FT integrals given (Fig. 3-34d) is the frequency spectrum with two peaks ω_0 and $10\omega_0$, where the low frequency ω_0 is the same as Fig. 3-34b and $10\omega_0$ is 10 times larger in frequency. The distinctive feature of the FT technique, like FT-IR, is that is sees all of the wavelengths at all times. This provides improved resolution, spectral acquisition time, and S/N ratios over conventional dispersive Raman spectroscopy.

3.8.2 INSTRUMENTATION

As stated earlier, FT-Raman instruments employ a CW Nd:YAG laser with an excitation at 1.064 μm (9,395 cm^{-1}). The use of such a near-infrared laser (NIR) suffers from a 16-fold reduction in signal as compared to a visible laser lasing at 514.5 nm, since the cross-section of Raman scattering follows the v^4 relationship (Section 2.6). The maximum power of the laser comes as high as ~ 5 watts, although less power (~ 1 watt) is generally used.

Several detectors are available for use in the NIR. Cooled PbS, Ge, InGaAs, InSb and platinum silicide have been investigated as detectors for use in FT-Raman. Present commercial instrumentation uses an InGaAs detector. For a discussion on detectors for use in FT-Raman, see Refs. 100 and 102.

An important aspect of FT-Raman instrumentation is the necessity for optical filtering. The first task is to eliminate the stray light at the laser

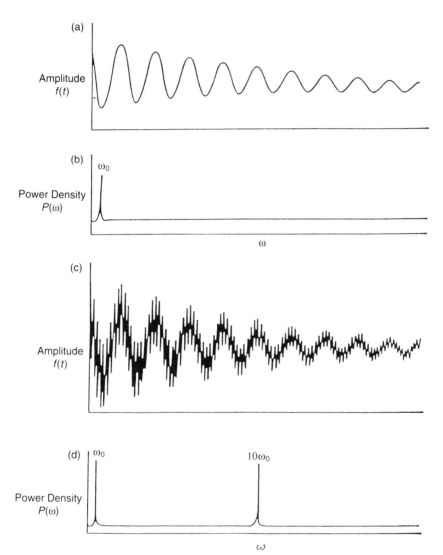

Figure 3-34 FT-Raman is a time-domain spectroscopy that is connected mathematically to a Fourier transform and then into a conventional frequency-domain spectrum using a computer program.

excitation, as it will saturate the detector and electronics. The filtering must be capable of reducing the Rayleigh line, which is 10^6 stronger than the Stokes-shifted lines in the Raman spectrum. In order to be sufficiently efficient and to obtain maximum information in the Raman experiment, the reduction of the Rayleigh line should be comparable to the strongest Raman line.

Dielectric long pass filters are used in series. Several other types of filters are used. These are the Chevron filters (also called the Raman notch filters). The Chevron filter is a set of four sequential narrow-band, high efficiency, laser-line filters. Other types of filters are discussed by Chase (98, 100). Filters are also necessary to remove the optical output of the He–Ne laser (used for referencing) because the laser has optical axes co-linear with the main laser source. Since the detectors used in FT-Raman are sensitive to the He–Ne wavelength, the laser is a source of interference. Here plasma emission filters can be used. The white light of the instrument is filtered with a NIR cutoff filter. A final filter may be used in front of the detector. Figure 3-35 illustrates the optical diagram of a typical FT-Raman spectrometer. It may be observed that the laser radiation is directed to the sample by means of a lens and a parabolic mirror, and the scattered light from the sample is collected and passed to a beam splitter and to the moving and fixed mirrors (the interferometric part). It is then passed through a series of dielectric filters and focused onto a liquid-nitrogen-cooled detector (Ge). Table 3-3 compares the various FT-Raman commercial instruments in regard to spectral range. Table 3-4 compares the advantages and disadvantages of the FT-Raman system. It may be observed that the advantages far outweigh the disadvantages.

2.8.3 APPLICATIONS

The interest in FT-Raman has increased significantly. The capabilities of FT-Raman have been demonstrated in many applications. High-resolution

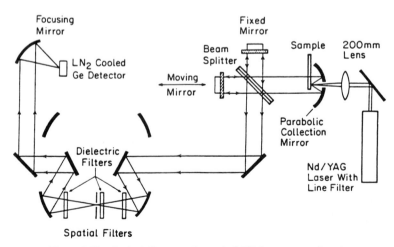

Figure 3-35 Optical diagram of a typical FT-Raman experiment.

Table 3-3 Comparison of Spectral Ranges of Commercial FT-Raman Instrumentation

FT-Raman Instrument[a]	Spectral Ranges[b] (cm^{-1})
Bomem (DA8)	3,500–100
Bruker (IFS 66V, 113V, 120 HR)	3,500–50
Digilab (FTS-40N, FTS-60AN)	3,200–220
Nicolet (800, 60SXR)	3,800–75 or 3,800–300
Perkin-Elmer (1700 X)	3,600–250
Spex (FTR/IR, FT/IR)	3,800–300

[a] Instruments to which Raman modules can be interfaced. Detector is InGaAs and laser is Nd:YAG for all instruments.

[b] High-frequency ranges are dependent on detector used. Low-frequency limit depends on filters. Numbers in tables are taken from instrument company brochures and are subject to changes with time. By the time this book is published the lower limit of the spectral range will probably be close to 40 cm^{-1}.

Table 3-4 Advantages and Disadvantages of FT-Raman System

Advantages	Disadvantages
Reduction or elimination of fluorescence	Absorption in the NIR
High resolution	Low sensitivity
Throughput	Does not allow detection of ppm impurities through spectrum subtraction
Spectral subtraction (can see 1% components) non-resonant scattering	v^4 dependence of intensity
Low-frequency capability[a]	Difficulty in studying samples at temperatures greater than 150°C due to thermal blackbody emission from the sample
Experimental flexibility	
Overheating not a problem in most cases	
Stokes and anti-Stokes Raman collected simultaneously, enabling one to obtain the spectroscopic temperature	
Use of both IR and Raman capabilities with one instrument	

[a] At present, commercial instruments reach 220 cm^{-1}. However, reports prevail that with proper filters ~40 cm^{-1} has been attained.

gas phase FT-Raman spectroscopy has also been demonstrated; this is not possible with conventional Raman scattering. However, Chase (101) has discussed certain limits of the capability of FT-Raman; some of these conclusions are:

(1) FT-Raman will not completely eliminate fluorescence background. Materials that absorb strongly in NIR will present problems (e.g., transition series complexes, transition series doped polymers, charge-transfer conductors, polycyclic aromatic compounds). Other methods of combating fluorescence were discussed in Section 2.7.

(2) It will not displace dispersive visible laser Raman spectroscopy.

(3) It will not detect ppm impurities through spectral subtraction.

(4) A serious problem of using a Nd:YAG laser to excite FT-Raman is the difficulty of attempting to study samples at temperatures greater than 150°C. The thermal blackbody emission from the sample becomes more intense (broad background) than the Raman signal. The S/N ratio is lowered and the detector becomes saturated. These types of studies might be accomplished by the use of a pulsed laser.

(a) Spectrum of Rhodamine (Dye, Control of Fluorescence Problem)

Fluorescence and thermal decomposition problems are extremely severe in conventional Raman spectroscopy whenever one studies laser dyes. Figure 3-36 shows the spectrum of rhodamine, which shows the suppression of fluorescence and which was not obtainable in conventional Raman spectroscopy (98, 102).

(b) Adsorbed Species

Previous studies of adsorbed species utilize infrared spectroscopy. Because of experimental difficulties with conventional Raman, studies of adsorbed species have lagged. Developments in FT-Raman now indicate that the technique offers major advantages in studying adsorbed species. Figure 3-37 shows FT-Raman spectra of pyridine adsorbed on ZY zeolite (103).[6] Pyridine is absorbed at high concentrations initially, and the spectrum shows only physisorbed pyridine species. As the pyridine desorbs, the 1,001 cm^{-1} band appears. The band at 1,030 shifts to 1,035 cm^{-1}, and the CH stretching band at 3,056 cm^{-1} moves to 3,067 cm^{-1}. These frequencies correspond to what is observed for aqueous pyridine. No bands could be attributed to coordinated pyridine although this may be due to a lack of penetration of the laser light.

---

[6]Zeolites are three-dimensional channeled networks of interconnected cavities of different sizes containing AlO_4 and SiO_4 tetrahedra joined by shared oxygen.

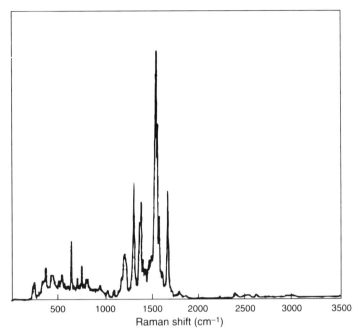

Figure 3-36 FT-Raman spectrum of rhodamine-6G. (Reproduced with permission from Ref. 102. Copyright 1986 American Chemical Society.)

(c) Spectra of Biologically Important Molecules

Biological materials have always posed a problem in conventional Raman spectroscopy. Mainly, the problem arises from the extreme fluorescence these materials possess, as well as the fragile nature and high elastically scattered background of such samples. Figure 3-38 shows the FT-Raman spectrum of bovine serum albumin (101). The incident power was kept at 200 mW to avoid thermal damage to the sample. Whereas biological materials were difficult or impossible to measure by conventional Raman spectroscopy, they now can be accomplished by FT-Raman spectroscopy.

(d) Forensic Identification of Illicit Drugs and Explosives

Conventional Raman spectroscopy has not been routinely used in forensic laboratories for the identification of illicit drugs and explosives because of the high background and time-consuming sample alignment. FT-Raman has overcome some of these problems. Figure 3-39 shows the FT-Raman spectra of the plastic Semtex[7] and 1,3,5-trinitro-1,3,5-triazene (RDX). It is seen that the chief explosive component of Semtex is RDX (104).

---

[7] Semtex is a plastic explosive containing several ingredients, chief of which is RDX.

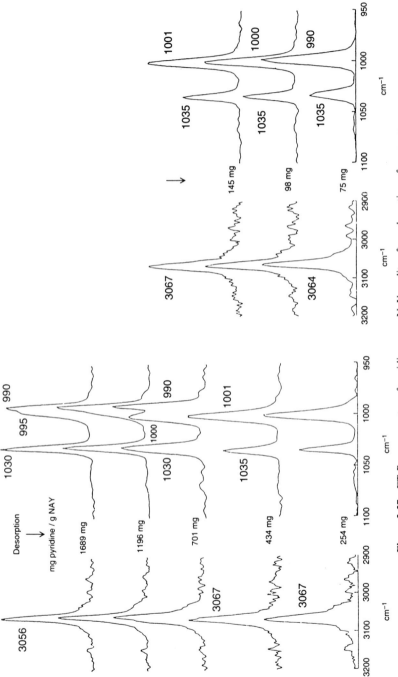

Figure 3-37 FT-Raman spectra of pyridine on a NaY zeolite after adsorption of an excess, followed by gradual removal as indicated by the mass of remaining adsorbed pyridine. (Reproduced from R. Burch, C. Passingham, G. M. Warnes, and D. J. Rawlence, *Spectrochim Acta* **46A**, 243, Copyright 1990, with permission from Pergamon Press Ltd., Headington Hill Hall, Oxford OX3 0BW, UK.)

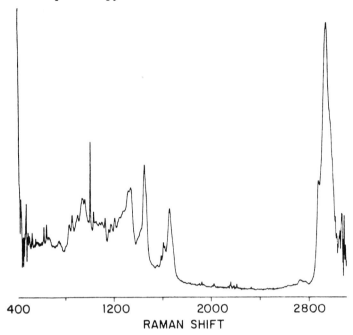

Figure 3-38   FT-Raman spectrum, 200 mW, bovine serum albumin (powder)—5,000 scans (101).

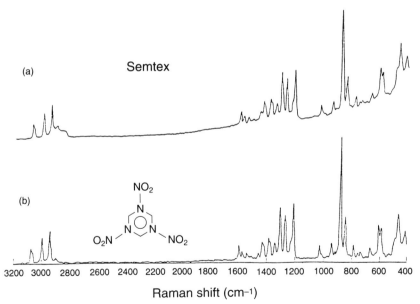

Figure 3-39   FT-Raman spectra of (a) Semtex; (b) the high explosive 1,3,5-trinitro-1,3,5-triazine (RDX). (Reproduced with permission from C. M. Hodges and H. Akhavan, *Spectrochim Acta* **46A**, 303, Copyright 1990, with permission from Pergamon Press Ltd., Headington Hill Hall, Oxford OX3 0BW, UK.)

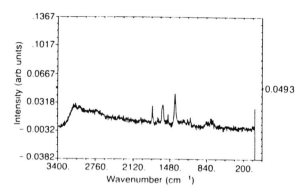

Figure 3-40 FT-Raman spectrum of a polyimide acquired in 60 min with 1.0 W of power and 500 scans. Both spectra 200 scans, 6 cm^{-1} resolution, scanning time 12 min. (Reproduced with permission from Ref. 105.)

(e)   Polymer Samples

Fluorescence problems have been a nuisance in studying various polymers. This situation has benefited from the advent of FT-Raman. Figure 3-40 shows the spectrum of a polyimide, where it is demonstrated that fluorescence is eliminated (105).

For further reading on this subject, see Refs. 106–111.

3.9   New Raman Imaging Spectrometry

A new technique for spectroscopic imaging microscopy has been introduced (112, 113). The technique involves an acoustic-optic tunable filter (AOTF) and a charge-coupled device (CCD) detector with an infinity-corrected microscope, and can be operated for chemical imaging for large areas (>20 km^2)—e.g., airborne and spaceborne remote-sensing imaging spectrometers— and on a micro-scale level—e.g., Raman (114, 115) and infrared (116) microscopy.

For measurements involving Raman scattering, a broad-band source is allowed to strike the sample, and the AOTF unit is positioned in front of the imaging detector. Figure 3-41 shows a schematic of the acoustic-optic filtered Raman imaging spectrometer. It consists of image collection and transfer optics, a TeO$_2$, AOTF crystal, and a slow scan, liquid-nitrogen-cooled silicon CCD detector.

The AOTF Raman imaging technique has several advantages in that no moving parts are involved. For Raman measurements Raman images are collected in several seconds and discrimination between multiple components

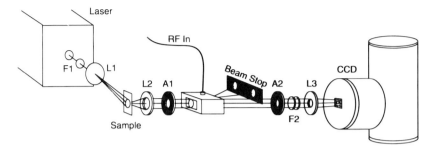

Figure 3-41 Schematic representation of the acousto-optic filtered Raman imaging spectrometer. The optical train shows the path of the incident laser beam and Raman-scattered light. The lipid/amino acid sample is mounted on a microscope slide positioned 45° relative to the incident laser; 90° Raman scattering is collected and spectrally filtered with the AOTF. Holographic Raman filters are placed after the AOTF to eliminate intense Rayleigh scatter before the image is focused onto a liquid-nitrogen-cooled CCD. (Reproduced with permission from Ref. 115.)

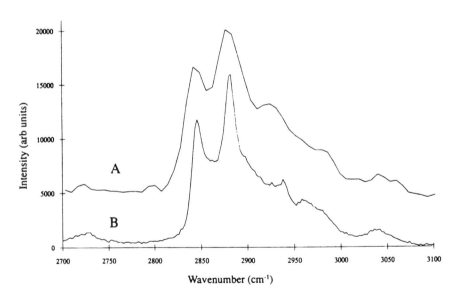

Figure 3-42 CH stretching mode region (2,700–3,100 cm^{-1}) Raman spectra of DPPC collected with the acousto-optic filtered imaging spectrometer (a) a dispersive scanning monochromator at 5 cm^{-1} resolution (b). Fourier self-deconvolution is applied to spectrum (a). The two spectra show similar features due to methyl and methylene vibrations arising from both the hydrocarbon chain and headgroup portions of the lipid. (Reproduced with permission from Ref. 116.)

in complex matrices is possible. High-quality Raman spectra of spatially resolved microscopic regions are easily obtained. This is an improvement over the conventional Raman microprobe, which provides limited imaging capabilities. The advantage of source filtering by AOTF for absorption studies is the greatly reduced optical power densities incident on the sample. These reduced power densities minimize thermal and photodecomposition of the sample and are particularly useful for sensitive biological samples.

Figure 3-42 illustrates the Raman spectra of dipalmtolylphosphatidylcholine (DPPC) in the CH stretching region as obtained by the AOTF method, and compared to the spectra obtained by a dispersive scanning monochromator. The results are similar, but certain features are enhanced in the AOTF technique. At this writing, the application of the technique is in its infancy. However, the technique has great potential as a tool for studies, especially in biochemistry and biophysics, and many applications will be forthcoming. For further details see References 112–116.

References

1. D. A. Long, "Raman Spectroscopy," pp. 219–242. McGraw-Hill, New York, 1977.
2. P. Esherick and A. Owyoung, "High resolution stimulated Raman spectroscopy," *in* "Advances in Infrared and Raman Spectroscopy" (R. J. H. Clark and R. E. Hester, eds.), Vol. 9, pp. 130–187. John Wiley, New York, 1982.
3. J. W. Nibler, W. M. Schaub, J. R. McDonald, and A. B. Harvey, "Coherent anti-Stokes Raman spectroscopy," *in* "Vibrational Spectra and Structure" (J. Durig, ed.), Vol. 6, pp. 173–225, Elsevier, Amsterdam 1977.
4. J. J. Barrett, *Appl. Spectrosc. Rev.* **21**, 419 (1985).
5. "Advances in Spectroscopy" (R. J. H. Clark and R. E. Hester, eds.), Vol. 15, John Wiley, New York, 1988. Contains several review articles on CARS.
6. M. D. Levenson and S. S. Kano, "Introduction to non-linear laser spectroscopy" (Revised edition). Academic Press, New York, 1988.
7. A. B. Harvey, ed., "Chemical Applications of Nonlinear Raman Spectroscopy," pp. 1–383. Academic Press, New York, 1981.
8. H. Hamaguchi, "Transient and time-resolved Raman spectroscopy of short-lived intermediate species," *in* "Vibrational Spectra and Structure" (J. Durig, ed.), Vol. 16, p. 227. Elsevier, Amsterdam, 1987.
9. D. E. Morris and W. H. Woodruff, "Vibrational spectra and the structure of electronically excited molecules in solution," *in* "Advances in Spectroscopy" (R. J. H. Clark and R. E. Hester, eds.), Vol. 14, p. 285. John Wiley, New York, 1987.
10. "Time resolved spectroscopy," *in* "Advances in Spectroscopy" (R. J. H. Clark and R. E. Hester, eds.), Vol. 18, pp. 1–406. John Wiley, New York, 1989.
11. G. H. Atkinson, "Time-resolved Raman spectroscopy," *in* "Advances in Infrared and Raman Spectroscopy" (R. J. H. Clark and R. E. Hester, eds.), Vol. 9. John Wiley, New York, 1982.
12. R. F. Dallinger, S. Farquharson, W. H. Woodruff, and M. A. J. Rodgers, *J. Am. Chem. Soc.* **103**, 7433 (1981).
13. P. G. Bradley, N. Kress, B. A. Hornberger, R. F. Dallinger, and W. H. Woodruff, *J. Am. Chem. Soc.* **103**, 7441 (1981).

14. P. K. Mallick, D. P. Strommen, and J. R. Kincaid, *J. Am. Chem. Soc.* **112**, 1686 (1990).
15. E. Whittle, D. A. Dows, and G. C. Pimentel, *J. Chem. Phys.* **22**, 1943 (1954).
16. H. E. Hallam, ed., "Vibrational Spectroscopy of Trapped Species." John Wiley, New York, 1973.
17. M. Moskovits and G. A. Ozin, eds., "Cryochemistry." John Wiley, New York, 1976.
18. A. J. Downs and M. Hawkins, "Raman studies of molecules in matrices," *in* "Advances in Infrared and Raman Spectroscopy" (R. J. H. Clark and R. E. Hester, eds.), Vol. 10, pp. 1–109. John Wiley, New York, 1983.
19. M. J. Almond and A. J. Downs, "Spectroscopy of matrix isolated species," *in* "Advances in Spectroscopy" (R. J. H. Clark and R. E. Hester, eds.), Vol. 17, pp. 1–511. John Wiley, New York, 1989.
20. H. J. Jodl, "Raman spectroscopy on matrix isolated species," *in* "Vibrational Spectra and Structure" (J. R. Durig, ed.), Vol. 13, pp. 285–426. Elsevier, Amsterdam, 1984.
21. A. J. Barnes, J. C. Bignall, and C. J. Purnell, *J. Raman Spectrosc.* **4**, 159 (1975).
22. J. S. Shirk and H. H. Claassen, *J. Chem. Phys.* **54**, 3237 (1971).
23. L. M. Proniewicz, I. R. Paeng, and K. Nakamoto, *J. Am. Chem. Soc.* **113**, 3294 (1991).
24. W. Scheuermann and K. Nakamoto, *Appl. Spectrosc.* **32**, 251 and 302 (1978).
25. J. W. Nibler and D. A. Coe, *J. Chem. Phys.* **55**, 5133 (1971).
26. W. F. Howard, Jr., and L. Andrews, *Inorg. Chem.* **14**, 767 (1975); L. Andrews, "Spectroscopy of molecular ions in noble gas matrices," *in* "Advances in Infrared and Raman Spectroscopy" (R. J. H. Clark and R. E. Hester, eds.), Vol. 7. John Wiley, New York, 1980.
27. J. R. Ferraro, "Vibrational Spectroscopy at High Pressures—The Diamond Anvil Cell" pp. 1–264. Academic Press, New York, 1984.
28. J. A. Xu, H. K. Mao, and P. M. Bell, *Science* **232**, 1404 (1986); R. J. Hemley, P. M. Bell, and H. K. Mao, *Science* **237**, 605 (1987).
29. C. E. Weir, E. R. Lippencott, A. Van Valkenburg, and E. N. Bunting, *J. Res. Natl. Bur. Stand. Sec. A* **63**, 55 (1959).
30. R. A. Forman, G. J. Piermarini, J. D. Barnett, and S. Block, *Science* **176**, 284 (1972).
31. A. Jayaraman, *Science* **250**, 54 (1984).
32. M. G. Gonikberg, K. H. E. Stein, S. A. Unkholm, A. A. Opekerinov, and V. T. Aleksanias, *Opt. Spectrosc.* (Engl. Transl.) **6**, 166 (1959).
33. C. Postmus, V. A. Maroni, and J. R. Ferraro, *Inorg. Nucl. Chem. Lett.* **4**, 269 (1968).
34. J. W. Brasch and E. R. Lippencott, *Chem. Phys. Lett.* **2**, 99 (1968).
35. R. Pucci and G. Piccitto, "Molecular systems under high pressure," *in* "Proceedings of the II Archimedes Workshop on Molecular Solids under Pressure" (R. Pucci and G. Piccitto, eds.), Vol. 61, pp. 1–388. Elsevier, Amsterdam, 1991.
36. D. M. Adams, P. D. Hatton, A. C. Shaw, and T. K. Tan, *J. Chem. Soc. Chem. Commun.* 226 (1981).
37. E. Wigner and H. B. Huntington, *J. Chem. Phys.* **3**, 764 (1935).
38. F. V. Gigor'yev, S. B. Kormer, D. L. Mikhaylova, A. P. Tolochko, and V. D. Urlin, *Zh. Eksp. Teor. Fiz. Pisma Red.* **5**, 286 (1972).
39. H. K. Mao and R. J. Hemley, *Science* **244**, 1462 (1989); R. J. Hemley and H. K. Mao, *Science* **249**, 391 1990).
40. J. H. Eggert, F. Moshary, W. J. Evans, H. E. Lorenzana, K. A. Goettel, J. F. Silvera, and W. C. Moss, *Phys. Rev. Lett.* **66**, 193 (1991).
41. A. L. Ruoff and C. A. Vanderborgh, *Chem. & Eng. News*, March 11, 1991, pp. 5–6.
42. L. G. Liu, *Phys. Earth Planet Index* **11**, 281 (1976).
43. T. Yagi, H. K Mao, and P. M. Bell, *Phys. Chem. Minerals* **3**, 97 (1978); *in* "Physical Geochemistry" (S. K. Sakema, ed.), Vol. 2, p. 317. Springer-Verlag, Berlin, 1982.
44. E. Knittle and R. Jeanloz, *Science* **235**, 668 (1987).
45. R. D. Hemley, R. E. Cohen, A. Yeganch-Haeri, H. K. Mao, D. J. Weidner, and E. Ito,

"Raman spectroscopy and lattice dynamics of $MgSiO_3$. Perovskite at high pressure," *in* "Perovskite: Structure of Great Interest to Geophysics and Materials Science" (A. Navrotsky and D. J. Weidner, eds.), pp. 35–44. American Geophysical Union, Washington, D.C., 1989.

46. J. R. Ferraro, *Coord. Chem. Rev.* **29**, 1 (1979).
47. J. R. Ferraro, *Coord. Chem. Rev.* **43**, 205 (1982).
48. J. R. Ferraro and L. J. Basile, *Appl. Spectroc.* **28**, 505 (1974).
49. A. Jayaraman, *Rev. Mod. Phys.* **55**, 65 (1983).
50. A. Jayaraman, *Science* **250**, 54 (1984).
51. W. F. Sherman, *Bull. Soc. Chim. Fr.* **9–10**, 347 (1982).
52. W. F. Sherman, "Raman and infrared spectroscopy of crystals at various pressures and temperatures," *in* "Advances in Infrared and Raman Spectroscopy" (R. J. H. Clark and R. J. Hester, eds.), Vol. 6, Chapter 4, pp. 158–336. Heyden, London, 1980.
53. H. G. Drickamer and C. W. Frank, "Electronic Transitions and the High Pressure Chemistry and Physics of Solids," pp. 1–220. Chapman and Hall, London, 1973.
54. A. J. Melvegar, J. W. Brasch and E. R. Lippencott, "High pressure vibrational spectroscopy," *in* "Vibrational Spectra and Structure" (J. R. Durig, ed.), pp. 51–72. Marcel Dekker, New York, 1972.
55. M. Fleischman, P. J. Hendra, and J. McQuillan, *Chem. Phys. Lett.* **26**, 163 (1974).
56. D. L. Jeanmaire and R. P. Van Duyne, *J. Electroanal. Chem.* **84**, 1 (1977).
57. M. G. Albrecht and J. A. Creighton, *J. Am. Chem. Soc.* **99**, 5215 (1977).
58. R. L. Garrell, *Anal. Chem.* **61**, 401A (1989), and references therein.
59. B. Chase and B. A. Parkinson, *Appl. Spectrosc.* **42**, 1186 (1988).
60. I. Pochrand, "Surface enhanced Raman vibrational studies at solid/gas interfaces," *in* "Springer Tracts in Modern Physics," Vol. 104, pp. 1–164, and references therein. Springer-Verlag, Berlin, 1984.
61. M. Moskovits, *Rev. of Mod. Phys.* **57**, 783 (1985).
62. D. A. Weitz, M. Moskovits, and J. A. Creighton, *in* "Chemical Structure at Interfaces" (R. B. Hall and A. B. Ellis, eds.), Chapter 5, p. 197. VCH, Deerfield Beach, Florida, 1986.
63. X. Jiang and A. Campion, *Chem. Phys. Lett.* **140**, 95 (1987).
64. P. Hildebrandt, K. A. Macor, and R. S. Czernuszewicz, *J. Raman Spectrosc.* **19**, 65 (1988).
65. A. M. Alak and T. Vo. Dinh, *Anal. Chem.* **59**, 2149 (1987).
66. M. M. Carabba, R. B. Edmonds, and R. D. Rauk, *Anal. Chem.* **59**, 2559 (1987).
67. R. L. Garrell, *Anal. Chem.* **61**, 401A (1989).
68. R. K. Chang and B. L. Laube, *CRC Critical Review in Solid State and Materials Sciences* **12**, 1 (1984).
69. P. B. Dorain, *Phys. Chem.* **90**, 5808 (1986).
70. D. L. Jeanmaire and R. P. Van Duyne, *J. Electroanal. Chem.* **84**, 1 (1977).
71. H. Yamada, *Appl. Spectrosc. Rev.* **17**, 227 (1986).
72. E. Koglin and J. M. Séquaris, "Surface enhanced Raman scattering of biomolecules," *Topics in Current Chemistry* **134**, 1 (1986).
73. T. M. Cotton, "The applications of surface-enhanced Raman scattering to biochemical systems," *in* "Spectroscopy of Surfaces" (R. J. H. Clark and R. E. Hester, eds.), Vol. 16. Chapter 3, p. 91, and references therein. John Wiley, New York, 1988.
74. J. Creighton, "The selection rules for surface-enhanced Raman spectroscopy," *in* "Spectroscopy of Surfaces" (R. J. H. Clark and R. E. Hester, eds.), Vol. 16, Chapter 2, p. 37, and references therein. John Wiley, New York, 1988.
75. R. Aroca and G. J. Kouracs, "Surface enhanced Raman spectroscopy," *in* "Vibrational Spectra and Structure" (J. R. Durig, ed.), Vol. 19, pp. 55–112. Elsevier, New York, 1991.
76. T. Takenaka and J. Uemura, "Application of infrared and Raman spectroscopy to the study of surface chemistry," *in* "Vibrational Spectra and Structure" (J. R. Durig, ed.), Vol. 19, pp. 215–314. Elsevier, New York, 1991.

77. R. J. H. Clark and R. E. Hester, eds., "Advances in Spectroscopy," Vol. 16. John Wiley, New York, 1988. This volume contains several other Reviews related to SERS.

78. P. P. Van Duyne, *J. Phys. (Paris)* **38**, C5-239 (1977).

79. R. P. Cooney, M. R. Mahoney, and A. J. McQuillan, "Raman spectroelectrochemistry," *in* "Advances in Infrared and Raman Spectroscopy" (R. J. H. Clark and R. E. Hester, eds.), Vol. 9, pp. 188–278. Heyden, London, 1982.

80. D. L. Jeanmaire and R. P. Van Duyne, *J. Am. Chem. Soc.* **97**, 1699 (1975).

81. R. S. Czernuszcewicz and K. A. Macor, *J. Raman Spectrosc.* **19**, 553 (1988).

82. D. L. Jeanmaire and R. P. Van Duyne, *J. Am. Chem. Soc.* **98**, 4029 (1976).

83. D. L. Jeanmaire and R. P. Van Duyne, *J. Am. Chem. Soc.* **98**, 4034 (1976).

84. D. L. Jeanmaire and R. P. Van Duyne, *J. Electroanal. Chem.* **66**, 235 (1975).

85. M. Delhaye and P. Dhamelincourt, *J. Raman Spectrosc.* **3**, 33 (1975).

86. G. J. Rosasco, E. S. Etz, and W. A. Cassatt, *Appl. Spectrosc.* **29**, 396 (1975).

87. G. J. Rosasco, *in* "Advances in Infrared and Raman Spectroscopy" (R. J. H. Clark and R. J. Hester, eds.), Vol. 7, Chapter 4, pp. 223–282, and references therein. Heyden, London, 1980.

88. F. Adar, *Microchemical Journal* **38**, 50 (1988).

89. J. M. Lerner and F. Adar, *Laser Focus World*, Pennwell Publishing Co., Feb. 1989.

90. S. W. Lee, R. A. Condrate, Sr., N. Malani, D. Tammaro, and F. E. Woolley, "Considerations concerning the application of the Raman microprobe technique to the systematic analysis of bubbles in glass," *Spectrosc. Lett.* **23**(7), 945–966 (1990).

91. S. W. Lee and R. A. Condrate, Sr., "Raman microprobe investigation of bubbles formed in $NaPO_3$ glass from nitriding with ammonia," *J. Solid State Chem.* **98**, 423–425 (1992).

92. R. G. Messerschmidt and D. B. Chase, *Appl. Spectrosc.* **43**, 11 (1989).

93. Bruker, personal communication.

94. B. Schrader, *in* "Practical Fourier Transform Infrared Spectroscopy, Industrial and Laboratory Chemical Analysis" (J. R. Ferraro and K. Krishnan, eds.), pp. 167–202. Academic Press, San Diego, 1990.

95. J. J. Blaha, "Raman microprobe spectroscopic analysis," *in* "Vibrational Spectra and Structure" (J. R. Durig, ed.), Vol. 10, pp. 227–267. Elsevier, New York, 1981.

96. G. W. Chantry, H. A. Gebbie, and C. Helsum, *Nature* **203**, 1052 (1964).

97. T. Hirschfeld and D. B. Chase, *Appl. Spectrosc.* **40**, 133 (1986).

98. D. B. Chase, *Anal. Chem.* **59**, 881A (1987).

99. T. Hirschfeld and E. R. Schildkraut, *in* "Laser Raman Gas Diagnostics" (M. Lapp, ed.), pp. 379–388. Plenum Press, New York, 1974.

100. T. Hirschfeld and D. B. Chase, *Appl. Spectrosc.* **40**, 133 (1986).

101. D. B. Chase, *Mikrochim Acta* **111**, 81, (1987).

102. D. B. Chase, *J. Am. Chem. Soc.* **108**, 7485 (1986).

103. R. Burch, C. Passingham, G. M. Warnes, and D. J. Rawlence, *Spectrochim. Acta* **46A**, 243 (1990).

104. C. M. Hodges and J. Akhavan, *Spectrochim. Acta* **46A**, 303 (1990).

105. F. J. Purcell, *Spectroscopy* **4**(2), 24 (1989).

106. D. B. Chase, "FT-Raman spectroscopy," *in* "Proceedings of Eleventh International Conference on Raman Spectroscopy" (R. J. H. Clark and D. A. Long, eds.), pp. 39–41. J. Wiley & Sons, New York, 1988.

107. B. Schrader, A. Simon, and M. Tischer, "NIR-FT-Raman spectrometry, new techniques and applications," *ibid.*, pp. 959–960.

108. D. J. Cutler, P. J. Hendra, H. M. Mould, R. A. Spragg, and A. J. Turner, "Practical Considerations in NIR FT-Raman Spectroscopy," *ibid.*, p. 964.

109. D. J. Cutler, *Spectrochima Acta* **46A**, 231 (1990).

110. P. J. Hendra and H. M. Mould, International Laboratory, April 1990. See also Refs. 98, 101, 103, and 106.

111. P. J. Hendra, C. Jones, and G. Warner, "Fourier Transform Raman Spectroscopy." Prentice Hall, New Jersey, 1992.
112. P. J. Treado, I. W. Levin, and E. N. Lewis, *Applied Spectrosc.* **46**, 553 (1992).
113. C. D. Tran, *Anal. Chem.* **64**, 971A (1992).
114. P. J. Treado and M. D. Morris, *Spec. Acta Rev.* **13**, 355 (1990).
115. P. J. Treado and I. W. Levin, *Applied Spectrosc.* **46**, 1211 (1992).
116. P. J. Treado and M. D. Morris, *in* "Spectroscopic and Microscopic Imaging of the Chemical State" (M. D. Morris, ed.). Marcel Dekker, New York, 1992.

Chapter 4

Applications

Raman spectroscopy has a variety of applications. As discussed in Section 1.8, Raman and IR spectroscopies are complementary, and both should be utilized whenever possible. Currently, IR is more common than Raman because IR instruments are more available in most laboratories and less expensive. However, Raman spectroscopy has a great potential to grow because it provides unique applications that cannot be obtained by IR spectroscopy (Chapter 3). In the following, we have chosen several examples for each application that we regard as important. To compensate for any unbalanced presentation on our part, the reader should consult many review articles and reference books cited in this and other chapters. For convenience, this chapter is divided into four sections: Applications to Structural Chemistry; Applications to Biochemistry, Biology, and Medicine; Solid State Applications; and Industrial Applications.

4.1 Applications to Structural Chemistry

Among a variety of spectroscopic methods, vibrational spectroscopy is most commonly used in structural chemistry. IR/Raman spectroscopy provides information about molecular symmetry of relatively small molecules and functional groups in large and complex molecules. Furthermore, Raman spectroscopy enables us to study the structures of electronically excited

Table 4-1 Number of Fundamentals for Tetrahedral XeF_4

T_d	Activity	Number of Fundamentals	$\nu(XeF)$ Stretching	$\delta(FXeF)$ Bending
A_1	R	1	1	0
A_2	ia[a]	0	0	0
E	R	1	0	1
F_1	ia	0	0	0
F_2	IR, R	2	1	1
Total	IR	2	1	1
	R	4	2	2

[a] ia = inactive.

molecules and unstable species produced by laser photolysis at low temperatures. Several other applications that are important in structural chemistry are also discussed in this section.

4.1.1 STRUCTURE DETERMINATION BY SYMMETRY SELECTION RULE

When a molecule is relatively small and/or belongs to a point group of relatively high symmetry, it is possible to elucidate the molecular structure by using the symmetry selection rules discussed in Section 1.14. Molecules of XY_2 (linear $\mathbf{D}_{\infty h}$ or bent $\mathbf{C}_{2v}$), XY_3 (planar $\mathbf{D}_{3h}$ or pyramidal $\mathbf{C}_{3v}$), XY_4 (square-planar $\mathbf{D}_{4h}$ or tetrahedral $\mathbf{T}_d$) and XY_5 (trigonal–bipyramidal $\mathbf{D}_{3h}$ or tetragonal–pyramidal $\mathbf{C}_{4v}$) types may take one of the structures indicated in parentheses. Since the number of IR/Raman-active vibrations is different for each structure, the most probable structure can be chosen by comparing the number of observed IR/Raman bands with that predicted for each structure by symmetry selection rules.

Vibrational spectroscopy played the major role in structure determination when XeF_4 was first prepared by Claassen et al. (1) in 1962. Tables 4-1 and 4-2 show the number and IR/Raman-activity of fundamental vibrations predicted for tetrahedral and square-planar XeF_4 molecules. (These results can be obtained via Appendices 1 and 2.) Group theory predicts that the tetrahedral structure should exhibit two $\nu(XeF)$ and two $\delta(FXeF)$, while the square-planar structure should exhibit two $\nu(XeF)$ and only one $\delta(FXeF)$ in the Raman spectrum.[1] The observed Raman spectrum (2) shows two $\nu(XeF)$ at 554 and 524 cm^{-1} and one $\delta(FXeF)$ at 218 cm^{-1} in agreement with the

---

[1] In general, nine ($3 \times 5 - 6$) normal vibrations are expected for XY_4-type molecules. As seen in Tables 4-1 and 4-2, this rule holds if we consider that vibrations belonging to E and F species are doubly and triply degenerate, respectively.

Table 4-2 Number of Fundamentals for Square-Planar XeF_4

D_{4h}	Activity	Number of Fundamentals	$\nu(XeF)$ Stretching	$\delta(FXeF)$ Bending
A_{1g}	R	1	1	0
A_{1u}	ia^a	0	0	0
A_{2g}	ia	0	0	0
A_{2u}	IR	1	0	1
B_{1g}	R	1	1	0
B_{1u}	ia	0	0	0
B_{2g}	R	1	0	1
B_{2u}	ia	1	0	1
E_g	R	0	0	0
E_u	IR	2	1	1
Total	IR	3	1	2
	R	3	2	1

^{a}ia = inactive.

square-planar structure. In the IR spectrum, one $\nu(XeF)$ and one $\delta(FXeF)$ are expected for the tetrahedral structure, while one $\nu(XeF)$ and two $\delta(FXeF)$ are predicted for the square-planar structure. The observed IR spectrum exhibits one $\nu(XeF)$ at 586 cm^{-1} and two $\delta(FXeF)$ at 291 and 161 cm^{-1}, again confirming the square-planar structure. The same conclusion can be obtained by simple application of the IR/Raman mutual exclusion principle (Section 1.7), since the point group D_{4h} has a center of symmetry that is lacking in the point group T_d.

Recently, Christe and co-workers (3) obtained the XeF_5^- ion as the tetramethylammonium salt at $-86°C$, and determined its structure by x-ray diffraction as well as vibrational spectroscopy. This anion takes a highly unusual D_{5h} structure, shown in Fig. 4-1, which can be derived from that of a pentagonal bipyramidal IF_7 in which the two axial fluorine ligands are replaced by two sterically active free valence electron pairs. The XeF_5^- anion has 12 $(3 \times 6 - 6)$ normal vibrations that are classified into $1A_1'(R) + 1A_2''(IR) + 2E_1'(IR) + 2E_2'(R) + E_2''$ (inactive) under D_{5h} symmetry. Thus, only three vibrations (A_1' and $2E_2'$) are Raman-active while three vibrations (A_2'' and $2E_1'$) are IR-active. The observed Raman spectrum shown in Fig. 4-2 clearly indicates the presence of three bands at 502, 422 and 377 cm^{-1}, in agreement with the D_{5h} structure.

Buckminsterfullerene (C_{60}) has attracted considerable attention in recent years. As expected from its extremely high symmetry (I_h point group), this molecule exhibits very small numbers of vibrations in IR and Raman spectra (see Section 4.3.6).

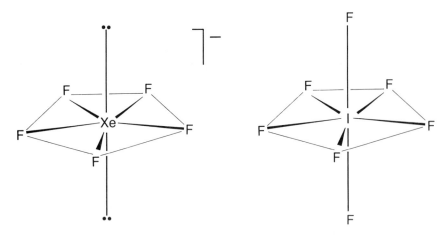

Figure 4-1 Structures of XeF_5^- and XeF_7. (Reproduced with permission from Ref. 3. Copyright 1991 American Chemical Society.)

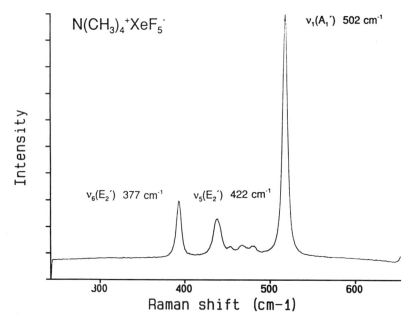

Figure 4-2 Single-crystal Raman spectrum of $N(CH_3)_4 \, XeF_5$ obtained by 514.5 nm excitation. (Reproduced with permission from Ref. 3. Copyright 1991 American Chemical Society.)

Another example is provided by a series of octahedral MX_nY_{6-n} type ions ($n = 0$–6), where M is Pt(IV), Os(IV), and Ir(IV), and X and Y are halogens. Preetz and co-workers (4) prepared these mixed-halogeno ions and assigned their IR/Raman bands based on point group symmetry. Table 4-3 shows the point group and classification of IR/Raman-active fundamental vibrations. Figure 4-3 shows the IR/Raman spectra and band assignments of the $[PtCl_nBr_{6-n}]^{2-}$ series. It should be noted that these ions exhibit $v(PtCl)$, $v(PtBr)$ and bending vibrations in 350–320, 220–190 and below 190 cm^{-1}, respectively. It is of particular interest to see whether symmetry selection rules are obeyed in pairs of stereoisomers. For $n = 3$, there are two isomers, fac (*cis*) and mer (*trans*).

According to Table 4-3, the former should exhibit two $v(PtCl)$ ($A_1(p)$ and $E(dp)$) and two $v(PtBr)$ ($A_1(p)$ and $E(dp)$) both in IR and in Raman spectra. Here, p and dp denote polarized and depolarized vibrations, respectively. This was found to be the case (trace 3a of Fig. 4-3). In contrast, the latter should exhibit three $v(PtCl)$ ($2A_1(p)$ and $B_1(dp)$) and three $v(PtBr)$ ($2A_1(p)$ and $B_2(dp)$) both in IR and in Raman spectra. Trace 3b of Fig. 4-3 shows that two polarized bands of A_1-type are observed for each vibration because B-type vibrations are weaker than A-type vibrations. Thus, comparison of Raman spectra of the two isomers should be made using A-type (polarized) vibrations.

As seen earlier, some fundamental vibrations are relatively weak. Furthermore, some overtone and combination bands become unusually strong when Fermi resonance (accidental degeneracy) occurs. A typical example is given by CO_2, where the frequency of the first overtone of the v_2(667 cm^{-1}) is very close to that of the v_1 fundamental (1,337 cm^{-1}). Since v_1 and $2v_2$ belong to the same symmetry species (Σ_g^+), they interact with each other to give rise to two strong Raman bands at 1,388 and 1,286 cm^{-1}. Finally, it should be noted that the point group symmetry in the crystalline state is not necessarily the same as that in the isolated state. Thus, this method must be applied with caution.

Table 4-3 Number of IR/Raman-Active Vibrations of Octahedral MX_nY_{6-n} Type Molecules (Ref. 4)

Spectrum[a] n	Ion	Point Group	$\nu(MY)$[b]	$\nu(MX)$[b]	Bending[b]
0	MY_6	O_h	$A_{1g} + E_g + F_{1u}$	—	$F_{1u} + F_{2g} + F_{2u}$
1	MXY_5	C_{4v}	$2A_1 + B_1 + E$	A_1	$A_1 + B_1 + B_2 + 3E$
2a	$cis\text{-}MX_2Y_4$	C_{2v}	$2A_1 + B_1 + B_2$	$A_1 + B_1$	$3A_1 + 2A_2 + 2B_1 + 2B_2$
2b	$trans\text{-}MX_2Y_4$	D_{4h}	$A_{1g} + B_{1g} + E_u$	$A_{1g} + A_{2u}$	$B_{2g} + E_g + A_{2u} + B_{2u} + 2E_u$
3a	$fac\text{-}MX_3Y_3$	C_{3v}	$A1 + E$	$A_1 + E$	$2A_1 + A_2 + 3E$
3b	$mer-MX_3Y_3$	C_{2v}	$2A_1 + B_1$	$2A_1 + B_2$	$2A_1 + A_2 + 3B_1 + 3B_2$
4a	$cis\text{-}MX_4Y_2$	C_{2v}	$A_1 + B_1$	$2A_1 + B_1 + B_2$	$3A_1 + 2A_2 + 2B_1 + 2B_2$
4b	$trans\text{-}MX_4Y_2$	D_{4h}	$A_{1g} + A_{2u}$	$A_{1g} + B_{1g} + E_u$	$B_{2g} + E_g + A_{2u} + B_{2u} + 2E_u$
5	MX_5Y	C_{4v}	A_1	$2A_1 + B_1 + E$	$A_1 + B_1 + B_2 + 3E$
6	MX_6	O_h	—	$A_{1g} + E_g + F_{1u}$	$F_{1u} + F_{2g} + F_{2u}$

[a] See Fig. 4-3.
[b] ... Raman-active, —— IR-active.

4.1.2 STRUCTURE-SENSITIVE VIBRATIONS

Although molecules and ions exhibit a number of normal vibrations, some vibrations are inherently sensitive to changes in electronic structure, molecular conformation and intermolecular interaction, while others are not. These structure-sensitive vibrations are known for a number of compounds. Here, we discuss three classes of compounds of biological interest.

(a) Iron porphyrin

Iron porphyrins shown in Fig. 4-4 (M = Fe) are highly important as active sites of biological functions of heme proteins (Section 4.2.1-2). This is largely due to the versatility of the Fe center, which can take a variety of oxidation states (II, III, IV, and V), spin states (high, low, and intermediate), and coordination numbers (four, five, and six). It is, therefore, important to find key bands that are sensitive to these parameters.

As shown in Fig. 1-32, metalloporphyrins exhibit a number of porphyrin core vibrations in which local modes such as $\nu(C{=}C)$ and $\nu(C{=}N)$ are strongly coupled (Section 1.21) due to its planar π-conjugated structure. Several groups of workers (5–8) have carried out normal coordinate analysis on metalloporphyrins. If we consider the simplest metalloporphyrin in which all the peripheral groups are the hydrogen atoms, it should have 105 $(3 \times 37 - 6)$ normal vibrations, which can be classified under $\mathbf{D}_{4h}$ symmetry as shown in Table 4-4. Table 4-5 shows major local coordinates that describe general characters of 35 Raman-active in-plane vibrations (8) together with observed frequencies for Ni(OEP) (Fig. 1-32). These normal mode descriptions are applicable to other metalloporphyrins with minor modifications.

Extensive RR studies on iron porphyrins and heme proteins (9) have shown that all porphyrin core frequencies give negative linear correlations with the core size of the Fe center, and the slopes of these correlations are roughly proportional to the degree of contribution of the methine stretch, $\nu(C_{\alpha}{-}C_m)$ to the normal mode. It is seen from Table 4-5 that ν_3, ν_{10}, ν_{19} and ν_{28} are core-size sensitive. High-spin Fe(III) has a larger core size than low-spin Fe(II) because in the former, the d_{z^2} and $d_{x^2-y^2}$ orbitals contain electrons. As a result, their $C_{\alpha}{-}C_m$ bonds are weaker and the frequencies of core-size marker bands are lower. The ν_3 near 1,500 cm^{-1} is used as a spin state marker band because it is strong under A-term resonance. The ν_3 for Fe(II) complexes are 1,488–1,476 cm^{-1} (high spin) < 1,511–1,493 cm^{-1} (low spin). Similarly, for Fe(III) complexes, the ν_3 are 1,495–1,493 cm^{-1} (high spin) < 1,513–1,504 cm^{-1} (low spin) for Fe(OEP)LL' type compounds.

Frequencies of intermediate spin complexes are close to those of low spin complexes because the d_{z^2} orbital is occupied but the $d_{x^2-y^2}$ orbital is empty. The core size is also influenced by the coordination number and the degree

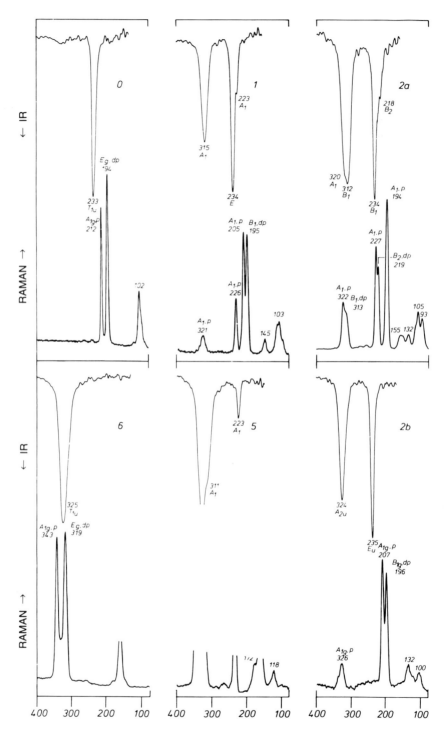

Figure 4-3 IR and Raman spectra of (TBA)$_2$ [PtCl$_n$Br$_{6-n}$], $n = 0$–6. Excitation wavelengths: 647.1 nm for $n = 0$, 2a, 3a, and 4a; 568.2 nm for $n = 1$, 2b, and 3b; and 514.5 nm for $n = 4$b, 5, and 6. (Reproduced with permission from Ref. 4.)

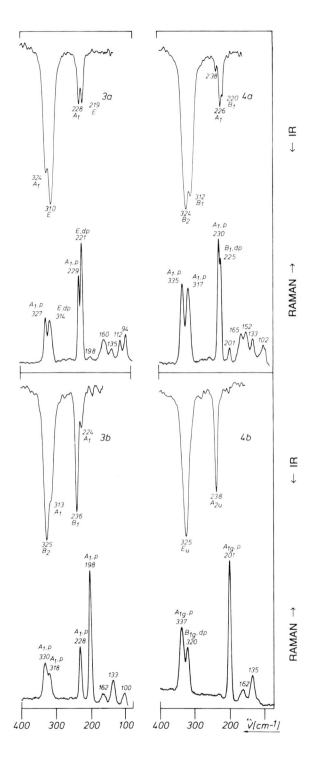

↓ IR

RAMAN →

↓ IR

RAMAN →

3a

324
A_1

310
E

228
A_1

219
E

E,dp
221

A_1,p
229

A_1,p
327

E,dp
314

198

160

135

112

94

4a

238

324
B_2

312
B_1

226
A_1

220
B_1

A_1,p
230

B_1,dp
225

A_1,p
335

A_1,p
317

201

165

152

133

102

3b

224
A_1

313
A_1

236
B_1

325
B_2

A_1,p
198

A_1,p
330

A_1,p
318

A_1,p
228

162

133

100

4b

238
A_{2u}

325
E_u

A_{1g},p
201

A_{1g},p
337

B_{1g},dp
320

162

135

400   300   200   100 $\tilde{v}[cm^{-1}]$

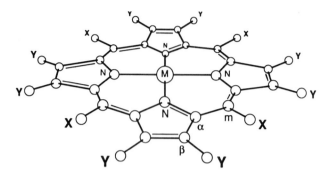

Figure 4-4 Structure of metalloporphyrin: octaethylporphyrin (OEP), X = H and Y = C_2H_5; tetraphenylporphyrin (TPP), X = C_6H_5 and Y = H.

Table 4-4 Classification of Normal Vibrations in Metalloporphyrin of D_{4h} Symmetry

In-Plane	Vibrations	Out-of-Plane	Vibrations
$A_{1g}(R)$	9	A_{1u}(ia)	3
A_{2g}(ia)^a	8	A_{2u}(IR)	6
$B_{1g}(R)$	9	B_{1u}(ia)	5
$B_{2g}(R)$	9	B_{2u}(ia)	4
E_u(IR)	18	$E_g(R)$	8
Total	71		34

$^a A_{2g}$ species become Raman-active under resonance conditions. ia: inactive.

of Fe($d\pi$)–porphyrin (π^*) back donation in Fe(II) complexes. Thus, v_3 must be compared with caution.

The v_4 is the strongest when RR spectra are obtained with Soret band excitation. As shown in Table 4-5, it is mainly due to $v(C_\alpha-N)$, a totally symmetric pyrrole ring breathing mode. This band is well known as an oxidation state marker. It is at $\sim 1,355$ cm^{-1} for Fe(II) and at ~ 1.370 cm^{-1} for Fe(III) with relatively small dependence on spin state. For example, the order of the v_4 are

$$\text{Fe(II)(TPP)(pip)} < \text{Fe(III)(TPP)Cl} < (\text{Fe(IV)(TPP)})_2\text{C}$$

l.s.   1,355   h.s.   1,367   l.s.   1,370 cm^{-1}

When π-acids such as CO and O_2 are bound to Fe(II), this frequency is upshifted to the Fe(III) region because the Fe($d\pi$)–porphyrin(π^*) back donation is reduced by the Fe($d\pi$)–axial ligand(π^*) back donation. Since v_4 is sensitive to the occupancy of the porphyrin(π^*) orbital, it is more generally

Table 4-5 Allocation of Ni(OEP) Skeletal Mode Frequencies (cm^{-1}) to Local Coordinates[a]

Local Mode[b]	A_{1g}	A_{2g}	B_{1g}	B_{2g}	E_u
$\nu(C_m{-}H)$	ν_1 [3,041][c]			ν_{27} [3,041]	ν_{36} [3,040]
$\nu(C_\alpha{-}C_m)_{asym}$		ν_{19} 1,603	ν_{10} 1,655		ν_{37} [1,637]
$\nu(C_\beta{-}C_\beta)$	ν_2 1,602		ν_{11} 1,577		ν_{38} 1,604
$\nu(C_\alpha{-}C_m)_{sym}$	ν_3 1,520			ν_{28} 1,483	ν_{39} 1,501
ν(Pyr. quarter-ring)		ν_{20} 1,393		ν_{29} 1,407	ν_{40} 1,396
ν(Pyr. half-ring)$_{sym}$	ν_4 1,383		ν_{12} 1,343[d]		ν_{41} [1,346]
$\delta(C_m{-}H)$		ν_{21} 1,307	ν_{13} 1,220		ν_{42} 1,231
$\nu(C_\beta{-}C_1)_{sym}$					ν_{44} 1,153
ν(Pyr. half-ring)$_{asym}$	ν_5 1,138	ν_{22} 1,121	ν_{14} 1,131	ν_{30} 1,159	ν_{43} 1,133
$\nu(C_\beta{-}C_1)_{asym}$		ν_{23} 1,058		ν_{31} 1,015	ν_{45} 996
δ(Pyr. def.)$_{asym}$				ν_{32} 938	ν_{46} 927
ν(Pyr. breathing)	ν_6 804		ν_{15} 751		ν_{47} 766[d]
δ(Pyr. def.)$_{sym}$	ν_7 674	ν_{24} 597	ν_{16} 746[e]		ν_{48} 605
δ(Pyr. rot.)		ν_{25} 551	ν_{18} 168	ν_{33} 493	ν_{49} 544
ν(NiN)	ν_8 360/343[f]		ν_{17} 305		ν_{50} [358]
$\delta(C_\beta{-}C_1)_{asym}$		ν_{26} [243]		ν_{34} 197	ν_{51} 328[e]
$\delta(C_\beta{-}C_1)_{sym}$	ν_9 263/274[f]				ν_{52} 263[e]
δ(Pyr. transl.)				ν_{35} 144	ν_{53} 212[e]

[a] Ref. 8) Observed values from CS$_2$ solution RR (A_g and B_g modes) and matrix-isolated IR (E_u modes) spectra.

[b] See Fig. 4-4 for illustration of the local coordinates.

[c] [ ] calculated frequencies; not observed.

[d] Observed only in the meso-d_4 isotopomer and its ^{15}N double isotopomer; not observed in the natural abundance species. Adding the calculated d_4 shift, 12 cm^{-1} to the 1.331 cm^{-1} meso-d_4 frequency gives 1,343 cm^{-1} as the assigned value for ν_{12}.

[e] These frequencies from 12 K RR spectra of tetragonal crystals.

[f] Pairs of frequencies attributed to ethyl orientational isomers.

called a π-electron density marker. For more detailed discussion of these structure-sensitive bands, the reader should consult Ref. 9.

(b)   Peptides and Proteins

The peptide (—CO—NH—) groups in proteins are nearly planar because of resonance stabilization involving the C=O and C—N bonds.

$$
\begin{array}{ccccccccc}
\text{H} & \text{H} & \text{O} & & \text{H} & \text{O} & & \text{H} & \text{O} \\
| & | & || & \psi\;|\;\phi & | & || & & | & || \\
-\text{N}- & \text{C} - & \text{C}-\text{N} & \text{C} & \text{C} & \text{C}- & \text{N}- & \text{C} - & \text{C}- \\
| & | & & | & | & & | & | \\
& \text{R}_1 & & \text{H} & \text{R}_2 & & \text{H} & \text{R}_3
\end{array}
$$

However, the torsional angles (ψ and ϕ) between two —CO—NH— groups can vary depending upon the amino acid residues involved. This, together with hydrogen bonding between the CO of one peptide and the NH of the other, produces several secondary structures of proteins such as α-helix, β-sheet and random coil (10).

   Raman spectroscopy provides structure-sensitive bands for distinguishing these secondary structures. Assignments of the —CO—NH— group vibrations were first made via normal coordinate analysis on N-methylacetamide by Miyazawa et al. (11).

$$
\begin{array}{ccc}
\text{CH}_3 & & \text{H} \\
\searrow & & \nearrow \\
& \text{C}-\text{N} & \\
\diagup\diagup & & \searrow \\
\text{O} & & \text{CH}_3
\end{array}
$$

Table 4-6 lists observed frequencies and band assignments of structure-sensitive amide vibrations. Here, we discuss only amide I and III bands for

Table 4-6   Band Assignments of Amide Vibrations in N-Methylacetamide (cm^{-1})

Raman	IR	Assignment
1,657	1,653	Amide I (80% C=O str.)
—	1,507	Amide II (60% NH in-pl. bend, 40% C—N str.)[a]
1,298	1,299	Amide III (40% C—N str., 30% NH in-pl. bend, 20% CH$_3$—C str.)
—	725	Amide V (NH out-of-pl. bend)[b]
628	627	Amide IV (40% O=C—N bend, 30% CH$_3$—C str.)
600	600	Amide VI (C=O out-of-pl. bend)

[a] in-pl. bend.—bending vibration in the molecular plane.
[b] out-of-pl. bend.—bending vibration in the plane perpendicular to the molecular plane.

Table 4-7 Key Raman Bands of Amino Acid Residues

Amino Acid Residue	Raman Band (cm^{-1})
Phenylalanine (Phe)	1,203 (w), 1,032 (w), 1,004 (s), 624 (w)
Tryptophan (Trp)	1,623 (w),[a] 1,555 (s), 1,436 (s), 1,016 (s), 882 (w), 762 (s)
Tyrosine (Tyr)	Doublet at 850 and 830[b]
Histidine (His)	1,408
Disulfide (S—S) bond	540–510

[a] Strong by 251 nm excitation (see Section 4.2.5).
[b] See Section 4.2.6.

which abundant data are available. The general trends shown in the table below were found by correlating x-ray structural data with Ramanfrequencies.

	Amide I	**Amide III**
α-Helix	1,645–1,600	1,300–1,260
β-Sheet	1,680–1,658	1,243–1,230
Random coil	1,665–1,660	1,243

In general, α-helix exhibits lower amide I and higher amide III than β-sheet and random coil. However, distinction of the latter two is not clear-cut. Amide III is more structure-sensitive than amide I. For example, cobramine B, a small basic protein from cobra venom, contains α-helix, β-sheet and random coil structures. As a result, three amide III bands are observed at 1,270 (α-helix), 1,254 (hydrogen-bonded random coil) and 1,235 cm (β-sheet). However, it exhibits a single strong band at 1,672 cm^{-1}, suggesting a large fraction of β-sheet structure (12). Table 4-7 lists Raman bands of amino acid residues containing the phenyl (Phe, Tyr), imidazole (His) and indole (Trp) rings, together with those of disulfide bonds. These vibrations are sensitive to the environment in which respective amino acid residues are buried in proteins. In some cases, it is possible to enhance vibrations of individual amino acid residues selectively by using UV resonance Raman (UVRR) techniques. Figure 4-5 shows the UVRR spectra of Tyr, Trp–Tyr and Trp obtained by Rava and Spiro (13). It is seen that Tyr and Trp vibrations of tryptophyltyrosine (Trp–Tyr) are selectively enhanced by using 200 and 218 nm excitation, respectively. Recently, Asher (14) reviewed the applications of UVRR spectroscopy to analytical, physical and biophysical chemistry.

More discussions on vibrational spectra of peptides and proteins are found in Refs. 15–18.

(c) Nucleic Acids

According to x-ray analysis, double-stranded DNA can take the A, B and Z forms shown in Fig. 4-6. The A and B forms are found in low salt fiber

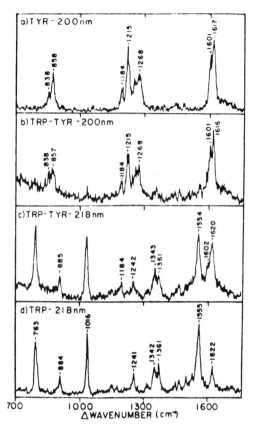

Figure 4-5 Raman spectra in the ring-mode region for aqueous solutions of (a) tyrosine, (b) and (c) tryptophyltyrosine, and (d) tryptophan (all in 10^{-3} M). (Reproduced with permission from Ref. 13. Copyright 1985 American Chemical Society.)

sodium DNA at 75% and 98% relative humidity, respectively. DNA in aqueous solution contains largely the B form. The Z form is produced when the solution contains high concentrations of $MgCl_2$ and NaCl, etc. Table 4-8 lists the parameters that characterize these three forms. The phosphodiester stretching vibration observed in Raman spectra is useful in distinguishing these forms. The A and B forms exhibit this vibration at 810 and 835 cm^{-1}, respectively, while the Z form does not show it in this region. Figure 4-7 compares the Raman spectra of DNA (B form) and transfer-RNA (A form) obtained by Peticolas et al. (19). In contrast, the phosphoionic stretch near 1,100 cm^{-1} is not structure-sensitive. These forms can also be distinguished by the guanine vibrations: 665(A), 682(B) and 625(Z) cm^{-1}.

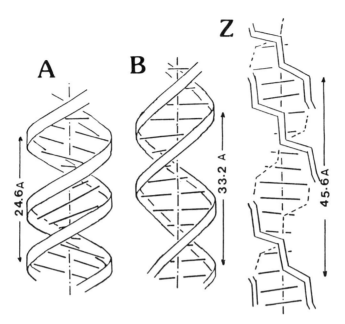

Figure 4-6 A, B and Z forms of DNA. (Provided by M. Tsuboi.)

For more detailed conformational analysis, the reader should consult a review article by Nishimura and Tsuboi (20).

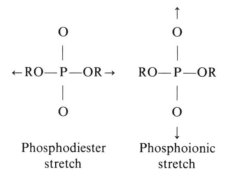

Phosphodiester stretch · Phosphoionic stretch

Table 4-9 lists some prominent bands that are useful in identifying nucleic acid bases: guanine (G), cytosine (C), adenine (A), thymine (T) and uracil (U). Some of these bands are assigned in Fig. 4-7. It should be noted that most of these vibrations originate in the purine or pyrimidine rings of these bases. UVRR studies (excitation wavelength ranging from 200 to 300 nm) by Kubasek *et al.* (21) show that deoxyribonucleotides such as GMP (guanine

Table 4-8 Structural Parameters of A, B and Z forms of DNA

	Helical Sense	Pitch (Å) per Turn	Base Pairs per Turn	Base Pair Distances (Å)	Base Inclination (Degrees)[a]
A form	Right	24.6	10.7	2.3	19
B form	Right	33.2	10	3.3	−1
Z form	Left	45.6	12	3.8	−9

[a] From the normal to the helical axis.

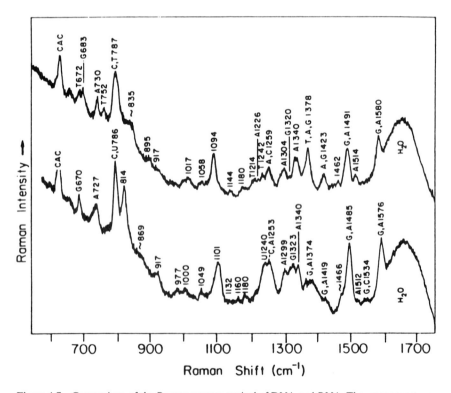

Figure 4-7 Comparison of the Raman spectra typical of DNA and RNA. The upper spectrum of a 2.5% aqueous solution of calf thymus DNA is representative of the B-form of DNA in aqueous solution. The lower spectrum of yeast RNA in a 2.5% aqueous solution at pH 7 is illustrative of the A-form structure adopted by RNA. (Reproduced with permission from Ref. 19. Copyright © 1987 John Wiley & Sons, Inc.)

Table 4-9 Prominent Raman Bands of Nucleic Acid Bases $(cm^{-1})^a$

Adenine (A)—1,580 (s), 1,510 (m), 1,484 (m), 1,379 (m), 1,340 (s), 1,310 (s), 1,255 (w), 729 (s)
Guanine (G)—1,582 (s), 1,487 (s), 1,375 (m), 1,328 (m), 670 (s)
Cytosine (C)—1,657 (m), 1,607 (m), 1,528 (m), 1,292 (s), 1,240 (s), 782 (s)
Uracil (U)—1,680 (s), 1,634 (m), 1,400 (m), 1,235 (s), 785 (s)

^{a}Ref. 22.

monophosphate), CMP, AMP and UMP can be distinguished based on their excitation profiles in the UV region.

4.1.3 CHARACTERIZATION OF UNSTABLE SPECIES PRODUCED BY LASER PHOTOLYSIS

In Section 3.3, we have described the method to produce unstable species in inert gas matrices by laser-photolysis and simultaneously determine their structures by RR spectroscopy. The example given was oxyferryl(IV) porphyrins produced by laser photolysis of the corresponding oxyiron(III) porphyrins:

$$Fe(III)(por)O_2 \xrightarrow[406.7\,nm]{hv} O{=}Fe(IV)(por).$$

Using similar procedures, Wagner and Nakamoto (23) prepared nitridoiron(V) porphyrins from laser photolysis of the corresponding azidoiron(III) porphyrins:

$$Fe(III)(por)N_3 \xrightarrow[488\,nm]{hv} N{\equiv}Fe(V)(por).$$

These nitrodoiron porphyrins provided rare examples of Fe(V) compounds.

Figure 4-8 shows RR spectra of thin films of $N{\equiv}Fe(OEP)$ obtained by gradually increasing laser power from 5 mW (trace a) to 100 mW (trace f). In trace a, three bands characteristic of the $Fe-N_3$ group are readily assigned as shown in the figure. These three bands together with porphyrin bands at 710 and 341 cm^{-1} become weaker, whereas the bands at 876, 752 and 673 cm^{-1} become stronger as the laser power is increased and/or as irradiation time is lengthened. Figure 4-9 shows the RR spectra of photolysis products of $^{54}Fe(OEP)N_3$ and its ^{NA}Fe, $^{15}NN_2$ and $^{15}N_3$ derivatives, respectively. (^{NA}Fe(Fe in natural abundance) contains 91.7% ^{56}Fe.) The upshift of the 876 cm^{-1} band in going from trace B to trace A indicates that this vibration involves the motion of the Fe atom. Furthermore, this band is shifted to 854 cm^{-1} by $^{14}N/^{15}N$ substitution (trace D), and the photolysis product of the $^{15}NN_2$ derivative exhibits two bands of equal intensity at 854 and 876 cm^{-1}

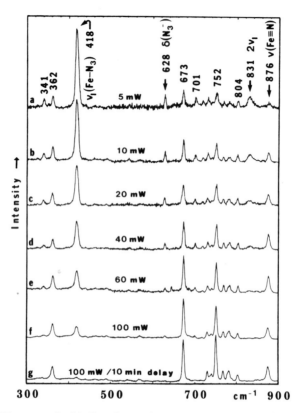

Figure 4-8 RR spectra of a thin film of $N_3Fe(OEP)$ at ~30 K, 488 nm excitation with different excitation power as indicated. (g) was obtained after 10-min pre-irradiation with 488 nm, 100 mW.

(trace C). It should be noted that the reaction of the $^{15}NN_2$ ion with Fe(OEP)Cl produces an equimolar mixture of

$$Fe—^{15}N\!\!=\!\!N\!\!\equiv\!\!N \qquad\qquad Fe—N\!\!=\!\!N\!\!\equiv\!\!^{15}N$$

so that laser photolysis yields an equimolar mixture of the $Fe\!\!\equiv\!\!^{15}N$ and $Fe\!\!\equiv\!\!^{14}N$ bonds. The observed shifts ($+3$ cm^{-1} by $^{56}Fe/^{54}Fe$ substitution and -23 cm^{-1} by $^{14}N/^{15}N$ substitution) are in perfect agreement with theoretical values expected for a $Fe\!\!\equiv\!\!N$ diatomic harmonic oscillator. The bands at 752 and 673 cm^{-1} are porphyrin modes of $N\!\!\equiv\!\!Fe(OEP)$.

As stated in the preceding section, the ν_4 vibration of the porphyrin core is the best oxidation state marker. In $N\!\!\equiv\!\!Fe(OEP)$, this band appears at 1,384 cm^{-1} which is higher than that of the Fe(IV) state (1,379 cm^{-1} for $O\!\!=\!\!Fe(OEP)$). Thus, the Fe(V) state is suggested. However, the spin state

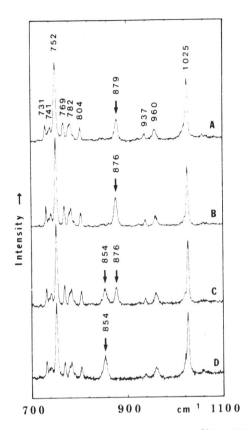

Figure 4-9 RR spectra of the photolysis products of (a) $N_3{}^{54}Fe(OEP)$, (b) $N_3Fe(OEP)$, (c) $^{15}NN_2Fe(OEP) + N_2{}^{15}NFe(OEP)$ in a 1:1 ratio, and (d) $^{15}N_3Fe(OEP)$, thin film at ~ 30 K, 488 nm excitation, 60 mW.

of $N{\equiv}Fe(OEP)$ could not be determined by core size sensitive bands since a relatively small core size is expected from either the low spin $(d_{xy}^2 d_{xz})$ or the high spin $(d_{xy}d_{xz}d_{yz})$ configuration. As seen in Fig. 4-10 $N{\equiv}Fe(V)(OEP)$ is isoelectronic with $O{=}Mn(IV)(OEP)$, which is known to be high spin. Therefore, the nitrido complex is probably high spin. It should be noted that, in the Cr, Mn and Fe series, abrupt drops in the $\nu(M{=}O)$ and $\nu(M{\equiv}N)$ frequencies occur when the electrons occupy anti-bonding orbitals (23).

Finally, the $\nu(Fe{\equiv}N)$ at 876 cm^{-1} disappears completely and a new set of bands appear at 798, 438 and 341 cm^{-1} when $N{\equiv}Fe(OEP)$ is irradiated by the 413.1 nm line of a Kr-ion laser. Based on $^{54}Fe/^{56}Fe$ and $^{14}N/^{15}N$ isotope shift data, the 438 cm^{-1} band is assigned to the symmetric stretching mode

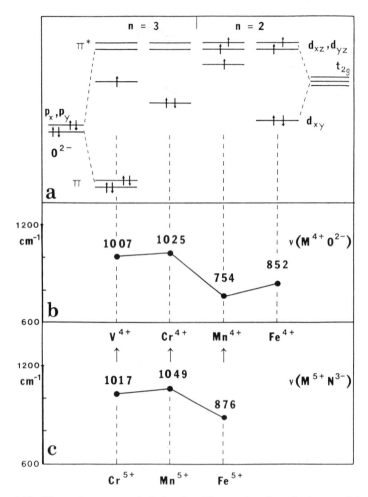

Figure 4-10 Electronic structures and vibrational frequencies of porphyrins containing M=O and M≡N groups. For the sources of vibrational frequencies, see Ref. 23.

of a linear Fe—N—Fe bridge of the $[Fe(OEP)]_2N$ dimer. The remaining bands are attributed to porphyrin core vibrations of the dimer. These results suggest that the $[Fe(OEP)]_2N$ dimer is formed via N≡Fe(OEP) as an intermediate species.

4.1.4 METAL–METAL BONDS AT ELECTRONIC GROUND AND EXCITED STATES

A number of compounds containing centrosymmetric metal–metal (M–M) bonds are known, and their totally symmetric v(M–M) vibrations appear strongly in Raman/RR spectra because of large changes in polarizability (24). As stated in Section 1.15, these vibrations are expected to show a series of overtones under resonance conditions (A-term resonance). As an example, Fig. 4-11 shows the RR spectrum (530.9 nm excitation) of the $[Re_2F_8]^{2-}$ ion obtained by Peters and Preetz (25). The strongest band observed at 320 cm^{-1} (v_1) is the totally symmetric $v(Re-Re)$. The next strongest band at 625 cm^{-1} (v_2) is the totally symmetric $v(Re-F)$. The remaining weak bands are assigned to overtones of v_1 and their combination bands as indicated in the figure.

Using Eq. (1-30) of Chapter 1, we can express the frequency of the nth overtone of a totally symmetric vibration as

$$v(n) = n\omega_1 - X_{11}(n^2 + n) + \text{higher terms.}$$

Here, ω_1 is the wavenumber corrected for anharmonicity, and X_{11} indicates the magnitude of anharmonicity. If we plot $v(n)/n$ against n, we obtain a straight line, and the slope and the intercept of such a plot give X_{11} and ω_1, respectively. In the case of $Re_2Fe_8^{2-}$, these values were found to be -0.45 ± 0.05 cm^{-1} and 319.6 ± 0.6 cm^{-1}, respectively.

The Re—Re bond contains one σ-bond, two π-bonds and one δ-bond (26). The lowest electronic transition occurs from the $(\sigma)^2(\pi)^4(\delta)^2(^1A_{1g})$ to the $(\sigma)^2(\pi)^4(\delta)(\delta^*)(^1A_{2u})$, producing a strong electronic band near 19,000 cm^{-1}. The RR spectrum of $Re_2Cl_8^{2-}$ obtained by 647.1 nm excitation exhibits the

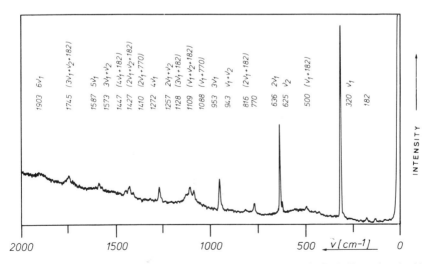

Figure 4-11 RR spectrum of $(TBA)_2(Re_2F_8)\cdot 4H_2O$ (530.9 nm excitation). (Reproduced with permission from Ref. 25.)

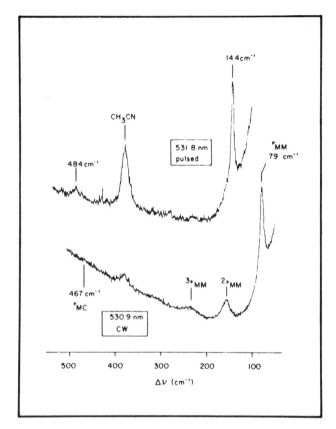

Figure 4-12 Lower trace: ground-state RR spectrum of $Rh_2b_4^{2+}$ obtained by CW excitation at 530.9 nm. Upper trace: excited-state RR spectra of $Rh_2b_4^{2+}$ obtained by pulsed laser excitation with the second harmonic of the Q-switched Nd:YAG laser (531.8 nm). (Reproduced with permission from Ref. 28. Copyright 1981 American Chemical Society.)

v(Re—Re) at 272 cm^{-1}. Dallinger (27) first observed the RR spectrum at the $\delta\delta^*$ excited state by using two-color pump (640 nm)/probe (355 nm) techniques described in Section 3.2 (TR^3 spectroscopy). The $\delta\delta^*$ state exhibited three bands at 138, 204 and 366 cm^{-1}. The last band was assigned to the v(Re—Cl) (ground state, 359 cm^{-1}), while the first band was attributed to the δ(ClReRe) (not observed in the ground state). The remaining band at 204 cm^{-1} was assigned to the v(Re—Re) of the $\delta\delta^*$ state. The large frequency decrease in going from the ground state (272 cm^{-1}) to the excited state (204 cm^{-1}) was attributed to the decrease in bond order from 4 to 3.

For other systems, the v(M—M) becomes higher in going from the ground to the excited state. For example, Fig. 4-12 shows the RR spectra of the

$[Rh_2b_4]^{2+}$ (b = 1,3-diisocyanopropane) ion obtained by Dallinger *et al.* (28). The ground state spectrum shown by the lower trace exhibits the strong $v(Rh—Rh)$ band at 79 cm^{-1} together with its overtones and the $v(Rh—C)$ band at 467 cm^{-1}. The upper trace shows the excited state spectrum obtained by pulsed laser excitation. It is seen that the $v(Rh—Rh)$ is now upshifted to 144 cm^{-1}. This upshift is attributed to the change in electronic structure from the $(d\sigma)^2(d\sigma^*)^2(^1A_{1g})$ ground state to the $(d\sigma)^2(d\sigma^*p\sigma)(^3A_{2u})$ excited state. The Rh—Rh bond order is much higher in the latter because one electron is promoted from the $d\sigma^*$ to the $p\sigma$ orbital. A review by Morris and Woodruff (29) provides more information on this and other subjects on TR^2 spectroscopy.

4.1.5 POTENTIAL ENERGIES OF VIBRATIONS OF LARGE AMPLITUDES

Ring molecules exhibit ring puckering vibrations that are anharmonic and have large amplitudes of vibration. These vibrations are observed in far-IR and low-frequency Raman spectra. As the first approximation, the potential energy of a ring puckering vibration of a four-membered ring is expressed by an anharmonic potential such as (30)

$$V(a) = ax^4 + bx^2,$$

where a and b are constants, and x is the ring puckering coordinate shown in Fig. 4-13. Planar ring compounds have a single potential minimum at

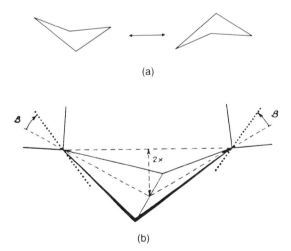

(a)

(b)

Figure 4-13 (a) Ring-puckering vibration of cyclopentene. (b) Definition of the ring-puckering coordinates for a four-membered ring molecule. (Reproduced with permission from Ref. 30.).

$x = 0$, whereas non-planar ring compounds have two minima at positive and negative values of x.

If the Schrödinger equation (Section 1.3) is solved using this potential, the resulting eigenvalues are expressed as a function of a quantum number, n. Selection rules for transitions are: $\Delta n = 1$ for IR and $\Delta n = 2$ for Raman. The values of a and b must be chosen so that calculated IR and Raman frequencies agree with those observed. As an example, consider 1,3-disilacyclobutane:

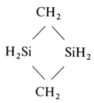

Figures 4-14 and 4-15 show the gas-phase IR and Raman spectra, respectively, of this compound obtained by Irwin et al. (31). Using the foregoing procedures, these workers obtained the potential energy curve shown in Fig. 4-16. The values of a, b and the barrier were found to be 2.3×10^5 cm^{-1}/Å^4, -9.0×10^3 cm^{-1}/Å^2 and 87 cm^{-1}, respectively. For more details on ring puckering vibrations, see a review by Laane (32).

Another example of large amplitude of vibration is a torsional mode around the C—C single bond. The potential energy of an internal rotation of a CH$_3$ group (local symmetry, $\mathbf{C}_{3v}$) relative to a reference framework can be expressed by a cosine function:

$$V(\phi) = V_3(1 - \cos 3\phi),$$

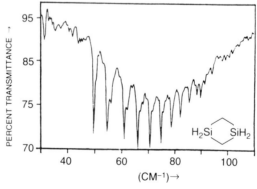

Figure 4-14 Far-IR spectrum of 1,3-disilacyclobutane. (Reproduced with permission from Ref. 31. Copyright 1977 American Chemical Society.)

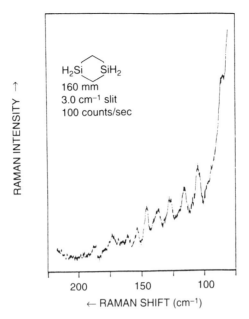

Figure 4-15 Raman spectrum of 1,3-disilacyclobutane. (Reproduced with permission from Ref. 31. Copyright 1977 American Chemical Society.)

where ϕ is the angle of internal rotation, and V_3 is the height of the barrier of internal rotation. $V(\phi)$ becomes zero at staggered configurations ($\phi = 0°$, $-60°$ and $+60°$). Through quantum mechanical treatments (33), one can calculate eigenvalues of such a rotator as a function of v. Figure 4-17 shows .a potential function of such a rotator obtained by Fateley and Miller (34). Although the energy levels are triply degenerate, they are split into the A_1 (or A_2) and E levels by the quantum mechanical tunnel effect, and the magnitude of splitting becomes larger as the levels approach the top of the barrier. At low energies, the CH_3 group performs torsional oscillation relative to its framework. At high energies, however, its amplitude becomes large enough to cross over the barrier (free rotation). In the former case, transitions between the levels of different v values are observed either in IR and Raman spectra. In Raman, $\Delta v = 1$ transitions may be weak even if symmetry-allowed. However, $\Delta v = 2$ transitions are relatively strong since they are totally symmetric. As an example, Fig. 4-18 shows the gas-phase Raman spectrum of ethylchloride(CH_3–CH_2Cl) obtained by Durig et al. (35). The three bands at 488, 455 and 418 cm^{-1} were assigned to 2–0, 3–1 and 4–2 transitions, respectively. Using these assignments, the parameters involved in theoretical

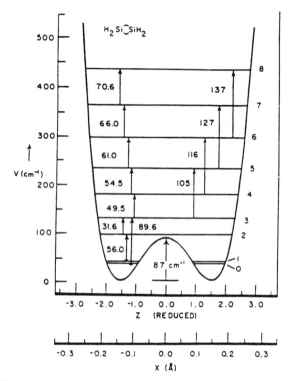

Figure 4-16 Ring-puckering potential energy function for 1,3-disilacyclobutane. (Reproduced with permission from Ref. 31. Copyright 1977 American Chemical Society.)

expressions of torsional frequencies were determined. This led to the V_3 value of 3.72 kcal/mole (or 1,302 cm^{-1})^2 for ethylchloride.

4.1.6 INTERMOLECULAR INTERACTIONS

(a) Donor–acceptor interaction

Molecular complexes are formed when donor and acceptor molecules are mixed in solution. Small shifts resulting from donor–acceptor interactions can be measured with high accuracy by using the difference Raman techniques described in Section 2.9. Shelnutt (36) studied the RR spectra of Cu(II) uroporphyrin I (CuURO, M = Cu(II), X = H and Y = CH_2COO^- in Fig. 4-4) mixed with a variety of 1,10-phenanthroline derivatives. As an example, Fig. 4-19 shows the RR spectra of CuURO and CuURO of the

^{2} 1 cm^{-1} = 2.858 cal/mole.

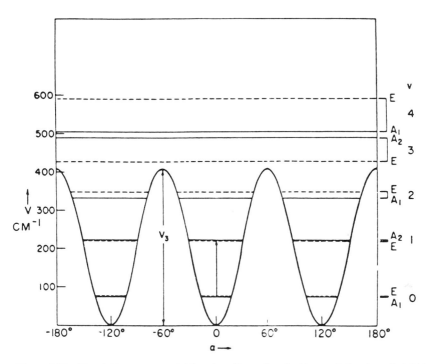

Figure 4-17 Threefold rotor potential function. (Reproduced with permission from Ref. 34.)

same concentration saturated with 5-methylphenanthroline. It should be noted that phenanthroline vibrations do not appear in the latter since only the vibrations of CuURO are in resonance at 514.5 nm excitation. Although both spectra appear to be similar, the difference spectrum shown in the upper trace clearly indicates small shifts that can be determined with high accuracy (± 0.1 cm^{-1} in this case). It was found that $v_3(\sim 1,500$ cm^{-1}), v_{19} ($\sim 1,582$ cm^{-1}) and v_{10} ($\sim 1,637$ cm^{-1}) give relatively large upshifts (2.7–0.1 cm^{-1}), and v_4 ($\sim 1,379$ cm^{-1}) gives much smaller shifts (0.8–0.1 cm^{-1}) when CuURO forms molecular complexes with phenanthroline derivatives. (For normal modes of these vibrations, see Table 4-5.) These upshifts have been attributed to the decrease of the π-electron density of CuURO as a result of electron donation to phenanthroline derivatives. In fact, these shifts are linearly related to the acceptor ability of phenanthroline derivatives and the known free energy of association. Steric properties of some phenanthroline derivatives suggest that the plane of phenanthroline is parallel to that of CuURO in these molecular complexes. Finally, all the electron-density-sensitive bands

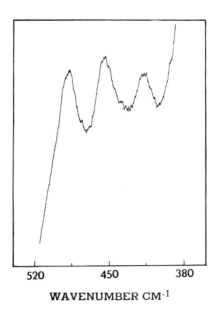

WAVENUMBER CM^{-1}

Figure 4-18   Raman torsional overtones of ethyl chloride, recorded at room temperature. (Reproduced with permission from Ref. 35. Copyright 1977 American Chemical Society.)

mentioned earlier are downshifted when CuURO is mixed with the viologen dication:

$$H_3C - \overset{+}{N} \text{〈pyridyl ring〉} - \text{〈pyridyl ring〉} \overset{+}{N} - H_3C$$

Thus, CuURO acts as an acceptor in this case.

(b)   Solute–Solvent Interaction

As shown in Section 1.15, porphyrin core vibrations are resonance-enhanced when the exciting line is chosen within the $\pi-\pi^*$ transitions of the porphyrin (α, β, and Soret bands). The middle trace of Fig. 4-20 shows the RR spectrum (406.7 nm excitation) of $[Co(TPP\text{-}d_8)(pyridine)^{18}O_2]$, which was produced by dissolving $Co(TPP\text{-}d_8)$ in toluene containing 3% pyridine ($-85°C$, ~ 4 atm $^{18}O_2$ pressure) (37). Here, $Co(TPP\text{-}d_8)$ is the d_8 analogue of $Co(TPP)$ ($X = C_6H_5$ and $Y = D$ in Fig. 4-4). The strong band at $1,094 \text{ cm}^{-1}$ is assigned to the $\nu(^{18}O_2)$ of $[Co(TPP\text{-}d_8)\text{-}(pyridine)^{18}O_2]$, and the remaining strong bands are attributed to porphyrin core vibrations. The d_8 analogue was used in this case because Co(TPP) exhibits an internal mode at 1,080 cm^{-1} that partly overlaps on the $v(^{18}O_2)$ band. The $v(^{18}O_2)$ as well as porphyrin vibrations are resonance-enhanced since the Co—O$_2$ CT transition is located near the Soret band.

The upper trace of Fig. 4-20 shows the RR spectrum obtained by using $^{16}O_2$ instead of $^{18}O_2$. If the bound O$_2$ is regarded as a diatomic harmonic oscillator, the $^{16}O_2$–$^{18}O_2$ isotope shift is calculated to be ~ 64 cm^{-1}. Thus, the $v(^{16}O_2)$ band is expected to appear as a single band near 1,158 cm^{-1}. Instead, two strong, unresolved bands are observed at 1,160 and 1,151 cm^{-1}. The latter at 1,151 cm^{-1} must originate in toluene because (1) toluene alone exhibits a weak band at 1,155 cm^{-1} and (2) only a single peak is observed at 1,159 cm^{-1} when toluene-d_8 is used (bottom trace). Strong enhancement of the solvent vibration is attributed to vibrational coupling between the $v(^{16}O_2)$ and the internal mode of toluene at 1,155 cm^{-1}. Namely, resonant vibrational energy has been transferred from the $^{16}O_2$ moiety of [Co(TPP-d_8)(pyridine)$^{16}O_2$] to toluene. This is possible because these two frequencies are close to each other (energy matching) and these two moieties are closely associated in time scale of vibrational transitions. In fact, such a coupling

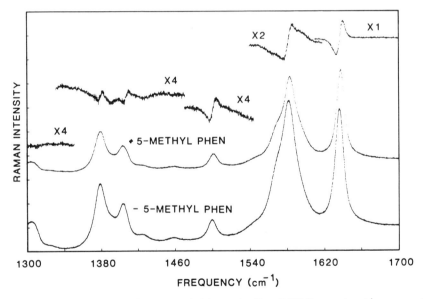

Figure 4-19 RR spectra of the 5-methylphenanthroline–CuURO complex (upper curve), CuURO (lower curve), and the Raman difference spectra. For the latter, the CuURO spectrum has been multiplied by a constant (shown) in order to balance the intensity of the Raman line in each spectrum. 514.5 nm excitation, 300 mW. (Reproduced with permission from Ref. 36. Copyright 1983 American Chemical Society.)

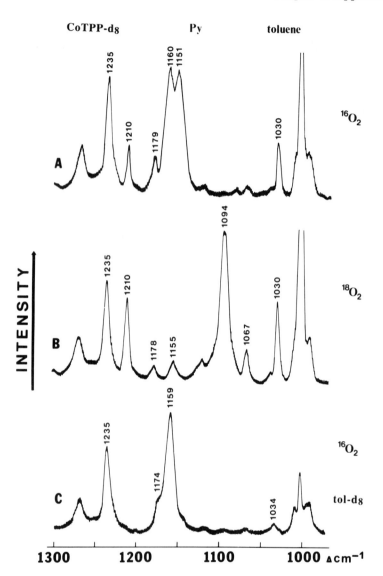

Figure 4-20   RR spectra of Co(TPP-d_8) in toluene containing 3% pyridine at $-85°C$ saturated with $^{16}O_2$ (trace A), $^{18}O_2$ (trace B) and $^{16}O_2$ in toluene-d_8 (406.7 nm excitation).

disappears if a picket-fence porphyrin is used instead of an open porphyrin such as Co(TPP). It is well known that the former prevents the access of a large molecule such as toluene to the bound O_2 inside the fence. The preceding phenomenon was first observed by Kincaid et al. (37). These workers have also shown that internal modes of chlorobenzene (solent) and pyridine (axial ligand) are also resonance-enhanced via similar mechanisms. In fact, the 1,067 cm^{-1} band in the middle trace of Fig. 4-20 is due to an internal mode of pyridine.

4.1.7 MOLECULAR ORIENTATION ON ELECTRODE SURFACE

As discussed in Section 3.5, surface-enhanced Raman spectroscopy (SERS) has been widely used to study the structures of a variety of compounds at surfaces. Since biological applications of SERS/SERRS have been reviewed extensively (38–41), we focus our attention on SERS dependence on the chromophore–surface distance. In 1980, Cotton et al. (42) first obtained the SERRS of cytochrome c and myoglobin on a silver electrode. Although the heme chromophores of these proteins are buried in the protein matrix (see Section 4.2.1), they exhibit a number of strong porphyrin ring vibrations. Later, Cotton et al. (43) studied the relationship between SERRS intensity and the chromophore–surface distance and showed that enhancement caused by the electromagnetic effect due to the surface can occur at distances greater than a few angstroms.

Figure 4-21 shows the SERS of cytidine-3′-monophosphate as a function of the applied electrode potential obtained by Koglin et al. (44). When the electrode potential was −0.1 V (vs. Ag/AgCl reference electrode), only a single band was observed at 236 cm^{-1}. This result indicates that the phosphate group is directly attached on the Ag surface, resulting in the Ag–OPO$_3$ vibration at 236 cm^{-1}. No enhancement of the cytosine base is observed because it is too far from the surface. When the electrode potential was changed to −0.6V, the 236 cm^{-1} band became markedly weaker, and a series of cytosine vibrations appeared in the 1,700–750 cm^{-1} region. The latter indicates that the pyrimidine ring is in direct contact with the electrode, as shown in Fig. 4-21.

Watanabe et al. (45) studied the SERS of adenine, adenosine, cytidine, etc., on silver electrode surfaces. They made an interesting observation that the degree of roughening of the electrode surface influences the spectra. The upper trace of Fig. 4-22 shows the normal Raman spectrum of adenosine in aqueous solution. The bottom trace shows a SERS of adenine (at −6 V vs. SCE) obtained by using a mildly roughened surface. It shows two prominent adenine vibrations at ~730 (ring breathing) and 1,320–1,330 cm^{-1} (ν(N$_7$—C$_5$), not shown). This result suggests that adenosine is adsorbed preferentially via the N$_7$ atom with the rest of the molecule directed outward as shown. The middle trace shows a SERS obtained after heavy roughening. It exhibits a series of weak bands in the 1,100–750 cm^{-1} region together with the adenine bands. Since most of the weak bands originate in the ribose

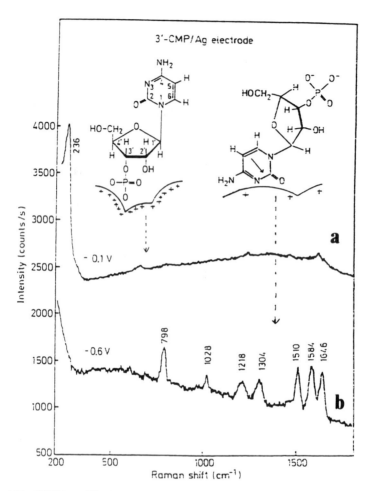

Figure 4-21 SERS in cytidine-3′-monophosphate (3′-CMP). 3′-CMP concentration 2 × 10⁻³ M, 0.15 M KCl, 2 × 10⁻³ M Tris buffer (pH 7.2). 514.5 nm excitation, laser power at the electrode 10 mW, prior activation of the Ag electrode: 1 × 5 mCb between −0.1 V and +0.2 V. (a) Adsorption potential E, −0.1 V vs. Ag/AgCl reference electrode. (b) Adsorption potential E, −0.6 V vs. Ag/AgCl reference electrode. (Reproduced with permission from Ref. 44.)

moiety, both the base and sugar moieties are in close contact with the silver electrode surface as indicated.

4.2 Applications to Biochemistry, Biology and Medicine

As stated in Section 1.8, Raman spectroscopy is ideal for studies of biological systems mainly for two reasons: (1) Since water is a weak Raman scatterer,

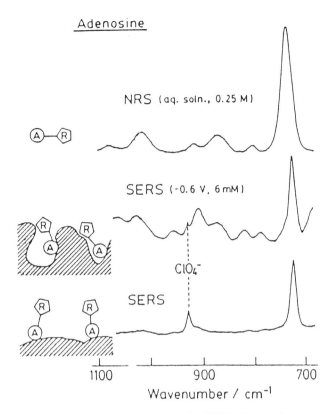

Figure 4-22 Effect of surface roughening on the SERS of adenosine; top, normal Raman spectrum; middle, SERS obtained by heavy roughening; bottom, SERS obtained by mild roughening. (Reproduced with permission from Ref. 45.)

it does not interfere with Raman spectra of solutes in aqueous solution. (2) By taking advantage of resonance Raman scattering, it is possible to selectively enhance particular chromophoric vibrations using a small quantity of biological samples. As a result, biological applications of Raman spectroscopy have increased explosively in recent years, and a number of review articles and monographs have already provided comprehensive coverage of this field. (See General References at the end of Chapter 1.) Here, we demonstrate its utility using selected examples. Only oxygen-binding proteins are covered in the first three subsections.

4.2.1   HEMOGLOBIN AND MYOGLOBIN

Heme proteins such as hemoglobin (Hb), myoglobin (Mb) and cytochromes contain the heme group (iron protoporphyrin, Fig. 4-23) as the active site of

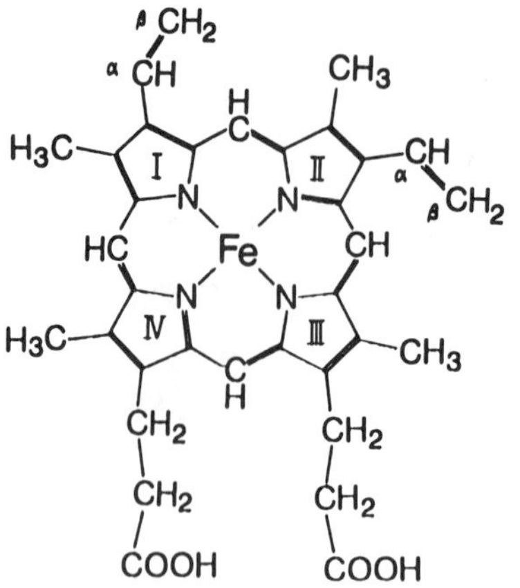

Figure 4-23 Structure of iron protoporphyrin IX.

their biological functions. As discussed in Section 1.15, porphyrin rings are ideal for RR studies because strong resonance enhancement is produced when the laser wavelength is chosen to coincide with $\pi-\pi^*$ transitions of the porphyrin ring. Since many review articles (46) are available on RR spectra of heme proteins, only brief discussions on RR spectra of Hb—O_2 and TR^3 spectra of the Hb—Co photoproduct are presented in this subsection.

Myoglobin (MW $\sim$ 16,000) is an oxygen storage protein in animal muscles. Figure 4-24 shows the well-known crystal structure of Mb as determined by x-ray analysis. It is a monomer containing 153 amino acids and an iron protoporphyrin that is linked to the proximal histidine (F8) of the peptide chain. Figure 4-25 illustrates the structural changes caused by oxygenation. In the deoxy state, the iron is divalent and high spin, and the iron atom is out of the porphyrin plane ($\sim$0.6 Å). Upon oxygenation, the O_2 molecule coordinates to the vacant axial position, and the heme plane becomes planar. The iron in the oxy state is low spin and its oxidation state is close to Fe(III). Hemoglobin (MW $\sim$ 64,000) is an oxygen transport protein in animal blood. It consists of four subunits (α_1, α_2, β_1 and β_2), each of which takes a structure similar to that of Mb. However, they are not completely independent of each other; the oxygen affinity of each subunit depends upon the number of subunits that are already oxygenated (cooperativity).

Figure 4-26 shows the RR spectra of oxy- and deoxy-Hb obtained by Spiro and Strekas (47). It is seen that the bands at 1,358 (Band I), 1,473 (Band II), 1,552 (Band III) and 1,607 (Band IV) cm^{-1} of the deoxy state are shifted to 1,374, 1,506, 1,586 and 1,640 cm^{-1}, respectively, upon oxygenation. These bands correspond to the v_4, v_3, v_{19} and v_{10} of Ni(OEP) discussed in Section 4.1.2. The v_4 (Band I) is an oxidation state marker, and its upshift

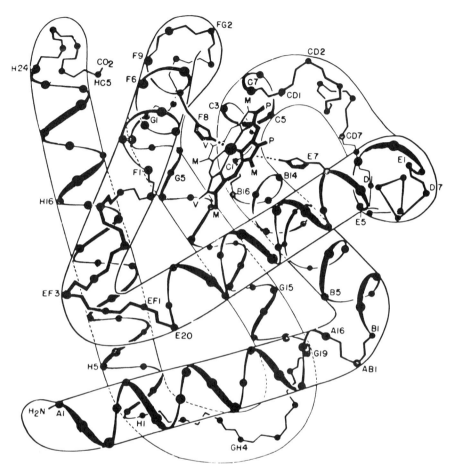

Figure 4-24 Structure of sperm-whale myoglobin. (Reproduced with permission from Dickerson, "The Proteins," Vol. 2, 2nd Ed. Academic Press New York, 1964.)

from 1,358 to 1,374 cm^{-1} indicates oxidation from Fe(II) to Fe(III). The v_3 (Band II), v_{19} (Band III) and v_{10} (Band IV) are core-size–sensitive, and their upshifts upon oxygenation support high to low spin conversion. Although the $v(O_2)$ of Hb—O_2 has not been observed in RR spectra, IR studies have shown it in the superoxo (O_2^-) region from 1,160 to 1,100 cm^{-1}. (For complexity of IR spectra, see Ref. 48.) Thus, the best formulation of the Fe—O_2 bond is Fe(III)—O_2^-.

As discussed in Section 3.3, the Fe—O_2 moiety can take either end-on or side-on geometry. Duff *et al.* (49) observed two $v(Fe—O_2)$ of Hb—O_2 at 567

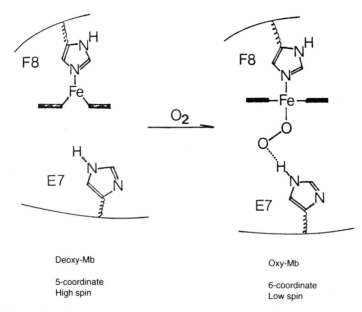

Deoxy-Mb

5-coordinate
High spin

Oxy-Mb

6-coordinate
Low spin

Figure 4-25 Schematic diagram of deoxy- and oxy-myoglobin near the active site.

and 540 cm^{-1} when Hb was oxygenated by ^{16}O—^{18}O. These frequencies are exactly the same as those of Hb—$^{16}O_2$ and Hb—$^{18}O_2$, respectively. Thus, their results provide definitive evidence to support the end-on structure. According to x-ray analysis, the Fe—O_2 bonding is stabilized by a hydrogen bond between the bound O_2 and the N—H group of distal histidine (E7) as shown in Fig. 4-25. The presence of such hydrogen bonding is also supported by RR studies of Kitagawa et al. (50), who observed a small upshift (2 cm^{-1}) of the $\nu(O_2)$ of Co-substituted Mb—O_2 at 1,134 cm^{-1} by H_2O—D_2O exchange.

In normal Hb, the $\nu(Fe$—$N)$ of the proximal histidine (F8) is near 220 cm^{-1} (51). In mutant Hb such as Hb M Iwate and Hb M Boston, F8 histidine and E7 histidine are replaced by tyrosine residues, respectively. In five-coordinate ferric α subunits of these compounds, the $\nu(Fe$—O^- (phenolate)) bands are observed at 589 and 603 cm^{-1}, respectively (52).

Raman spectra of short-lived species of heme proteins can be obtained by using TR^2/TR^3 techniques (Section 3.2). Terner et al. (53) employed this method to monitor structural changes of Hb—CO (low spin) following the photolysis. Figure 4-27 shows the TR^3 spectra obtained by 576 nm pulse excitation of deoxy Hb (trace c) and the Hb—CO photoproduct with ~30 ps (trace a) and ~20 ns (trace b) pulses. It is seen that the spectra of the photoproduct are almost identical to that of deoxy Hb (high spin). Since

ν_{10}, ν_{19} and ν_{11} are all spin state sensitive, these observations suggest that a high spin Fe(II) species has been produced by the photolysis of Hb—CO within ~ 30 ps. This spin conversion is ~ 10^3 times faster than typical spin conversion rates in the ground state of Fe(II) complexes. Terner *et al.* suggest that the photolysis pathway involves intersystem crossing for the initially

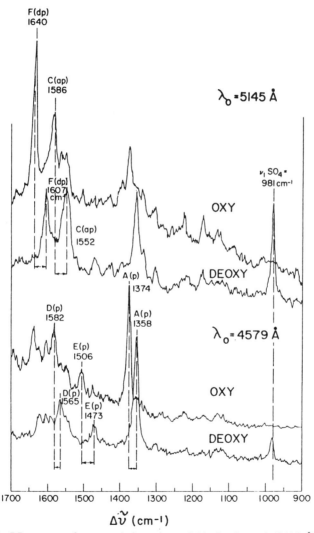

Figure 4-26 RR spectra of oxy- and deoxy-hemoglobin in the α–β (5,145 Å) and Soret (4,579 Å) scattering regions. Frequency shifts for corresponding bands are marked by the arrows between vertical broken lines. (Reproduced with permission from Ref. 47. Copyright 1974 American Chemical Society.)

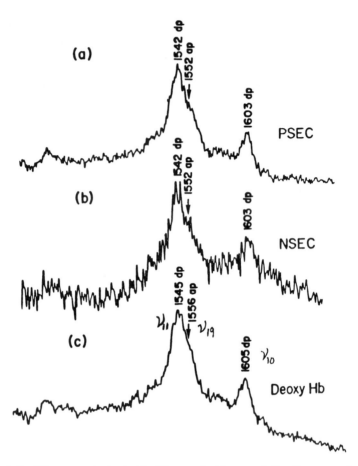

Figure 4-27 RR spectra obtained with 576 nm pulsed laser excitation from a synchronously pumped dye laser. (a) and (b), the Hb–Co photoproduct obtained with ~30 ps and ~20 ns pulses, respectively. (c) Deoxy-Hb. (Reproduced with permission from Ref. 53.)

excited singlet π–π^* state to a low-lying excited state of Hb—CO. The observed small downshifts in going from deoxy Hb to the photoproduct indicate a slightly larger core-size of the latter relative to the former. (The Fe atom in the photoproduct is closer to the heme plane than in deoxy Hb (~06 Å).) These downshifts are seen even when the laser pulses are lengthened to ~20 ns. This may suggest that the slow relaxation to the structure of deoxy Hb is associated with changes in the globin tertiary structure.

4.2.2 CYTOCHROMES AND PEROXIDASES

Another class of heme proteins containing iron protoporphyrin as the active center includes enzymes such as cytochrome P-450 and horseradish peroxidase (HRP). The former is a monooxygenase enzyme (MW $\sim$ 50,000) that catalyzes hydroxylation reaction of substrates such as drugs, steroids and carcinogens:

$$R\text{—H} + O_2 + 2H^+ + 2e^- \rightarrow R\text{—OH} + H_2O.$$

Figure 4-28 shows a proposed reaction cycle of cytochrome P-450 (54). In contrast to Hb and Mb, its Fe center is axially bound to a mercaptide sulfur

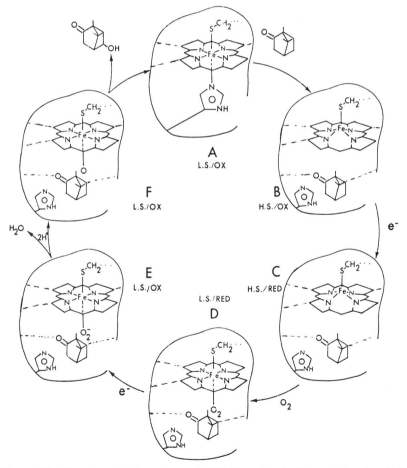

Figure 4-28 Reaction cycle of cytochrome P-450. (Reproduced with permission from Ref. 54.)

(RS$^-$) of a cysteinyl residue. In fact, the v(Fe—S$^-$) vibration of cytochrome P-450 cam (camphor as the substrate) (B-state) has been observed at 351 cm^{-1} by Champion et al. (55). The oxidation state marker band of C-state was observed at 1,346 cm^{-1} by Ozaki et al. (56). It is much lower than the corresponding band of deoxy-Hb at 1,356 cm^{-1}. This marked lowering has been attributed to the strong π-basicity of the thiolate ligand, which donates electrons via the Fe ($d\pi$)–porphyrin ($p\pi^*$) overlap. As stated in the preceding section, the v(O$_2$) of oxy-Hb has not been observed by Raman spectroscopy. However, Bangcharoenpaurpong et al. (57) were able to observe the v(O$_2$) of cytochrome P-450 cam (D-state) at 1,140 cm^{-1}.

According to the reaction cycle shown in Fig. 4-28, oxoferryl (O=Fe(IV)) porphyrin is formed in F-state via the O—O bond breaking. This cleavage is partially facilitated by the weakening of the O—O bond due to the thiolate ligand in the trans position. The marked difference in biological function between Hb and Mb (reversible O$_2$ binding) and cytochrome P-450 (O—O bond cleavage) is largely attributed to the difference in the axial ligand (imidazole nitrogen vs. thiolate sulfur).

Horseradish peroxidase (MW $\sim$ 40,000) catalyzes the oxidation of organic and inorganic compounds by H$_2$O$_2$:

$$AH_2(\text{substrate}) + H_2O_2 \xrightarrow{\text{HRP}} A + 2H_2O.$$

The reaction cycle of HRP involves two intermediates, HRP-I and HRP-II:

$$\text{HRP(ferric)} + H_2O_2 \rightarrow \text{HRP-I} + H_2O,$$

$$\text{HRP-I} + AH_2 \rightarrow \text{HRP-II} + AH,$$

$$\text{HRP-II} + AH \rightarrow \text{HRP(ferric)} + A + H_2O,$$

Thus, HRP-I (green) and HRP-II (red) have oxidation states higher than the native Fe(III) state by two and one, respectively. It has been found that both intermediates are oxoferryl (Fe(IV)) porphyrins and that HRP-II is low spin Fe(IV), whereas HRP-I is its π-cation radical, which is one electron deficient in the porphyrin π-orbital of HRP-II.

As expected from its high oxidation state, HRP-II exhibits the v_4 at 1,377 cm^{-1}, which is the highest among heme proteins (58). The v(Fe=O) vibrations of HRP-II were first reported by Hashimoto et al. (58) and Terner et al. (59) almost simultaneously. Figure 4-29 shows the RR spectra of HRP-II obtained by the former workers. Upon reacting HRP with H$_2$O$_2$ at alkaline pH, a new band appears at 787 cm^{-1} that is shifted to 790 cm^{-1} by ^{56}Fe/^{54}Fe substitution, and to 753 cm^{-1} by H$_2$^{16}O$_2$/H$_2$^{18}O$_2$ substitution. Thus, this band was assigned to the v(Fe=O) of HRP-II. In neutral solution, the corresponding band was observed at 774 cm^{-1}, which was shifted to 740

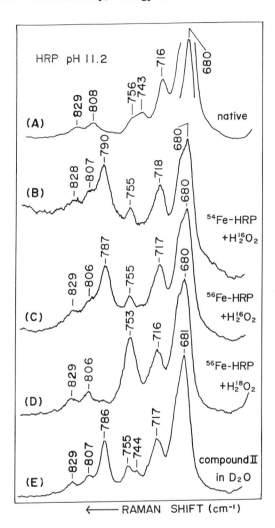

Figure 4-29 RR spectra of HRP-II at pH = 11.2 (406.7 nm excitation). (Reproduced with permission from Ref. 58.)

cm^{-1} by H$_2$^{16}O$_2$/H$_2$^{18}O$_2$ substitution. The observed downshift of the v(Fe=O) in going from alkaline to neutral solution has been attributed to the formation of a hydrogen bond in neutral solution as depicted in Fig. 4-30 (58). These v(Fe=O) frequencies are much lower than those of O=Fe (TPP-d_8) (853 cm^{-1} (Section 3.3) because HRP-II is six-coordinate. Kitagawa's review (60) provides more information on RR spectra of reaction intermediates of heme proteins.

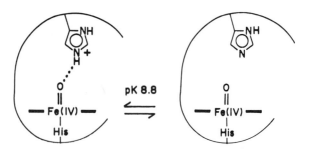

Figure 4-30 Equilibrium between hydrogen-bonded and non-hydrogen-bonded structures of HRP-II. (Reproduced with permission from Ref. 58.)

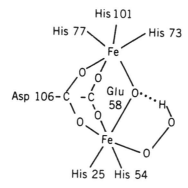

Figure 4-31 Active site structure of oxy-hemerythrin.

4.2.3 Non-heme Respiratory Proteins

Hemerythrin (Hr) is a non-heme oxygen carrier found in invertebrate phyla. Hr isolated from a sipunculan worm (MW $\sim 108,000$) consists of eight identical subunits, and each contains 113 amino acids and two Fe atoms (61). The deoxyform (colorless) turns to pink upon oxygenation ("pink blood"), and one molecule of O_2 binds to a pair of Fe atoms. Originally, Kurtz *et al.* (62) measured the RR spectra of oxy-Hr with isotopically scrambled dioxygen ($^{16}O_2/^{16}O^{18}O/^{18}O_2 \cong 1:2:1$) and found that $v(^{16}O^{18}O)$ near 820 cm^{-1} splits into two peaks, indicating non-equivalence of the two oxygen atoms. This finding was also supported by the RR spectra of the $v(Fe-O_2)$ region (510–470 cm^{-1}). However, the structure of the active site was not clear. Later, Stenkamp *et al.* (63) proposed the structure shown in Fig. 4-31 based on x-ray studies; the coordinated dioxygen is protonated and forms

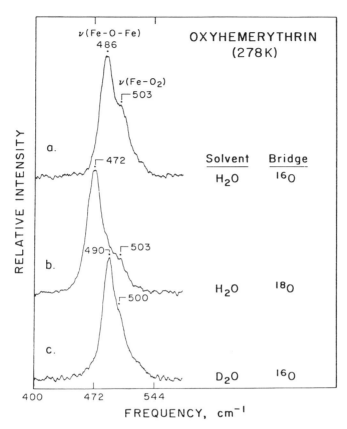

Figure 4-32   RR spectra of oxy-hemerythrin (excitation at 363.8 nm near $O_2^{2-} \rightarrow$ Fe(III) CT transition). (Reproduced with permission from Ref. 64. Copyright 1986 American Chemical Society.)

an intramolecular hydrogen bond with the μ-oxo bridge oxygen. This structure was supported by RR studies by Shiemke *et al.* (64). Figure 4-32 shows the low-frequency RR spectra of oxy-Hr obtained by these workers. The bands at 753 (not shown), 503 and 486 cm^{-1} were assigned to the ν_a (FeOFe), ν (Fe—O_2) and ν_s (FeOFe), respectively. The fact that the latter two bands are shifted to 500 and 490 cm^{-1}, respectively, in D_2O solution provided definitive evidence for the intramolecularly hydrogen-bonded structure.

Hemocyanins (Hc) are oxygen-transport proteins found in the blood of insects, crustacea and other invertebrates (MW $\sim 10^5$–10^7) (65). One of the smallest Hc (MW $\sim$ 450,000) extracted from spiny lobster consists of six subunits, each containing two Cu atoms. Upon oxygenation, the deoxy-form

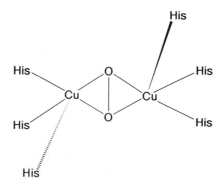

Figure 4-33   Structure of oxy-hemocyanin. (Reproduced with permission from Ref. 69.)

(Cu(I), colorless) turns to blue (Cu(II), "blue blood") by binding one molecule
of O_2 per two Cu atoms. Freedman *et al.* (66) measured the RR spectra of
oxy-Hc extracted from crustacea and observed the $v(O_2)$ near 750 cm^{-1},
which shifts to ~705 cm^{-1} upon $^{16}O_2/^{18}O_2$ substitution. In contrast to
oxy-Hr, the two O atoms of the coordinated O_2 were found to be equivalent
since its $^{16}O^{18}O$ adduct exhibited a single $v(^{16}O^{18}O)$ band at 728 cm^{-1} (67).
Using $^{63}Cu/^{65}Cu$ isotopic techniques, Larrabee and Spiro (68) assigned the
$v(Cu-N(Im))$ at 267 and 226 cm^{-1} (363.8 nm excitation). Recent studies
support the active site structure shown in Fig. 4-33 (69).

4.2.4   DRUG–DNA INTERACTIONS

Basically, three types of interaction are involved in drug–DNA complexes;
intercalation, groove binding and covalent bonding, which are often
reinforced by hydrogen-bonding and/or coulombic interaction. Recently,
Raman spectroscopy has become a powerful technique in elucidating the
mode of interaction. In particular, RR spectroscopy has the advantages that
drug vibrations can be selectively resonance-enhanced if the drug has a strong
absorption in the visible region.

Both aclacinomycin (ACM) and adriamycin (ADM) are antitumor and
antibiotic drugs that bind to DNA. Figure 4-34 shows their structures, and
Fig. 4-35 shows the RR spectra of these drugs mixed with poly(dA-dT) and
poly(dG-dC) obtained by Nonaka *et al.* (70). It is seen that the fluorescence
background is prominent in ADM-poly(dA-dT) but is quenched in
ADM-poly(dG-dC). On the other hand, a strong fluorescence background
is observed for ACM-poly(dG-dC) but is quenched for ACM-poly(dA-dT).
These results suggest that ADM is intercalated between the G–C/C–G

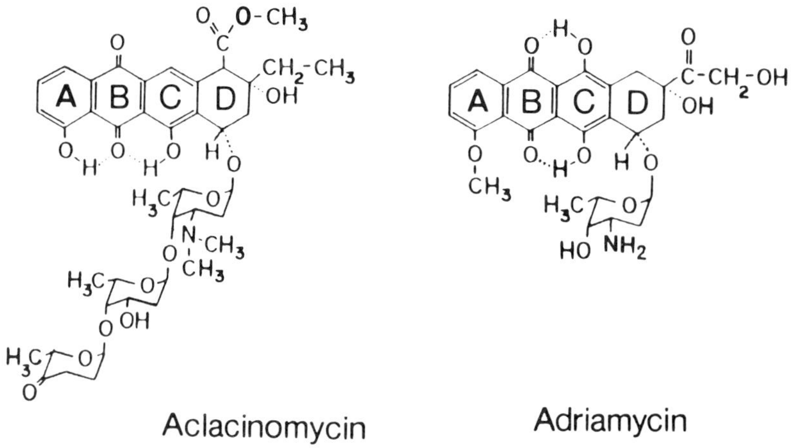

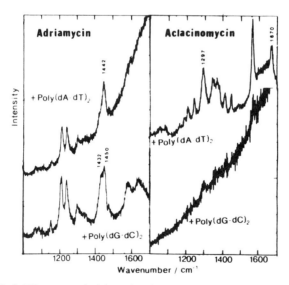

Figure 4-34   Structures of aclacinomycin and adriamycin.

Figure 4-35   Left: RR spectra of adriamycin mixed with poly(dA-dT) and poly(dG-dC) (457.9 nm excitation). Right: RR spectra of aclacinomycin mixed with poly(dA-dT) and poly(dG-dC) (406.7 nm excitation).

sequence, whereas ACM is intercalated between the A–T/T–A sequence of DNA.

A long, flexible molecule such as distamycin (Fig. 4-36) binds to A,T-rich regions in the minor groove of DNA via hydrogen bonding. Figure 4-37

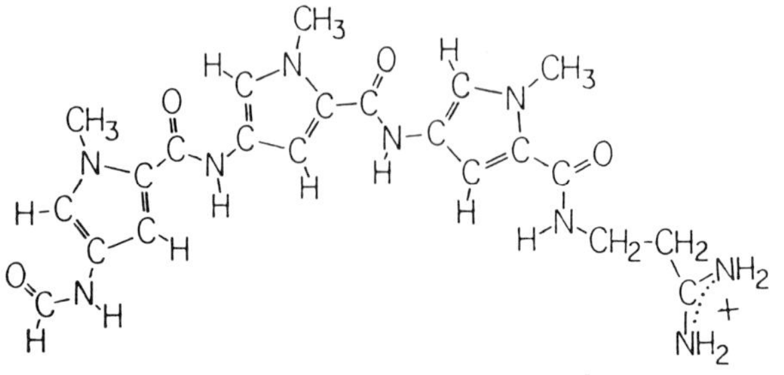

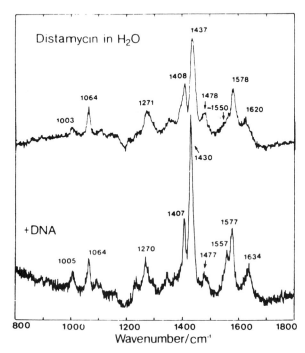

Figure 4-36 Structure of distamycin.

Figure 4-37 Raman spectra of distamycin (upper trace) and distamycin mixed with DNA (lower trace).

shows the Raman spectra of distamycin alone and its mixture with DNA obtained by Lu *et al.* (71). It is seen that the amide I band at 1,620 cm^{-1} is upshifted to 1,634 cm^{-1}, whereas the pyrrole ring mode at 1,437 cm^{-1} is downshifted to 1,430 cm^{-1} when distamycin is mixed with DNA. These changes suggest that the pyrrole ring and the peptide group are nearly coplanar in the free state, and that this coplanarity is destroyed when distamycin is bound inside the minor groove of DNA. When these groups are coplanar, a considerable amount of electron migration is expected to occur from the pyrrole ring to the peptide group via resonance:

When this resonance is disrupted on DNA binding, the amide I band would shift to higher and the pyrrole ring band to lower frequency. Such conformational change also accounts for the observed sharpening of Raman bands upon DNA binding. In the free state, all the internal rotational angles around the C—C and C—N bonds connecting the pyrrole ring and the peptide group fluctuate in a narrow range around $0°$. However, this fluctuation causes broadening of the vibrational bands. Upon binding to DNA, the conformation of distamycin is fixed by the steric requirements in the minor groove, and the bands become sharper.

Recently, Strahan et al. (72) discovered a novel phenomenon that water-soluble copper porphyrin, CuP4 (M = Cu(II)) (X = $\langle \rangle$N$^+$–CH$_3$) and Y=H in Fig. 4-4, which is intercalated between GC/CG sequence of DNA, is translocated to the ATAT site upon electronic excitation of CuP4 by a pulsed laser. As shown in Fig. 4-38, the RR spectrum of CuP4–DNA obtained by high-power pulsed laser exhibits new bands at 1,550 and 1,346 cm^{-1} (trace B) that are not observed by CW laser excitation (trace A). These new bands do not appear with low-power pulsed laser excitation, and they are observed with poly(dA-dT) (trace C) but not with poly(dG-dC) (trace E). They have been attributed to an electronically excited CuP4 that was stabilized by forming a π-cation radical exciplex, (CuP4)$^+$*(AT)$^-$, at an AT site. If oligonucleotides contain GC/CG as well as ATAT or a longer A/T sequence, the exciplex bands are observed as seen in traces G and H. More elaborate experiments show that, in these cases, some of the intercalated porphyrin at the GC/CG site is translocated to the ATAT site (major groove binding) by electronic excitation within 35 ns (Fig. 4-39). Vibrational studies on drug–nucleic acid interactions have been reviewed by Manfait and Theophanides (73).

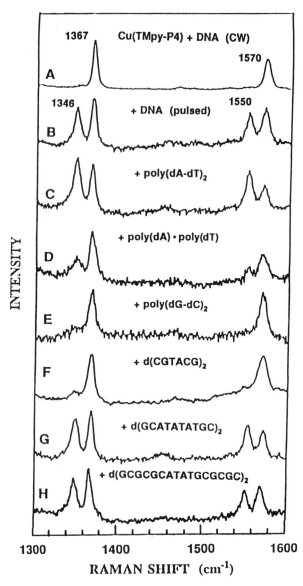

Figure 4-38   RR spectra of Cu(TMpy-P4)–nucleic acid complexes. All the spectra were obtained by using pulsed laser excitation at 416 nm except for the top spectrum (CW, 406.7 nm excitation).

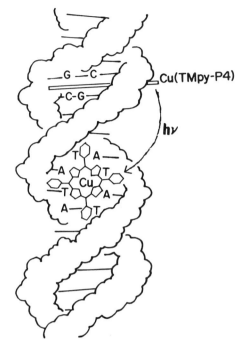

Figure 4-39 Schematic diagram showing translocation of Cu(TMpy-P4) from GC to ATAT site upon electronic excitation.

4.2.5. PLANTS, ANIMALS AND BACTERIA

Carotenoids are widely distributed among plants and animals and are ideal for RR studies because their vibrations can be selectively enhanced by choosing the exciting laser wavelength in the strong $\pi-\pi^*$ transition of a carotenoid pigment. Figure 4-40 shows the RR spectra (488 nm excitation) of β-carotene in live carrot root, canned carrot juice and of pure all-trans β-carotene in n-hexane obtained by Gill et $al.$ (74). The intense peaks at 1,527 and near 1,160 cm^{-1} are due to the $v(C=C)$ and $v(C-C)$ of β-carotene.

Carotenoids are also detected in human blood plasma. Figure 4-41a shows the RR spectrum (514.5 nm exc.) of blood plasma taken from a healthy woman by Larsson and Hellgreen (75). The strong bands at 1,520 and 1,160 cm^{-1} are due to carotenoids. This is contrasted with the RR spectrum of blood plasma taken from a woman with hepatitis shown in Fig. 4-41b. In the latter, the fluorescence background continuously increases with increasing frequency and shows a maximum near 3,400 cm^{-1}. When this patient recovered, her blood plasma gave a normal spectrum similar to that of

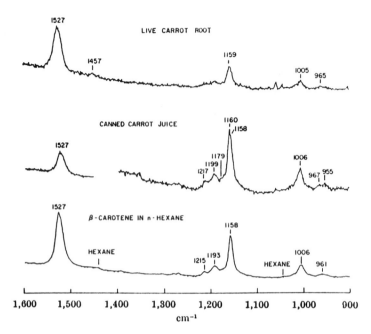

Figure 4-40 RR spectra of β-carotene in live carrot root (top), canned carrot juice (middle), and *n*-hexane (bottom; pure all trans form, 488 nm excitation). (Reproduced with permission from *Nature* from Ref. 74. Copyright 1970 Macmillan Magazines Limited.)

Fig. 4-41a. Thus, a combination of Raman and fluorescence scattering can be used for blood testing.

In contrast to the preceding, bacteria such as *E. coli* are colorless, and their chromophores such as proteins and nucleic acids absorb below 300 nm. Thus, RR spectra of bacteria can be obtained only by using UV laser excitation. Figure 4-42 shows the UVRR spectra of *E. coli* obtained by Britton *et al.* (76). The observed peaks have been assigned based on UV excitation profiles of individual amino acids and nucleotides. It was found that the 222.5 nm excitation spectrum is dominated by vibrations due to aromatic amino acids; 1,614 (Tyr), 1,558 (Trp), 1,178 (Tyr) and 1,008 cm^{-1} (Tyr), whereas the 250.9 nm excitation spectrum is dominated by vibrations of nucleic acid bases: 1,623 (U, Trp, Tyr), 1,580 (A, G), 1,486 (A, G), 1,335 (A, G) and 1,242 cm^{-1} (U). Such selective enhancement is possible because aromatic amino acids absorb in the 220–190 nm region while nucleic acid bases absorb in the 260–240 nm region.

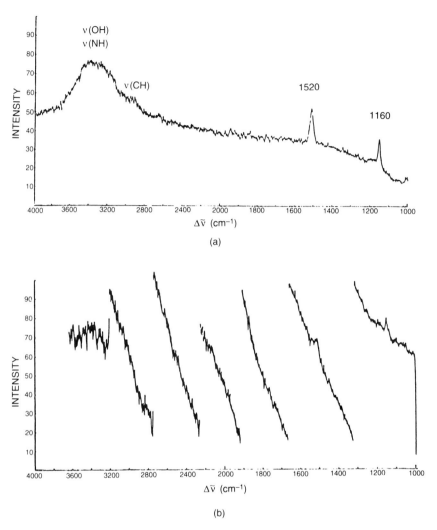

Figure 4-41 RR spectra obtained from human blood plasma. (a) A healthy woman, age 23. No history of any disease. (b) A 20-year-old woman with homologous serum hepatitis. (Reproduced with permission from Ref. 75.)

4.2.6 LENS PROTEINS

Lens aging and opacification can be monitored *in situ* via structural changes in lens proteins observed in Raman spectroscopy. Ozaki *et al.* (77) have carried out an extensive study on mouse lens proteins. Figure 4-43 shows the Raman spectrum and band assignments of a rat lens nucleus. It was

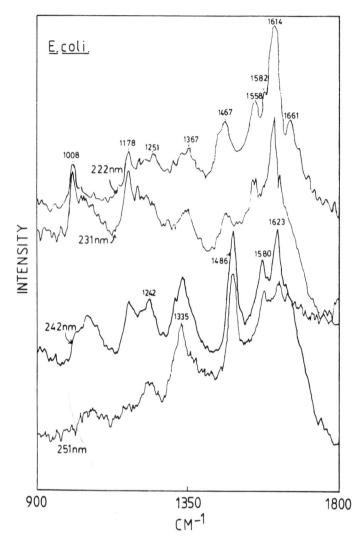

Figure 4-42 UVRR spectra of *E. coli* excited at 222.5, 230.6, 242.4, and 251.0 nm. (Reproduced with permission from Ref. 76.)

found that relative intensities of $\nu(OH)$ (lens water) at 3,390 cm^{-1} and $\nu(SH)$ (lens protein) at 2,579 cm^{-1} decrease markedly during the first four months of aging, and that these changes are parallel to the intensity decrease in the Trp band at 880 cm^{-1} and the intensity increase in the $\nu(S—S)$ band at 510 cm^{-1}. These observations suggest that the aging process involves lens

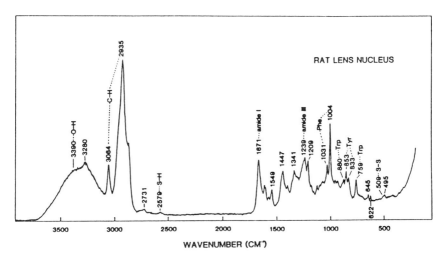

Figure 4-43 Raman spectrum of a SD-strain rat lens nucleus (5 months old). 488 nm excitation, 120 mW. (Reproduced with permission from Ref. 77.)

dehydration and the $2SH \rightarrow S{-}S$ conversion caused by an environmental change of the Trp residue (78).

In contrast to aging, lens opacification (cataract formation) is characterized by (1) the intensity increase of the $\nu(OH)$ at 3,390 cm^{-1}, and (2) the change in relative intensity of the Tyr doublet near 840 cm^{-1}. (1) is a better marker of opacification because the change is larger and observable even in precataractous stage (79).

Figure 4-44 shows the fluorescence and Raman spectra of a human lens of 14 years of age obtained by Yu *et al.* (80, 81). It is seen that fluorescence dominates the spectra when the exciting wavelength is shorter than 514.5 nm, while Raman bands are observed when it is longer than 514.5 nm. Thus, 514.5 nm is regarded as the critical wavelength (λ_C) of this particular lens. The λ_C of a normal lens increases with age; the λ_C is near 680 nm for a normal lens of a 78-year-old human. A plot of λ_C vs. age for normal lens has been obtained using 11 normal lenses. Any deviation of λ_C from such a plot may be regarded as a sign of deterioration of a lens.

4.2.7 MICROSTRUCTURES OF ALGAE CELLS AND GALLSTONES

As described in Section 3.7, Raman microscopy is a powerful technique for the studies of morphology and composition of samples in the micrometer size. Using the MOLE, LeFevre *et al.* (82) studied the compositions of a Triassic *Solenopora* algae cell. Figure 4-45 shows the Raman spectra of various

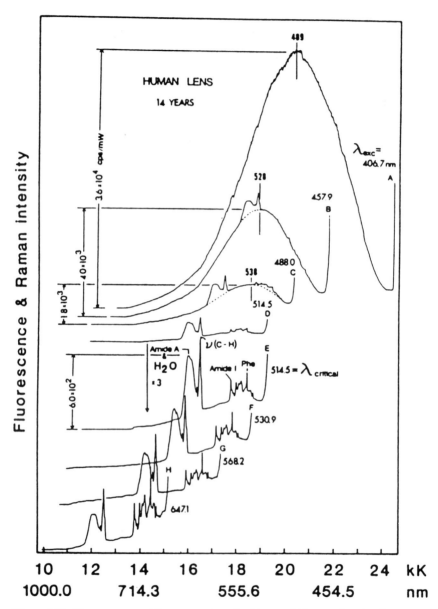

Figure 4-44 Fluorescence and Raman spectra of a 14-year-old human lens (nucleus center) obtained with excitation at various wavelengths indicated. (Reproduced with permission from Refs. 80, 81. Copyright © 1987 John Wiley & Sons, Ltd.)

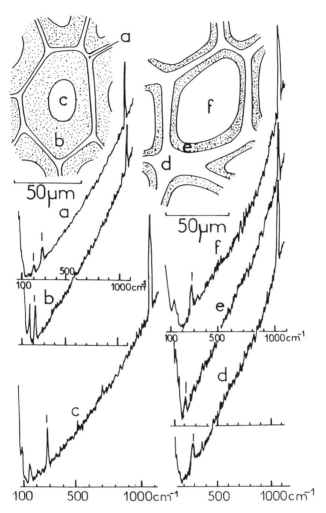

Figure 4-45 Raman spectra of cells of Triassic *Solenopora* algae at various locations indicated. (a) and (d): Peripheral cellar frame—calcite. (b) and (e): Diagenetic layer—aragonite. (c) and (f): Central part of the cells—calcite. (Reproduced with permission from Ref. 82.)

areas of the cell indicated. Although strong fluorescence backgrounds appear in the spectra, bands characteristic of calcite and aragonite structures (lattice vibrations) are clearly seen at 284 and 206 cm^{-1}, respectively, depending upon the location of the point focus (*cf.* Fig. 1-42).

Gallstones show complicated microstructures that reflect the history of their formation. Ishida and co-workers (83) studied the microstructure of a

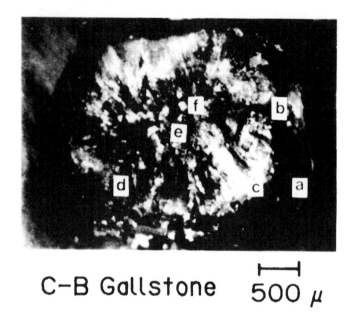

C-B Gallstone   500 μ

Figure 4-46 Optical micrograph of a cholesterol–bilirubin gallstone. (Reproduced with permission from Ref. 83.)

cholesterol–bilirubin gallstone by using the MOLE and FT-IR. Figure 4-46 is a micrograph showing its layered structure. The compositions at six points indicated were found to be:

Outer layer (a and b)—cholesterol
Midlayer (c and d)—$Ca_3(PO_4)_2$, cholesterol, bilirubin
Center (e)—bilirubin
White particle (f)—Ca palmitate

4.2.8   OTHER SYSTEMS

There are many other classes of compounds or topics of biochemical and biological importance. Since Raman spectroscopic studies on them have been reviewed extensively, only the references are cited: hydroporphyrins and chlorophylls (84, 85), flavins (86–88), iron–sulfur proteins (89, 90), metal–tyrosinate proteins (91), visual pigments and bacterial rhodopsin (92–94), photosynthesis (95), viruses and nucelproteins (96, 97), membranes (98, 99) enzyme–substrate reactions (100–102), and medical applications (103).

4.3 Solid State Applications

Numerous applications in the area of the solid state have utilized Raman spectroscopy as the diagnostic tool. In some cases, infrared spectroscopy has complemented the Raman technique. Examples will be presented to demonstrate typical types of problems in which Raman spectroscopy has played a significant role for their solution. Some of these examples illustrate phase transitions for a solid that occur at non-ambient conditions of temperature and/or pressure as determined by Raman spectroscopy. One example measures the non-stoichiometry and the point defect chemistry of the solid state, as followed by the changes in the Raman spectrum. This method has become the method of choice to determine stoichiometric information of solid materials. Another application involves the use of Raman spectroscopy to determine ZrO_2 concentrations in HfO_2–ZrO_2 solid solutions, which cannot be made easily by other techniques. Compressibilities of solids have been determined using Raman data. The latter three applications are also rather unusual and are included to illustrate the usefulness and versatility of Raman spectroscopy as a probe for changes in the solid state.

4.3.1 PHASE TRANSITIONS WITH TEMPERATURE VARIATIONS IN $UO_2(NO_3)_2 \cdot 6H_2O$

The Raman spectrum of uranyl nitrate hexahydrate was followed from 15 to 290 K (104). The study was made using a JEOL-400D double monochromator equipped with a RCA-31034a photomultiplier detector (see Section 2.5). The sample was cooled by placing it on a cold finger of a cryostat (Oxford, Helitran LT-3-110G). The temperature was accurate to ± 1 K. The study was repeated with a Spex Model 1403 and a Cary Model 82 spectrophotometer using an Ar^+ ion laser.

The changes taking place as a function of temperature were obtained by following the Raman spectra in the OH stretching region. The temperature-dependent Raman spectra are depicted in Fig. 4-47a, b. Figure 4-47a shows the spectra of H_2O-I (the water molecules in the inner coordination sphere) from 133–223 K. Figure 4-47b shows the spectra of H_2O-II (the water molecules in the outer sphere). The spectra above 223 K are not shown because of the overlap with fluorescence that is observed with the 514.5 nm excitation. Plots of the variations of band frequency with temperature are illustrated in Fig. 4-48a, b for H_2O-I and H_2O-II. Two discontinuities are observed at 195 ± 5 K and 140 ± 5 K, indicative of three distinct phases occurring in the temperature range studied, as indicated in Fig. 4-48a. The higher-frequency OH stretch region, as shown in Fig. 4-48b does not show

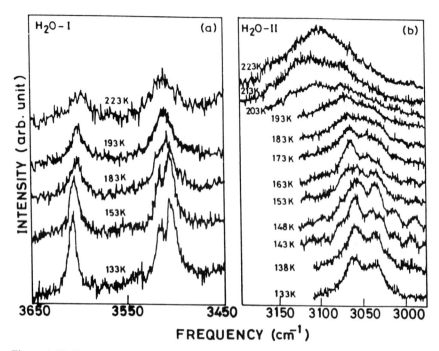

Figure 4-47 Temperature-dependent Raman spectra of uranyl nitrate hexahydrate for the O—H stretching mode of (a) H_2O-I and (b) H_2O-II (514.5 nm excitation). (Reproduced with permission from Ref. 104. Copyright © 1989 John Wiley & Sons, Ltd.)

any discontinuities for H_2O-I. A plot of full width at half maximum intensity (FWHM) vs. T for H_2O-I shows a discontinuity at ~140 K (Fig. 4-48c, d). Additional support for these phase transitions was found from the temperature dependences of the UO_2^{2-} vibrational mode, lattice vibrations and the NO_3^- ion vibrations (translations and rotations).

4.3.2 PHASE TRANSITIONS IN CsVO₃ OCCURRING WITH AN ELEVATION OF PRESSURE

Elevated pressure effects on solid $CsVO_3$ were followed by Raman spectroscopy (105). The study was made by using a gasketed diamond anvil cell with argon as the pressure medium to insure hydrostatic pressures. The pressure was determined using the ruby fluorescent method (see Section 3.4). Spectra were obtained in a backscattering geometry using a Spex double monochromator equipped with a photon-counting system and a 488 nm Ar⁺ laser at a power level of 20–40 mW. The spectra were collected at room

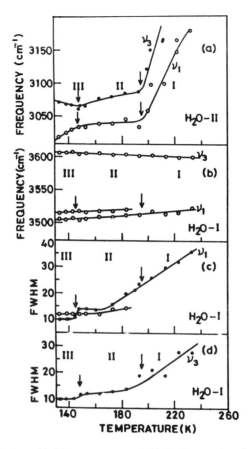

Figure 4-48 Variation with (a) temperature and (b) the frequencies of v_1 and v_2 modes for H_2O-II and H_2O-I and (c) and (d) the FWHM of the v_2 and v_3 modes of H_2O-I. (Reproduced with permission from Ref. 104. Copyright © 1989 John Wiley & Sons, Ltd.)

temperature. At room temperature, $CsVO_3$ has an orthorhombic structure, and based on the changes occurring in the Raman spectra as pressure is increased, new phases are also found at 10, 11.5 and 13 GPa (kbar). Plotting the Raman frequencies as a function of pressure shows discontinuities at the aforementioned pressures, indicative of phase transitions occurring. Figure 4-49a shows the pressure dependence in the region 400 cm^{-1} and between 500 and 1,000 cm^{-1}. Figure 4-49b shows the Raman spectra of phase I and of phases II, III and IV. The phase changes are attributed to a subtle structural change affecting the VO bond angle. The second and third transitions involve the twisting and rearrangement of the corneer-sharing geometry of the polyhedral units in the chains of $CsVO_3$.

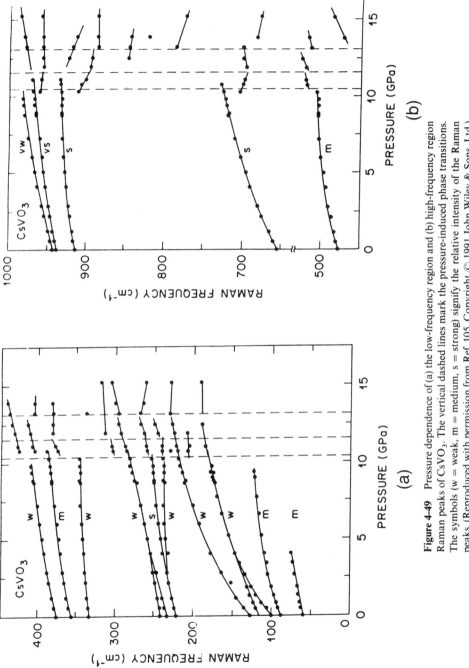

Figure 4-49 Pressure dependence of (a) the low-frequency region and (b) high-frequency region Raman peaks of $CsVO_3$. The vertical dashed lines mark the pressure-induced phase transitions. The symbols (w = weak, m = medium, s = strong) signify the relative intensity of the Raman peaks. (Reproduced with permission from **Ref. 105**. Copyright © 1991 John Wiley & Sons, Ltd.)

4.3.3 Measurement of Non-stoichiometry in Solids

The changes in non-stoichiometry and point defects of solid perovskite ($BaTiO_3$) at 900°C can be observed with Raman spectroscopy (106). The method is believed to be more sensitive than the neutron scattering technique and has become the standard in determining stoichiometric information for solid materials. The interest in perovskite-type materials stems from their use in solid-state capacitors.

Pressed pellets of $BaTiO_3$ were sintered in a platinum dish for six hours at 900°C in a controlled partial pressure of oxygen. The samples were quenched to room temperature, and the spectra recorded on a four-slit double-monochromator Raman spectrophotometer. An Ar^+ laser with excitation at 514.5 nm was the source. The spectra were recorded at room temperature. Figure 4-50 shows the spectrum of $BaTiO_3$ whose Ba/Ti ratio is equal to 0.9999. The Raman spectrum is sensitive to the Ba/Ti ratio and the oxygen non-stoichiometry. The half-band width is variable as well as the intensity ratio of the 525 and 713 cm^{-1} bands. The ratio ($I_{525/713}$) is at a minimum at the composition of 0.9999, and this can be observed in Fig. 4-51, which shows a plot of the intensity ratio (I_{525}/I_{713}) vs. the Ba/Ti composition. As the Ba/Ti ratio increases beyond 1.0, the intensity ratio I_{525}/I_{713} increases slightly and then levels off. When the Ba/Ti ratio decreases there is a significant rise in the I_{525}/I_{713} ratio. Figure 4-52 shows a plot of the 525 cm^{-1} band vs. the intensity ratio I_{525}/I_{713}. The two independent lines, Ba-rich and Ti-rich, have a common origin at Ba/Ti = 0.9999. This

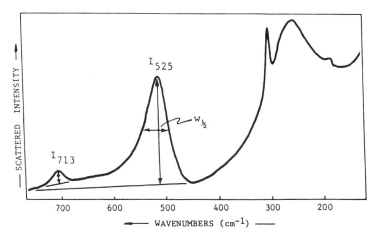

Figure 4-50 A Raman spectrum of $BaTiO_3$ for Ba/Ti = 0.9999 with spectral parameters (intensities and half-band width) defined. This sample was quenched after 3 hours in 1 atm oxygen. (Reproduced with permission from Ref. 106.)

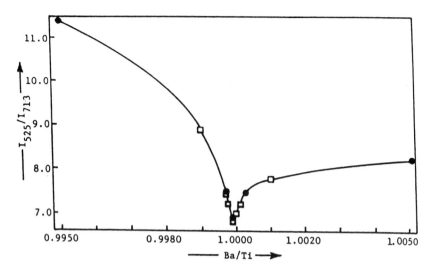

Figure 4-51 Dependence of the I_{525}/I_{713} ratio upon the Ba/Ti ratio for OGC samples ($\square$) and MTU samples ($\bullet$). The samples were quenched from 900°C in 1 atm. (Reproduced with permission from Ref. 106.)

OGC = Oregon Graduate Center samples.

MTU = Michigan Technological University samples

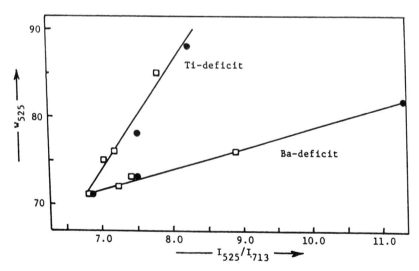

Figure 4-52 Correlation of the intensity ratio and half-band width for Ba-deficit and Ti-deficit samples (900°C, 1 atm O_2). (Reproduced with permission from Ref. 106.)

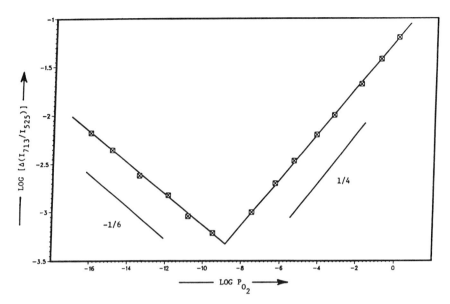

Figure 4-53 Oxygen partial pressure dependence of the change in Raman band intensity ratio for samples with Ba/Ti = 0.9999 ($R = I_{713}/I_{525}$, $\Delta R = R - R_0$). (Reproduced with permission from Ref. 106.)

allows one to determine quantitative information about stoichiometry at room temperature from Raman data. These results would be difficult to obtain by any other technique.

It has been possible to obtain intrinsic oxygen vacancies in solids of this type from Raman data. The dependence of the intensity ratio I_{713}/I_{525} upon the partial pressure of oxygen is non-linear, and the sign of the slope changes. The ratio first decreases and then increases as the partial pressure of oxygen decreases. This trend suggests that this ratio is sensitive to changes in the predominant type of defects occurring in the solid. This is illustrated in Fig. 4-53, where $\log[\Delta(I_{713}/I_{525})]$ is plotted vs. the oxygen partial pressure ($\log P_{O_2}$).

4.3.4 HfO$_2$–ZrO$_2$ SOLID SOLUTIONS

Hafnia and zirconia are known to form a continuous series of solid binary solutions. The use of Raman spectroscopy to provide information about the HfO_2/ZrO_2 ratios for the specific phases (107) (e.g., the HfO_2/ZrO_2 solid solutions), rather than for the total material, has proven to be more advantageous than the use of other traditional techniques such as x-ray

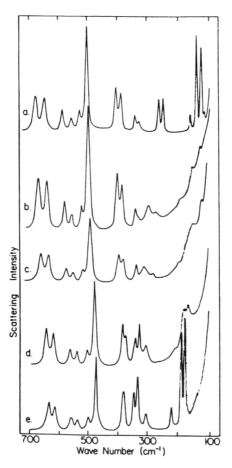

Figure 4-54 Raman spectra for precipitated and fired HfO_2–ZrO_2 solid solutions containing (a) 100% HfO_2, (b) 75%,HfO_2, (c) 50% HfO_2, (d) 25% HfO_2, and (e) 100% ZrO_2.
(Reproduced with permission of the American Ceramic Society from Ref. 107. Copyright 1982.)

diffraction whose peak locations do not show measurable shifts with changes in solid solution composition. Also, techniques such as fluorescence spectroscopy show relative Hf/Zr concentrations for total materials, but will not indicate whether or not these cations are occurring in solid solution phases.

Raman spectra of powdered samples in capillary tubes were obtained using a double monochromator spectrometer (Model 1401—Spex Industries, Inc.) with the blue laser line excitation (488 nm). The scattered radiation from the sample was taken at 90° to the incident beam.

Figure 4-54 shows the Raman spectra observed for the powders obtained for the various investigated HfO_2–ZrO_2 solid solutions. The observed bands do not split, but gradually shift between those found for the pure end members. This is illustrated in Fig. 4-55, which shows nearly linear changes in frequency with ZrO_2 content for the six Raman bands with the highest frequencies, except for the 189 cm^{-1} band, which shows a discontinuous non-linear change with the HfO_2 content (not shown in Fig. 4-55). Figure 4-54a is 100% HfO_2, b is 75% HfO_2, c is 50% HfO_2, d is 25% HfO_2, and e is 100% ZrO_2

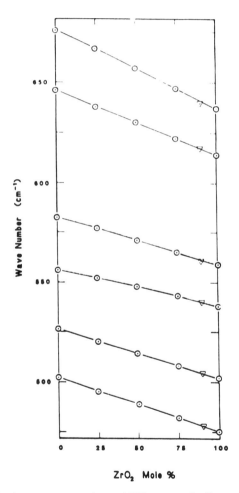

ZrO$_2$ Mole %

Figure 4-55 Relation between wavenumber and HfO_2 content for Raman bands of HfO_2–ZrO_2 solid solutions: (○) results for normally fired samples and (▽) results for plasma-fired samples. (Reproduced with permission of the American Ceramic Society from Ref. 107. Copyright 1982.)

Table 4-10   Correlation of Raman Bands of HfO_2–ZrO_2 Compositions

Wavenumber (cm^{-1})				
100% HfO_2	75% HfO_2	50% HfO_2	25% HfO_2	100% ZrO_2
676	667	657	647	637
646	638	630	622	614
582	577	571	565	559
556	552	548	543	538
526 (sh)[a]	520 (sh)	514 (sh)	508 (sh)	502 (sh)
502	495	489	482	475
403	398	394	387	382 (b)[a]
387	382	379	375 (sh)	
340	340	339	345	348
327	329 (sh)		330	334
262	296 (b)	311 (b)	309	310
246	274 (sh)	282 (b)		
162	185 (sh)	200 (sh)	212 (sh)	221 (sh)
141	155	155 (sh)	183	189
126	130	130 (sh)	172	177

[a] sh = shoulder and b = broad.

in mixtures of the solid solutions. The half-widths of the Raman bands of the pure end members and the solid solutions are smaller than the wavenumber separations between related Raman band shifts. For example, the half-width of the Raman band for HfO_2 at 502 cm^{-1}, which correlates to the band at 475 cm^{-1} for ZrO_2, is 8 cm^{-1}. Clear correlations can be made between the Raman bands of pure ZrO_2 and HfO_2 using those of the solid solutions since the Raman bands shift continuously with respect to the chemical compositions, and the relative Raman band intensities of the solid solutions correspond to those of the pure end members. Table 4-10 lists the wavenumbers of the correlated Raman bands from HfO_2, ZrO_2 and HfO_2–ZrO_2 solid solutions.

4.3.5   COMPRESSIBILITIES OF SOLIDS USING RAMAN DATA

A study was conducted to determine an indirect method for obtaining compressibilities of solids from Raman data (108). In this study, the solids used were $TbVO_4$ and $DyVO_4$. The method is based on determining the VO bond distance from Raman data involving the VO stretching frequency $(v_1(A_g))$. The bonds lengths are then correlated to the size of the unit cell of the crystal. This procedure allows one to determine the change in volume

Table 4-11 Relative Volumes of TbVO$_4$ and DyVO$_4$ at Different Pressures

Pressure (GPa)	TbVO$_4$ $V(P)/V(0)$	DyVO$_4$ $V(P)/V(0)$
0	1	1
0.5	0.9968	0.9976
1.0	0.9936	0.9941
1.5	0.9905	0.9911
2.0	0.9873	0.9882
2.5	0.9842	0.9853
3.0	0.9811	0.9824
3.5	0.9780	0.9795
4.0	0.9750	0.9766
4.5	0.9719	0.9738
5.0	0.9689	0.9709

of the crystals from the changes in the corresponding stretching VO frequency. For the lanthanide vanadates, a relationship between the volume of the unit cell (VO) and the VO bond distance [VO]

$$V(P) = C[d(VO)_P]^3. \tag{4-1}$$

The relative value $V(P)/V(0)$ is given by

$$V(P)/(V(0)) = [d(VO)_P]^3/[d(VO)_0]^3. \tag{4-2}$$

If the VO bond distances can be determined as a function of pressure from Raman spectroscopy, then the compressibility of the vanadates can be estimated. Bond distances can be determined from Raman data (109).

$$v_1(A_g) = 21,349 \exp(-1.9176R). \tag{4-3}$$

Utilizing Eqs. (4-2) and (4-3), $V(P)/V(0)$ can be calculated at different pressures as seen in Table 4-11, and in Fig. 4-56. The compressibilities are obtained from the solid lines in Fig. 4-56 by using a simple polynomial fit of the lines. Values of 6.42×10^{-3} GPa^{-1} for TbVO$_4$ and 6.07×10^{-3} GPa^{-1} for DyVO$_4$ were calculated. This study demonstrates a unique application of the Raman effect.

4.3.6 ELECTRICAL CONDUCTOR APPLICATIONS

Raman spectroscopy has played a significant role in the characterization of electrical conductors, some of which have become superconductors. There are three general classes of compounds that have been investigated using

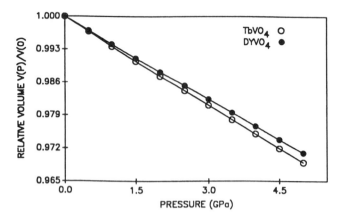

Figure 4-56   Volume compression of $TbVO_4$ and $DyVO_4$ calculated as a function of pressure. Solid lines are fits to a simple polynomial expression. (Reproduced with permission from Ref. 108.)

Raman spectroscopy. These are the high-T_c superconducting ceramics; the low-T_c superconducting charge-transfer (C-T organics), and the fullerenes (e.g., C_{60}), which form some superconducting salts with the alkali metals. Typical applications of Raman spectroscopy for each of these classes will be presented.

(a)   The Raman Spectra of C_{60} and Several of its Superconducting Salts

Since the discovery of the soccer-ball-like, 60-carbon buckminsterfullerene, scientists have been fascinated by this molecule. The molecule possesses a truncated icosahedral structure, of I_h symmetry, and is depicted in Fig. 4-57.

Raman spectroscopy has been an excellent diagnostic tool in the study of the C_{60} molecule and its salts, because of the high symmetry these substances possess. In the case of C_{60}. which has a center of symmetry, the gerade modes are observed only in the Raman spectrum. Furthermore, the materials are excellent Raman scatterers.

The early spectral work on C_{60} was suspect because of its impurities. The Raman spectrum of pure C_{60} was obtained using a Perkin-Elmer FT-IR/Raman Model 1720. To prevent overheating of the sample, very low values of laser power were used (~ 20 mW), at a resolution of 1 cm^{-1}. Figure 4-58 shows the Raman spectrum of pure C_{60}. Hendra and co-workers (110) observed 14 bands, although one was very weak, and two others were shoulders on strong bands and may be due to resonance effects. For a truncated icosahedral structure, 10 Raman-active modes are predicted. As shown in Fig. 4-58 the lowest-frequency band is the H_g "squashing" mode. The two A_g modes at 1,467 and 495 cm^{-1} are assigned to the cage breathing modes. For reference to the species of vibrations in an icosahedral (I_h)

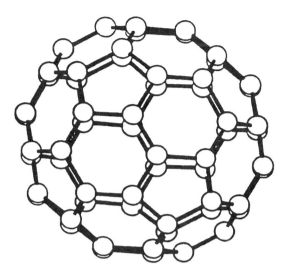

Figure 4-57 Structure of the C_{60} molecule.

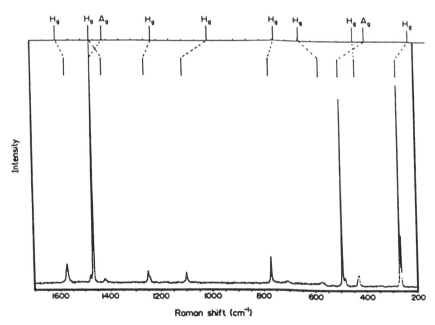

Figure 4-58 Raman spectrum of C_{60}. (Reprinted from T. J. Dennis *et al.*, *Spec. Acta* **47A**, 1289, Copyright 1991, with permission from Pergamon Press Ltd., Headington Hill Hall, Oxford OX3 0BW, UK.)

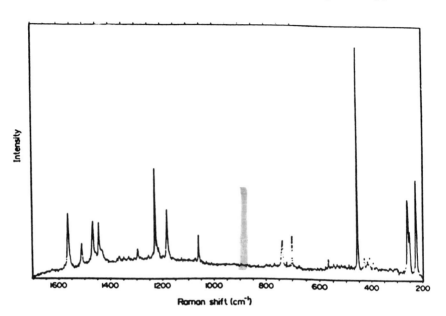

Figure 4-59   Raman spectrum of C_{70}. (Reprinted from T. J. Dennis *et al.*, *Spec. Acta* **47A**, 1289, Copyright 1991, with permission from Pergamon Press Ltd., Headington Hill Hall, Oxford OX3 0BW, UK.)

symmetry, see Table 1-5 and the character table for the I_h point group in Appendix 1.

It is possible to distinguish between C_{60} and C_{70} and higher carbon entities from their Raman spectra. Figure 4-59 shows the Raman spectrum of C_{70}. Comparison of the Raman spectrum of C_{60} with that of C_{70} (Figs. 4-58 and 4-59) illustrates the differences between the two molecules. In the C_{70} molecule the symmetry has decreased to D_{5h}, and thus more bands are expected to appear in the Raman spectrum, as is observed.

Upon salt formation with potassium and rubidium, C_{60} forms superconductive materials. Table 4-12 lists some doped fullerene salts that become superconductive along with their critical temperatures. In the salts of C_{60} with K^+ and Rb^+, the A_g breathing mode at $\sim 1,467$ cm^{-1} has been investigated by Raman spectroscopy. This mode shifts from 1,467 cm^{-1} in neat C_{60} to lower frequencies as one follows the doping process (salt formation). For the K_3C_{60} salt (111), the shift is from 1,467 cm^{-1} to 1,445 cm^{-1}. For the $Rb_3C_6^0$ salt (112), the shift is from 1,467 cm^{-1} to 1,447 cm^{-1}. If the doping is allowed to continue, the shift is lowered to 1,429 cm^{-1}. Figure 4-60 illustrates the Raman spectra of a C_{60} film during the rubidium doping process.

Table   4-12 Superconducting   Metal
Fullerides

Salt	T_c (K)
K_3C_{60}	18.0–19.6
K_2RbC_{60}	21.8–22.5
Rb_2KC_{60}	24.4–28.0
$Rb_{1.5}K_{1.5}C_{60}$	25.1
Rb_3C_{60}	28.0–29.8
Rb_2CsC_{60}	31.3
Cs_2RbC_{60}	33
Cs_xC_{60}[a]	29.5
Ca_5C_{60}[b]	8.4
Ba_6C_{60}[c]	7

[a] x not characterized

[b] From R. C. Haddon, G. P. Kochanski, A. F. Hebard, A. T. Fiory and R. C. Morris, *Science* **258**, 1636 (1992).

[c] From A. R. Kortan, N. Kopylov, S. Glarum, E. M. Gyorgy, A. P. Ramirez, R. M. Fleming, O. Zhou, F. A. Thiel, P. L. Trevor, and R. C. Haddon, *Nature* **360**, 566 (1992).

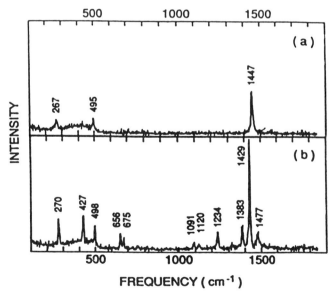

Figure 4-60 *In situ* Raman spectra of a C_{60} film taken during rubidium doping: (a) superconducting and (b) insulating. (Reproduced with permission from Ref. 112. Copyright 1992 American Chemical Society.)

(b) Charge-Transfer Organic Superconductors

Charge-transfer (C-T) reactions between an organic donor and an acceptor molecule and anion have, in some instances, produced superconductors. As in the case of the C_{60} molecule, Raman spectroscopy has played an important role in the characterization of the neat organic donor, as well as the C-T salts (113). Several examples of applications follow.

A successful donor has been bis(ethylenedithio)tetrathiafulvalene (BEDT-TTF or ET). The structure of ET is depicted in Fig. 4-61. The infrared and Raman frequencies of BEDT-TTF and BEDF-TTF-d_8 have been reported and assignments made (114). Figure 4-62 shows the spectra of several salts in the region of 800–50 cm^{-1}. Here, the research was done to determine the nature of the polyiodides in the molecule (115).

The Raman-active C=C stretching vibration in neat ET has been followed by Raman spectroscopy. Upon salt formation, the frequency lowers. Table 4-13 shows this result, comparing neat donors with their salts. The vibration becomes IR-active and combines with the conducting electrons in the salts (electron–molecular vibration, EMV) (116) and shifts to lower frequency. The extent of this shift is a measure of the electron–molecular vibration coupling in a given superconducting salt (116).

Limited Raman studies have been made on the ET salts because they are black and absorb the laser energy, causing them to decompose. This decomposition has occurred even at low laser power levels and with a near infrared laser.

These C-T salts are metallic at room temperature and are also two-dimensional. Upon lowering the temperature some of these salts become superconducting, provided that they can avoid the electronic instability toward a metal–insulator transition. The actual mechanism of superconductivity is not known at present.

(c) High-T_c Ceramic Superconductor Applications

In 1986, the synthesis of copper-oxide based ceramics with a superconducting temperature (T_c) exceeding the boiling point of liquid nitrogen (77 K) was a major contribution in the scientific field. To date, a number of superconductors of this type have been prepared, and the highest T_c reached is 122 K (117). Considerable vibrational studies have been made, with Raman spectroscopy

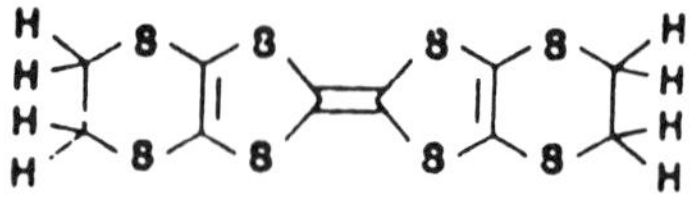

Figure 4-61 Structure of ET.

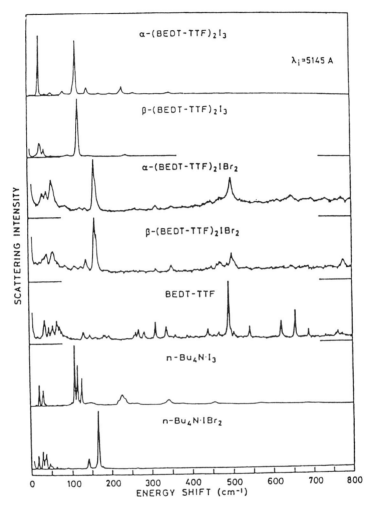

Figure 4-62 Spectra of several (ET) X salts. (Reproduced from S. Suga and G. Saito, *Solid State Commun.* **58**, 759, Copyright 1986, with permission from Pergamon Press Ltd., Headington Hill Hall, Oxford OX3 0BW, UK.)

being a major contributor. Most of the materials have a center of symmetry and thus the gerade, low-lying phonon modes are only observed in the Raman experiment (see Section 1.17). A typical application is presented herewith.

The original studies with the ceramic superconductors were conducted on powdered, impure phases, and therefore the early work reported bands belonging mainly to the impurities. Most of the definitive information that

Table 4-13   Comparison of the $v_{C=C}$ Frequencies of Neutral Donor Molecules with the Vibronic Frequencies of Their Salts (cm^{-1})

Donor	$v_{C=C}$	Salt	Vibronic Frequency
ET	1,511	κ-(ET)$_2$Cu(NCS)$_2$	1,290
TTF	1,518	(TTF)Br	1,368
TMTTF[a]	1,538	(TMTTF)Br	1,340
TMTSF[a]	1,539	(TMTSF)$_2$ReO$_4$	1,415

[a]TMTTF refers to tetramethyltetrathiafulvalene, and TMTSF to tetramethyltetraselenafulvalene.

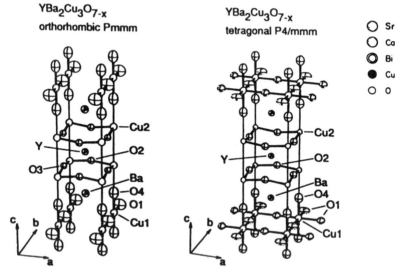

Figure 4-63   Structures of the orthorhombic and tetragonal forms of MBa$_2$Cu$_3$O$_7$. (Reproduced with permission from Ref. 117.)

has contributed to the identification and assignments of the Raman bands for these compounds has come from investigations made with single crystals.

This application will concentrate on the MBa$_2$Cu$_3$O$_{7-\delta}$ (123) system, where M = Y. Another designation for this system is YBCO. For this system, the optimum superconductivity is obtained for YBa$_2$Cu$_3$O$_{7-\delta}$, when $\delta = 0.3$. This formulation has an orthorhombic (O) structure. Upon loss of oxygen a phase transition takes place to an insulator with a tetragonal (T) structure. Figure 4-63 shows the structures for the O and T-phases.

Figure 4-64 shows the correlation diagram (see Appendix 6) for $y = 6$ to $y = 7$ for YBa$_2$Cu$_3$C$_y$. For the orthorhombic form, $5A_g$ Raman-active modes are predicted. In going from the O form to the T form, the vibrations change

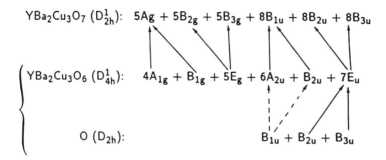

Figure 4-64 shows the correlation diagram content:

$YBa_2Cu_3O_7$ (D_{2h}^1): $5A_g + 5B_{2g} + 5B_{3g} + 8B_{1u} + 8B_{2u} + 8B_{3u}$

$YBa_2Cu_3O_6$ (D_{4h}^1): $4A_{1g} + B_{1g} + 5E_g + 6A_{2u} + B_{2u} + 7E_u$

O (D_{2h}): $B_{1u} + B_{2u} + B_{3u}$

Figure 4-64 Correlation diagram for $YBa_2Cu_3O_y$, $y = 6$ to $y = 7$. (Reproduced with permission from Ref. 118.)

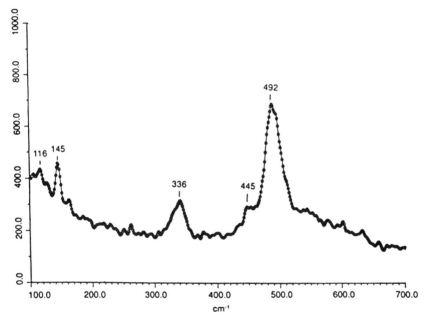

Figure 4-65 Raman spectrum of orthorhombic YBCO. (Reproduced with permission from Ref. 117.)

to $4A_g + B_{1g}$ modes. Figure 4-65 shows the Raman spectrum of the O-form. There is general agreement that $5A_g$ modes are observed for the O form of YBCO, as detected by polarization studies. These bands, located at 502, 436, 335, 146, and 115 cm^{-1}, are attributed to the following vibrations:

502 cm^{-1}: axial motion of the O(4) atoms,

$436 \, \text{cm}^{-1}$: Cu(2)–O(3) and –O(3) bond bending with the O(2) and O(3) atoms moving in phase,

$335 \, \text{cm}^{-1}$: Cu(2)–O(2) and –O(3) bond bending with the O(2) and O(3) atoms moving out of phase,

$146 \, \text{cm}^{-1}$: axial stretching of the Cu(2) atoms, and

$115 \, \text{cm}^{-1}$: axial stretching of the Ba atoms.

A major impurity found in these formulations has been $BaCuO_2$. This compound has an intense absorption at $639 \, \text{cm}^{-1}$. A second ubiquitous impurity was Y_2BaCuO_5. This phase shows Raman scattering at 380 and $595 \, \text{cm}^{-1}$. In a highly purified YBCO these bands are not present. Bands for the T-form of YBCO have been reported at 639, 480, 452, and $123 \, \text{cm}^{-1}$. Distinguishing between the (O)YBCO form and the T-form and the impurities is possible by Raman spectroscopy.

Raman spectroscopy has also contributed to the characterization of other ceramic superconductors (118). A normal coordinate analysis for Y_2BaCuO_5 has been made (119).

4.4   Industrial Applications

As mentioned in the introduction of this chapter, industrial applications of the Raman effect have garnered intense interest since the introduction of FT-Raman instrumentation. Concurrent with FT-Raman instrumentation development have been the fiber optics improvements and the advent of new detectors. These three factors have synergized and have led to the present interest in Raman spectroscopy. This has brought the Raman effect from the laboratory and into the plant, where *in-situ* measurements are now possible.

4.4.1   ENVIRONMENTAL APPLICATIONS

Several examples where Raman spectroscopy has played an important role in environmental analyses will be presented.

(a)   Determination of Ionic Species in Ground Water by SERS

Hazardous materials in trace amounts, such as metallic ions, find their way into the water system. It is necessary to be able to detect these materials *in situ* and in real time. Recently (120), such a method was found involving the SERS technique using optical fibers. The technique monitors the ions in aqueous solutions, measuring the changes in the Raman spectra of indicators, which form complexes with the metallic ions. The indicators used were

Eriochrome Black T (EBT), 4-(2-pyridylazo)resorcinol (PAR), cresol red, methyl red and 4-pyridinethiol.

Figure 4-66 illustrates the spectra of EBT–Cu^{2+}, of blank EBT, and the difference spectrum (traces a, b, c, respectively). The intensity of the band at 1,403 cm^{-1} in the difference spectrum corresponds to the Cu^{2+} concentration in solution. The 1,274 cm^{-1} band present in the uncomplexed EBT

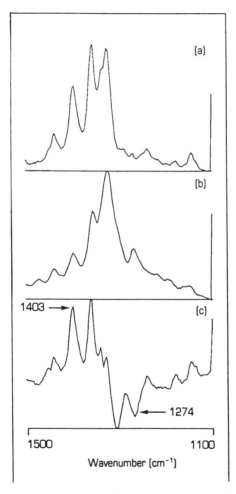

Figure 4-66 SERS spectra of (a) 1.8×10^{-4} M EBT and 1.8×10^{-4} Cu^{2+}; (b) 1.8×10^{-4} M EBT only; and (c) the difference spectrum of (a) − (b). (a) and (b) were obtained with 3-s integration times and 20 mW of 531 nm light for resonance excitation of EBT. (Reproduced with permission from Ref. 120.)

(trace b) disappears upon complexation. A calibration curve for detection of Cu^{2+} with EBT was made by plotting I_{1403}/I_{1274} vs. $pCu^{2+}(-\log Cu^{2+})$ concentration. This is illustrated in Fig. 4-67.

Figure 4-68a shows the spectrum of blank PAR. Figure 4-68b shows its spectrum in the presence of Pb^{2+}, and Fig. 4.-68c the spectrum in the presence of Fe^{3+}. Similar calibration curves could be formulated. For the Pb^{2+} concentration, plotting of I_{1323}/I_{1005} vs. $pPb^{2+}(-\log Pb^{2+})$ can be used for the detection of Pb^{2+}. Likewise, a plot of I_{1329}/I_{1362} vs. pFe^{3+} can be used for detection of the Fe^{3+} concentration in ground water by the SERS technique. The technique is particularly effective when used with optical fibers. Silver substrates provide the largest enhancements of the Raman signal, but rapidly degrade in air or water. A method of forming a durable, strongly enhancing SERS surface on silver is to roughen the fiber end followed by depositing a layer of silver, whereby some of degradation of the silver surface is avoided.

(b) Determination of Low Concentrations of the Nitrite Ion in Fresh Water and Seawater

Trace amounts of the nitrite ion (NO_2^-) are indicative of the extent of pollution and eutrophication. The multitude of methods that can measure nitrite ion concentrations, such as colorimetry, chemiluminescence or fluorimetry, are not capable of detecting subnanomole amounts of nitrite. These also suffer from interference problems. A highly sensitive and selective method for the determination of low concentrations of nitrite in aqueous solutions using surface-enhanced resonance Raman (SERRS) has been developed (121) (see Section 3.5).

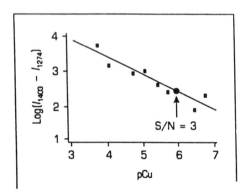

Figure 4-67 Calibration curve for detection of Cu^{2+} with EBT. The detection limit of 85 ppb is illustrated where $S/N = 3$. (Reproduced with permission from Ref. 120.)

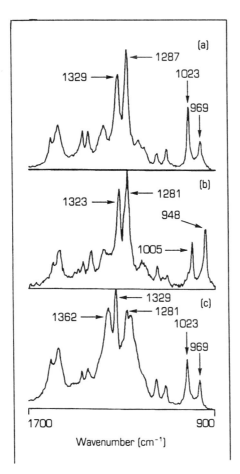

Figure 4-68 SERS spectra of etched silver foil that is coated with anchored thiolate of the indicator PAR disulfide (PARDS) immersed in (a) a blank solution showing PARDS only, (b) in the presence of Pb^{2+}, and (c) in the presence of Fe^{3+}. Spectra were obtained with 5-s integration times and 5 mW of 514.5 nm light for resonance excitation of PARDS. (Reproduced with permission from Ref. 120.)

The method uses a silver hydrosol active substrate. Raman spectra were obtained using a Raman micro-probe consisting of a Leitz Ortholux-1 microscope, optically coupled with a Spex Triplemate spectrograph (Model 1877). A charge-coupled device was used as the detector. All measurements were made with a 180° scattering geometry. The 507.1 nm line of an Ar laser (Spectra-Physics Model 2020) was used for excitation with a laser power of 6–8 mW at a resolution of 2 cm^{-1}. Prior to inducing the SERRS effect, the nitrite was transformed into a colored azo dye.

The dye has an azo form (Fig. 4-69, (1) and (2a)) and a hydrazone form (2b). The hydrazone form is the predominant form. The difference between the azo and hydrazone forms is that the N=N bond is associated with the azo forms (1) and (2), and the quinoid ring is associated with the hydrazone form

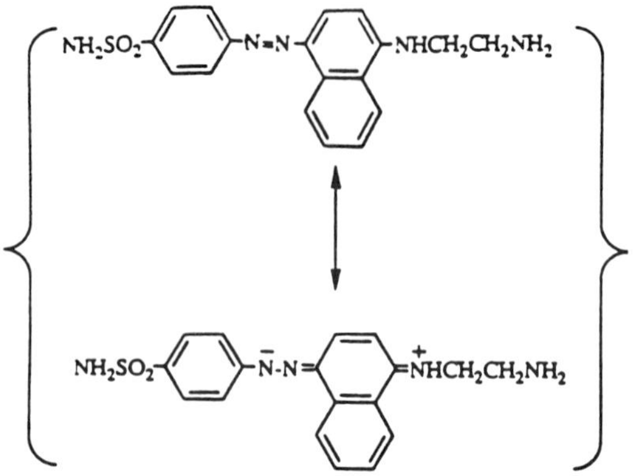

(1)

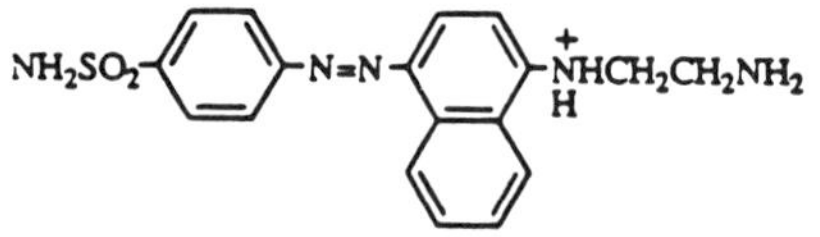

(2a)

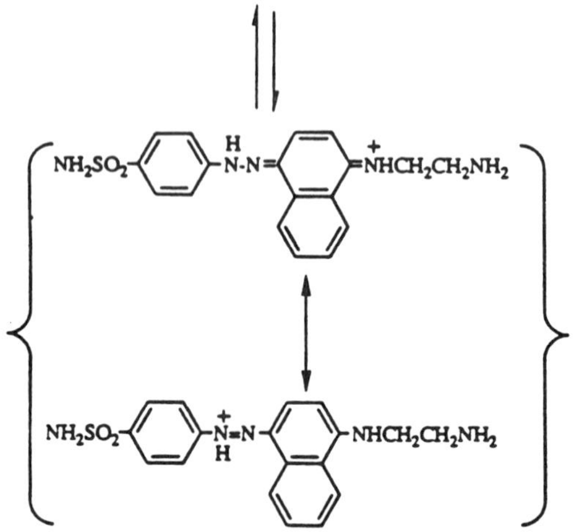

(2b)

Figure 4-69 Structures of (1) basic and (2) acidic forms of the azo dye.

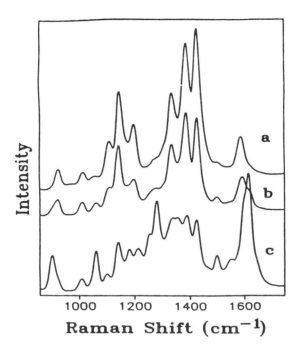

Figure 4-70   SERRS spectra of 1 μm azo dye in different pH solutions: (a) pH = 12; (b) pH = 7; (c) pH = 1. (Reproduced with permission from Ref. 121.)

(2b). The SERRS spectra of the azo dye at high, neutral and low pH solutions are shown in Fig. 4-70. Figure 4-71 shows the SERRS spectra compared to the resonance spectra (RR) at pH 2 and 12. The observed Raman bands and the tentative assignments are listed in Table 4-14. In basic solutions the SERRS spectrum of the azo dye looks the same as the RR spectrum (Fig. 4-71). In acidic solutions (Fig. 4-71), they differ, with two new bands at 1,328 and 1,283 cm^{-1} appearing in the SERRS spectrum. For the purpose of quantitative analyses the Raman measurements were made at high pH. The SERRS spectrum (Fig. 4-70a) in basic solution has the bands at 1,422 cm^{-1} (N=N band) and at 1,328 and 1,383 cm^{-1}, which show high selectivity and can be used for analytical purposes. Figure 4-72 shows analytical curves for the azo dye. It may be observed that linear relationships exist between the SERRS intensity and the dye concentration. For purposes of accuracy, an internal standard such as pyridine (10^{-3} M) was added to compensate for changes in excitation energy and for variations in sample positioning and optical alignment. Figure 4-73 shows the SERRS spectra of the azo dye in seawater; a was 35 m deep, and b, 500 m deep.

   Limits of detection for the nitrite ion are 0.02 nM. Reproducibility of the

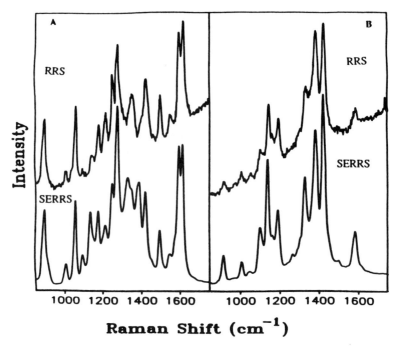

Figure 4-71 Comparison of SERRS spectra and RR spectra of 10 μM azo dye in different pH solutions: (a) pH = 2; (b) pH = 12. (Reproduced with permission from Ref. 121.)

method is satisfactory. The method is simple, is rapid, and demonstrates advantages over the high-sensitivity laser-induced fluorescent techniques, which require low temperatures and more complex sample preparation. The technique demonstrates the usefulness of the SERRS method for trace analysis in fresh water and seawater.

4.4.2 SURFACES (COATINGS)

(a) An Application in the Paint Industry

It has been advantageous to use the FT-Raman method to study various dynamic processes of interest to the paint industry. One such study was the study of an emulsion polymerization reaction whereby a FT-Raman system actually monitored the process (122).

Polymer latices[3] are of extreme technological importance in the development of water-borne paint systems. One method for the production

---

[3] The plural of latex, taken from *Webster's Collegiate Dictionary*.

Table 4-14 Tentative Assignment of Observed Raman Bands[a,b]

pH = 2		pH = 12	
1,620 vs	N'-ring C=C stretch		
1,597 vs	S-ring stretch		
		1,582 m	8a or 8b
1,546 vs	?		
1,496 m	S-ring 19a	1,499 vs	S-ring 19a
		1,450 vvw	N'-ring 19b
		1,422 vs	N=N stretch
1,425 sh	C—N' stretch		
		1,383 vs	S-ring 19b
1,351 sh	N'-ring C—C stretch		
		1,328 s	Ph—NMe$_2$
1,278 s	C—N stretch		
		1,266 vw	S-ring 14
1,253 s	S-ring Ph—NH—N		
1,214 m	?		
		1,193 m	9a
1,177 m	S-ring 9a		
		1,163 vw	9a coupled
1,145 w	C—N stretch	1,142 s	C—N stretch
		1,101 m	18b
		1,006 w	?
1,062 s	18a or b		
		915 m	5
897 m	3		

[a] Note: s = strong, w = weak, v = very, sh = shoulder.
[b] Using the Wilson mode numbering system see E. B. Wilson, *Phys. Rev.* **45**, 706 (1973).

of these latices involves emulsion polymerization, which allows careful control of particle size and morphology. Despite the fact that such polymerizations have been conducted for many years, they still are not too well understood. Spectroscopic techniques that can be used to study these reactions are hindered by the presence of water (e.g., infrared). In the case of Raman spectroscopy, the presence of water does not affect the quality of the spectrum.

This study illustrates a particular use of FT-Raman spectroscopy to monitor an emulsion polymerization of an acrylic/methacrylic copolymer. There are four reaction components to an emulsion polymerization: water-immiscible monomer, water, initiator, and emulsifier. During the reaction process, the monomers become solubilized by the emulsifier. Polymerization reactions were carried using three monomers: BA (butyl acrylate), MMA (methyl methacrylate), and AMA (allyl methacrylate). Figure 4-74 shows the FT-Raman spectra of the pure monomers, with the strong vC=C bands highlighted at 1,650 and 1,630 cm^{-1}. The reaction was

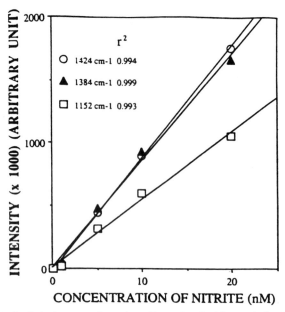

Figure 4-72   Analytical curves of azo dyes. (Reproduced with permission from Ref. 121.)

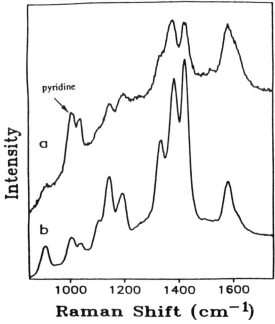

Figure 4-73   SERRS spectra of azo dye in seawater with internal standard (1 mM pyridine) at different depths: (a) 35 m; (b) 500 m. (Reproduced with permission from Ref. 121.)

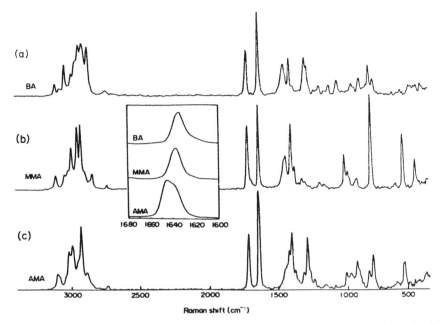

Figure 4-74 FT-Raman spectra of monomers: (a) BA, (b) MMA, (c) AMA (inset shows C=C stretching region). (Reproduced from G. Ellis, M. Claybourne, and S. E. Richards, *Spec. Acta.* **46A**, 227, Copyright 1990, with permission from Pergamon Press Ltd., Headington Hill Hall. Oxford OX3 0BW, UK.)

made at 74°C. As the polymerization proceeded, the disappearance of the C=C vibration could be followed, as illustrated in Fig. 4-75, which shows a plot of the concentration of the νC=C bands in the emulsion with reaction time. After two hours of the monomer feed, 5% of the unreacted double bonds remained. As the polymerization proceeded and the solids content increased, the S/N ratio (measured between the CH_2 deformation band at 1,450 cm^{-1} and the background at 2,500 cm^{-1}) of the spectrum improves from 10:1 at 7% solids to 70:1 at 35% solids. Figure 4-76 shows a plot of the ratio of the Raman bands at 1,450 and 3,450 cm^{-1} vs. the percent solids formed. This application illustrates the feasibility of monitoring a dynamic process by FT-Raman spectroscopy in the paint industry.

4.4.3 AN APPLICATION IN THE FOOD INDUSTRY

Fluorescence problems occurring with conventional Raman spectroscopy precluded the use of this technique in studying food and agricultural substances. However, with the advent of FT-Raman, renewed interest has

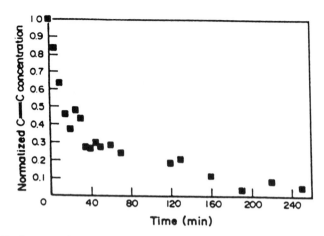

Figure 4-75 Concentration of C=C double bonds in the emulsion with reaction time. (Reproduced from G. Ellis, M. Claybourne, and S. E. Richards, *Spec. Acta.* **46A**, 227, Copyright 1990, with permission from Pergamon Press Ltd., Headington Hill Hall, Oxford OX3 0BW, UK.)

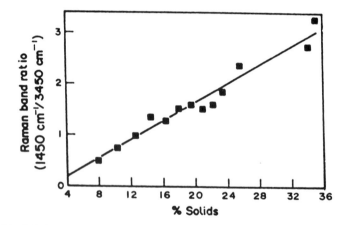

Figure 4-76 Relative Raman intensity of ratios of 1,450/3,450 cm^{-1} bands vs. solids content in the emulsion. (Reproduced from G. Ellis, M. Claybourne, and S. E. Richards, *Spec. Acta.* **46A**, 227, Copyright 1990, with permission from Pergamon Press Ltd., Headington Hill Hall, Oxford OX3 0BW, UK.)

arisen in these studies. The studies reported herein (123) were accomplished using a JEOL JRS-FT 6500N FT-Raman spectrometer equipped with an InGaAs detector. Excitation at 1,064 nm was provided by a CW Nd:YAG laser operating at 100–500 mW. Data were collected at 4 cm^{-1}, and 50–500 scans were accumulated.

The study was conducted on a series of lipids such as oils, tallow and butter. Figures 4-77 and 4-78 illustrate Raman spectra of sunflower, corn, sesame, rapeseed and olive oils and peanut, beef tallow and butter, respectively. The study determined that the iodine number the lipid containing foodstuffs could be estimated by measuring the FT-Raman spectra. The presence of double bonds in the unsaturated fatty acids in lipids provides a method of determining unsaturation. In treatment with iodine, two atoms of iodine are added per double bond of the unsaturated fatty acid. The related molar ratio that is measured with respect to these bonds is known as the

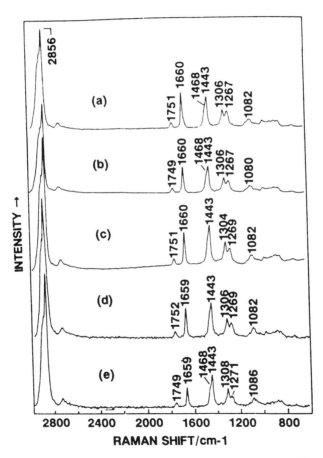

Figure 4-77   NIR FT Raman spectra of (a) sunflower, (b) corn, (c) sesame, (d) rapeseed, and (e) olive oils. (Reproduced from Y. Ozaki, R. Cho, K. Ikegaya, S. Muraishi and K. Kawauchi, *Applied Spectroscopy.* **46**, 1503 (1992), used with permission.

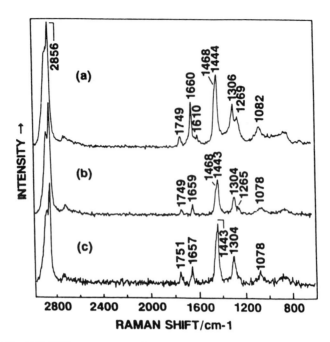

Figure 4-78 NIR FT Raman spectra of (a) peanut, (b) beef tallow, and (c) butter. (Reproduced with permission from Ref. 123.)

iodine number. It indicates the unsaturation level of the fat-containing food products. The higher the iodine value, the greater the unsaturation. In Fig. 4-77, the bands near 1,658 and 1,442 cm^{-1} are due to the unsaturated $v(C{=}C)$ stretching mode of the cis unsaturated fatty acid part, and the CH_2 scissoring mode of the saturated fatty acid part, respectively. The vC$=$C) stretching mode of the unsaturated fatty acids is very sensitive to the configuration around the C$=$C bond. For example, the trans unsaturated fatty acid shows the $v(C{=}C)$ stretching modes in the 1,670–1,680 cm^{-1} range, while the cis configuration shows the mode at 1,650–1,660 cm^{-1}.

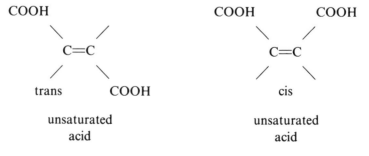

In Fig. 4-79 the $v(C{=}C)$ stretch is located at 1,660 cm^{-1}, indicating that most of the fatty acids studied are in the cis configuration around the $v(C{=}C)$ bond. Figure 4-79 plots the iodine number vs. the ratio of the intensities of I_{1658}/I_{1443}. As the iodine number increases the intensity ratio also increases, indicative of increasing cis-type unsaturated fatty acids in the lipids studied. Table 4-15 shows the iodine number and the percentage of fatty acids present in the fats described earlier.

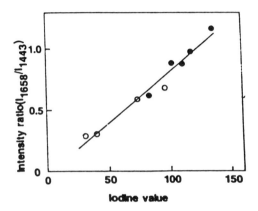

Figure 4-79 Iodine value vs. the intensity ratios of two bands at 1,658 and 1,443 cm^{-1} (I_{1658}/I_{1443}) for fat-containing foodstuffs investigated. (Reproduced with permission from Ref. 123.)

Table 4-15 Iodine Value (Number) and Percentages of Fatty Acids Constituting Lipids of Foods Investigated

	Iodine Value	Palmitic Acid	Stearic Acid	Oleic Acid	Linoleic[b] Acid	Linolenic[b] Acid
Sunflower oil	136	8%	3%	25%	59%	3%
Corn oil	118	10	6	34	48	1
Sesame oil	111	8	4	40	42	—
Rapeseed oil[a]	102	2	2	15	14	8
Olive oil	83	11	2	74	9	—
Peanut	96	9	5	52	23	—
Yolk	73	11	15	38	33	—
Beef tallow	40	28	24	44	3	—
Butter	31	25	9	33	5	2

[a]Rapeseed oil contains large amounts of erucine oil ($C_{21}H_{41}COOH$).
[b]Linolenic acid has one more $C{=}C$ in its chain than Linoleic acid.

4.4.4 APPLICATIONS IN THE FORENSIC AREA CONCERNING THE IDENTIFICATION OF ILLICIT DRUGS

Heretofore, Raman spectroscopy has not played a role in forensic science because of the fluorescent problems and the sample alignment, which is time-consuming. As a consequence, the technique was never seriously considered as a routine tool to study forensic materials. However, with the development of FT-Raman spectroscopy, the technique is now being reexamined. One such application in forensic science follows (124).

A FT-Raman spectrometer with excitation from a Nd:YAG laser at 1.064 μm and an GaInAs detector was used to collect the Raman data. The range

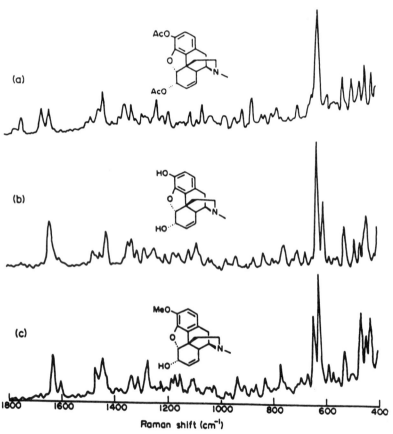

Figure 4-80 FT Raman spectra of pure alkaloids: (a) heroin, (b) morphine, and (c) codeine. 50 scans, 6 cm^{-1} resolution, incident laser power 200 mW. Scanning time 3 minutes. (Reproduced with permission from Ref. 124. Copyright © 1989 John Wiley & Sons, Ltd.)

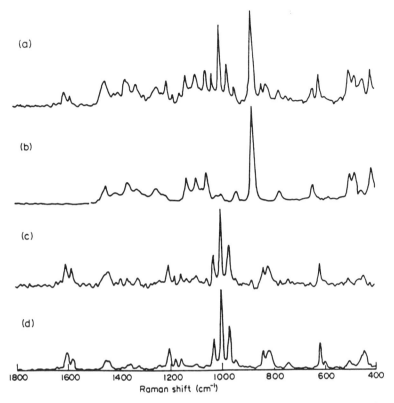

Figure 4-81 FT Raman spectra of (a) 23% amphetamine sulfate in sorbitol, and (b) pure sorbitol. Computer subtraction gives (c) the predicted spectrum of amphetamine sulfate while (d) shows the actual spectrum recorded for a sample of pure amphetamine sulfate. All spectra had 50 scans, 6 cm^{-1} resolution, and 200 mW of incident laser power. Scanning time was 3 minutes. (Reproduced with permission from Ref. 124. Copyright © 1989 John Wiley & Sons, Ltd.)

for the Raman spectra was 400–3,200 cm^{-1} at a resolution of 6 cm^{-1} and 200 mW power.

Figure 4-80 records the Raman spectra of three pure alkaloids (heroin, morphine and codeine) in the 400–1,800 cm^{-1} range. Figure 4-81 shows the spectra of a cut sample of amphetamine sulfate, the cutting agent, sorbitol, the subtraction spectrum, and pure amphetamine sulfate. Subtraction of the cutting agent from the cut sample gives the spectrum of amphetamine sulfate, which agrees with the spectrum of pure amphetamine sulfate. It appears that the technique is well suited for identification of illicit drugs. The technique utilizes small samples with no sample preparation and is non-destructive. The only possible problems that might be encountered are those where the

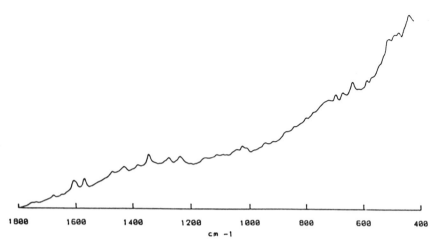

cm −1

Figure 4-82 Spectrum of cut heroin. (Reproduced with permission from Ref. 124. Copyright © 1989 John Wiley & Sons, Ltd.)

material is contaminated with a highly fluorescent compound (e.g., a fluorescent cutting agent), or a very dark material and a thermally sensitive substance. The effect of a fluorescent cutting agent is illustrated in Fig. 4-82, where fluorescence of the sample makes the identification of the drug virtually impossible. The capability of attaching a microscope to an FT-Raman, which is now commercially available, should make this technique even more attractive (see Section 3.7.4).

4.4.5   APPLICATION IN THE DYE INDUSTRY

It has been common knowledge that conventional Raman spectroscopy has failed in attempting to analyze dyes or dyestuffs. Most of the common dyes fluoresce intensely when excited in the visible. However, with the introduction of FT-Raman, the characterization of dyes has improved dramatically. Raman spectra are obtained, which are free of fluorescence and/or resonance effects. One such study involving the investigation of low levels of dyestuffs in acrylic fibers is presented (125).

A Perkin-Elmer 1710 interferometer using a Nd:YAG laser with an InGaAs detector was used in the study. The acrylic fibers studied are based on an acrylonitrile (94%), methacrylate (6%) copolymer with a diameter of 12–20 microns. The cell used for the measurement is illustrated in Fig. 4-83. Resolution was 3 cm^{-1}, and 50–150 scans were taken. Figure 4-84 shows the Raman spectra of a blue dye fiber, red dye fiber and an undyed methacrylic fiber. The dye vibrations can be readily observed. Figure 4-85 shows the subtraction spectrum (blue dyed fiber minus the undyed fiber). The subtracted spectrum (a) is compared to a blue cobalt dye (b), and the agreement is excellent.

The technique is more diagnostic in terms of detail, because the IR spectra are dominated by intense polymer absorptions, which cannot be eliminated completely by computer subtraction.

The results presented here illustrate that dye spectra may be obtained quickly and yield useful information. Small percentages of dye (1–2%) provide discernible bands. Computer subtraction can be used to remove the excess acrylic polymer bands. Here, also, the use of a micro FT-Raman instrument would be advantageous in lowering the sample size to be investigated.

4.4.6 APPLICATIONS IN METAL CORROSION STUDIES

Laser Raman spectroscopy has played a major role in the study of electrochemical systems (see Section 3.6). The technique provides molecular-specific information on the structure of the solid–solution interfaces *in situ* and is particularly suited for spectroelectrochemical studies of corrosion and surface film formation. Metals such as Pb, Ag, Fe, Ni, Co, Cu, Cr, Ti, Au and Sn, stainless steel and other alloys in various solutions have been studied by the technique.

A typical experimental layout for this technique is demonstrated in Fig. 4-86 (126). Generally Ar$^+$, Kr$^+$, and tunable dye lasers (see Section 2.2) are used, although NIR lasers can also be used. Various detectors have been used, such as photomultiplier tubes, diode arrays, silicon photon imaging

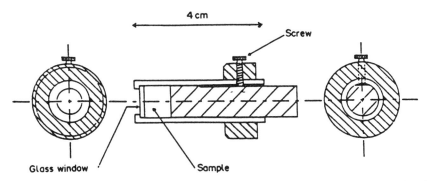

Figure 4-83 Fiber cell used for recording FT Raman spectra of fibers of dyes. (Reproduced from D. Bourgeois and S. P. Church, *Spec. Acta* **46A**, 295, Copyright 1990, with permission from Pergamon Press Ltd., Headington Hill Hall, Oxford OX3 0BW, UK.)

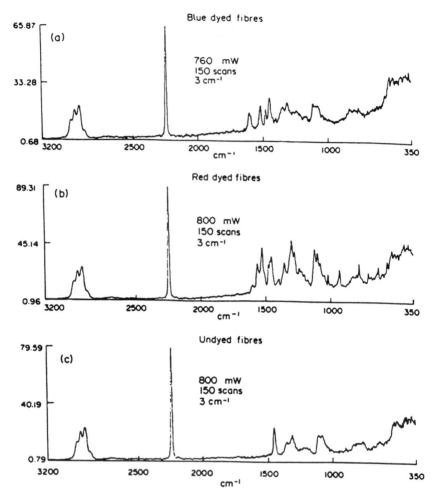

Figure 4-84 FT Raman spectra of acrylic fibers: (a) blue-dyed, (b) red-dyed, (c) undyed. (Reproduced from D. Bourgeois and S. P. Church, *Spec. Acta* **46A**, 295, Copyright 1990, with permission from Pergamon Press Ltd., Headington Hill Hall, Oxford OX3 0BW, UK.)

detectors, and silicon intensitied target. We shall describe the study of the corrosion of lead in dilute Na_2SO_4 solutions (127). The interest in lead corrosion is based on a better understanding of the deactivation mechanism in lead-acid batteries as well as simulating metallic corrosion in light water reactors. A Coherent Radiation Model CR 6 Ar^+ laser, incident at 60° to the surface normal, was used. The 5145 Å laser excitation was used at 100 mW.

In dilute solutions of Na_2SO_4 with an applied anodic potential of -0.85 V, the Raman spectrum was measured. The lead working electrode was made by melting lead shot onto a nickel support under a helium atmosphere, followed by filing to the desired shape and size. The external Ag/Ag_2SO_4 reference electrode was connected to the electrolytic cell via a long Teflon tube filled with zirconium oxide sand and saturated Na_2SO_4/Ag_2SO_4 solution. Figure 4-87 shows the temperature dependence of the Raman spectra observed. Bands are observed at 149 and 976 cm^{-1}, consistent with PbO

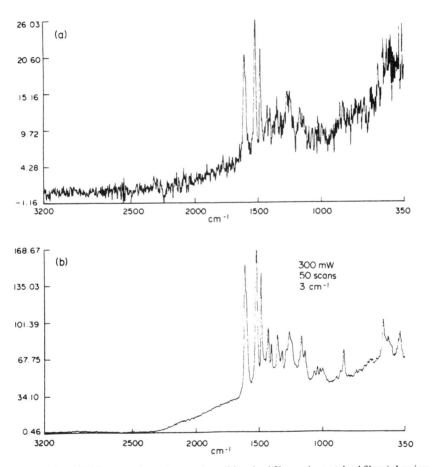

Figure 4-85   (a) FT Raman subtraction spectrum (blue-dyed fibers minus undyed fibers) showing the spectrum of the blue due after removal of acrylic polymer bands. (b) FT Raman spectrum of pure "Blue Cobalt" dye. (Reproduced from D. Bourgeois and S. P. Church, *Spec. Acta* **46A**, 295, Copyright 1990, with permission from Pergamon Press Ltd., Headington Hill Hall, Oxford OX3 0BW, UK.)

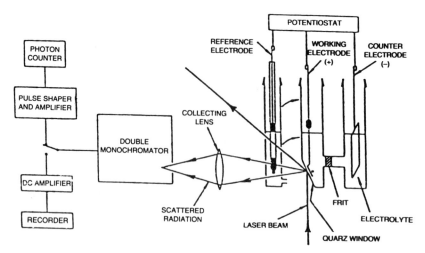

Figure 4-86 Experimental setup for laser Raman spectroelectrochemical studies. (Reprinted from Ref. 126 by permission of Kluwer Academic Publishers.)

and SO_4^{2-} vibrations, respectively, and the relative intensities were in accordance with an assignment of motions of PbO and SO_4^{2-} entities. The two phases were considered to be $PbOPbSO_4$ and $3PbOPbSO_4 \cdot H_2O$. The spectra persist with an increase in temperature to 553 K. This result agrees with x-ray diffraction data of the surface of the electrode, as well as the interpretation of cyclic voltammograms. At a higher voltage (-0.1) vs. Ag/Ag_2SO_4, the spectra observed as a function of temperature gave evidence for $PbSO_4$, which was confirmed by x-ray diffraction data. The results for both electrolytic oxidations of lead indicated three phases formed at the surface ($PbSO_4$, $PbOPbSO_4$, and $3PbOPbSO_4 \cdot H_2O$). Figure 4-88 shows the temperature dependency of the Raman spectra observed for lead in dilute Na_2SO_4 at -0.1 V.

4.4.7   FIBER-OPTICS RAMAN SPECTROSCOPY

One of the recent major developments has been the capability of making *in situ* and remote measurements of chemical processes in the laboratory or in the plant. These developments have occurred because of fiber-optics improvements (see Section 2.7), new detectors (array or CCDs) and the introduction of FT-Raman instrumentation.

To illustrate the new technique, a recent *in situ* measurement of an emulsion polymerization is reported. For commercial polymerizations of styrene and

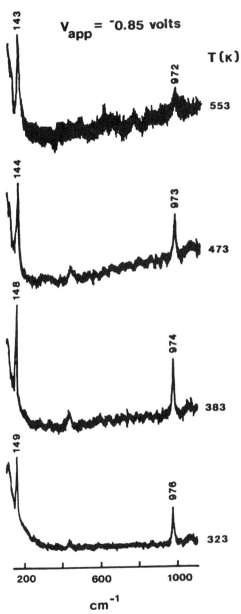

Figure 4-87 Temperature dependence of Raman spectra of lead in dilute solutions of Na_2SO_4 with applied voltage (anodic potential) of -0.85 V. (Reproduced with permission from Ref. 127.)

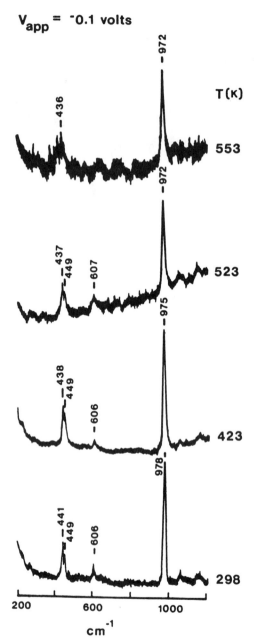

Figure 4-88 Temperature dependence of Raman spectra of lead in dilute solutions of Na_2SO_4 with applied anodic potential of -0.1 V. (Reproduced with permission from Ref. 127.)

other monomers, and for the copolymerizations of various materials, the method of emulsion polymerization has become the dominant process for commercial polymerization.

In this study (128) the emulsion polymerization of styrene was followed using fiber-optics Raman spectroscopy. Figure 4-89 shows the experimental arrangement. The styrene C=C band at 1,631 cm^{-1} was monitored with time. A band at 1,602 cm^{-1}, common to styrene monomer and polystyrene, was used as the pseudo-internal standard. In this region of the Raman the background of the SiO$_2$ fiber-optic probe was weak. The reaction was followed for close to one hour. Figure 4-90 shows the Raman spectra during the course of the polymerization. One may observe that the 1,631 cm^{-1} band disappears with time. The results were substantiated by UV-vis measurements. The method demonstrates the feasibility of making direct *in situ* analysis for systems of this type.

It should be cited that Garrison *et al.* (129) are studying the feasibility of using fiber-optics Raman spectroscopy to control distillation processes. This capability has important ramifications for the petroleum industry, as well as other industrial processes where distillation columns are involved. One of the systems that is being studied is a distillation column for the distillation of isopropanol/water.

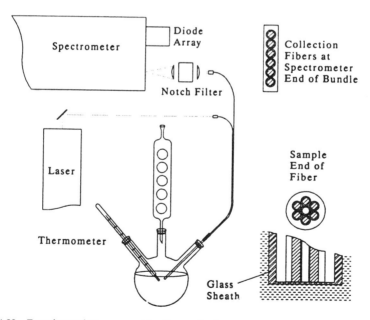

Figure 4-89 Experimental arrangement for fiber-optics Raman spectroscopy. (Reproduced with permission from Ref. 127.)

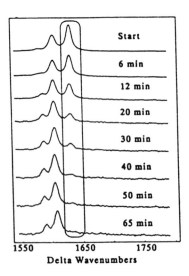

Figure 4-90 Raman spectra recorded during the course of a reaction. Spectra have been normalized. (Reproduced with permission from Ref. 128.)

The examples presented herein indicate the potential of using FT-Raman and fiber optics for *in situ* and remote applications, including environmental monitoring and process control.

For further reading on the subject of fiber-optics Raman spectroscopy, see Refs. 130–137, and for industrial applications of Raman spectroscopy, see Refs. 138–140.

4.4.8 APPLICATION IN THE PETROLEUM INDUSTRY

After Sir Raman's discovery of the Raman effect, petroleum scientists attempted to use the technique in order to study components of petroleum products. However, because of fluorescence problems, the interest waned. Today, in view of FT-Raman spectroscopy's (Section 3.8) capability to ameliorate fluorescence, renewed interest in the method has been generated. Recently, a new application for the study of blended gasolines, using FT-Raman spectroscopy, has been described (141).

After the addition of tetraethyllead to gasoline was prohibited, the oil companies were forced to make unleaded gasolines. In order to prevent engine knocking, the car manufacturers lowered the compression ratio of the engine, and oil companies changed the hydrocarbon composition of gasolines to incorporate more branched alkanes and aromatics to increase

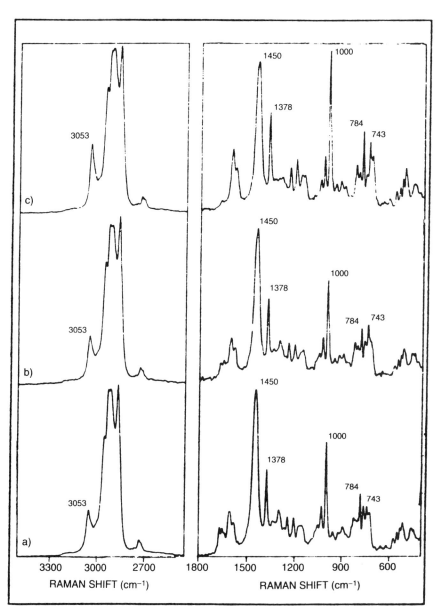

Figure 4-91 FT Raman spectra of blended gasolines: (a) 87 octane, (b) 89 octane, and (c) 93 octane. (reproduced with permission, Ref. 141).

the octane number. Benzene and toluene were some additives, as well as ethanol in some cases.

It was now necessary to be able to probe the composition of these blended gasolines for the additives present. The vibrational spectroscopies are excellent probes, and both IR and Raman spectroscopies can provide vital information concerning methyl/methylene ratios and identify the additives present in blended gasoline. This application pertains to the use of Raman spectroscopy to probe blended gasolines for additives.

Some of the frequencies that are useful as diagnostics for blended gasolines are listed as follows: For aliphatic groups:

(1) $2,960–2,870$ cn^{-1}, CH stretch in methyl groups;
(2) $2,925–2,850$ cm^{-1}, CH stretch in CH_2 groups;
(3) $2,890$ cm^{-1}, CH stretch in CH group.

For aromatic groups:

(1) $3,080–3,010$ cm^{-1}, CH stretch in benzene and derivatives and olefins;
(2) $1,000$ cm^{-1}, in substituted phenyl rings;
(3) $825–680$ cm^{-1}, bending modes in substituted phenyl rings.

For ethanol:

(1) 880 cm^{-1} diagnostic for ethanol.

Figure 4-91 shows the FT-Raman spectra of blended gasolines of various octane numbers. It may be observed that as the octane number increases there are increases in the methyl/methylene ratio ($3,053/2,870$ cm^{-1}) and $1,000/2,870$ cm^{-1} ratio, the latter indicative of aromatic additives. Table 4-16 shows the relative intensities of the aromatic bands with the grade of gasoline. The intensities at $3,053$ and $1,000$ cm^{-1} increase with octane number. The 780 cm^{-1} band increases, and this is indicative of the substituted phenyl range (e.g., toluene). The 743 cm^{-1} band increases as well, indicating a t-butyl group (iso-octane).

Figure 4-92 illustrates the spectra of ethanol and three gasolines with

Table 4-16   Relative Intensity of Aromatic Bands in FT-Raman Spectra of Gasolines

Gasoline Grade	Octane Number	$3,053/$ $2,870$ cm^{-1}	$1,000/$ $2,870$ cm^{-1}
Regular unleaded	Octane 87	0.264	0.158
Intermediate unleaded	Octane 89	0.284	0.161
Premium unleaded	Octane 93	0.387	0.236

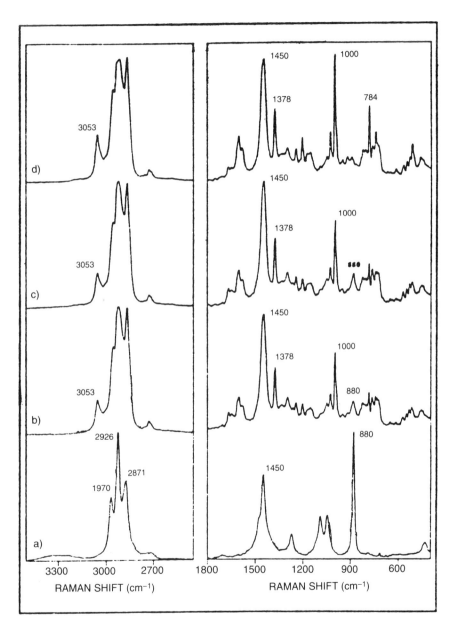

Figure 4-92 FT Raman spectra of (a) ethanol, (b) 87 octane gasohol, (c) 87 octane gasohol, and (d) 90 octane gasohol (reproduced with permission, Ref. 141).

varying octane numbers, all advertised as gasohol containing 10% ethanol. In these spectra a band at 880 cm^{-1} is diagnostic for alcohol. For the 90% octane gasoline, the 880 cm^{-1} band is not readily apparent, and it can be presumed that this gasoline is lacking ethanol. Additionally, the concentration of aromatic and branched hydrocarbons is significantly greater in the 90% octane gasoline, as compared to the lower octane gasolines.

It should be noted that IR could provide similar results. However, Raman spectroscopy is more sensitive to the concentration of unsaturated hydrocarbons in the fuels.

References

1. H. H. Claassen, H. Selig, and J. G. Malm, *J. Am. Chem. Soc.* **84**, 3593 (1962).
2. P. Tsao, C. C. Cobb, and H. H. Claassen, *J. Chem. Phys.* **54**, 5247 (1971).
3. K. O. Christe, E. C. Curtis, D. A. Dixon, H. P. Mercier, J. C. P. Sanders, and G. J. Schrobilgen, *J. Am. Chem. Soc.* **113**, 3351 (1991).
4. W. Preetz and G. Rimkus, *Z. Naturforsch.* **37b**, 579 (1982).
5. H. Ogoshi, Y. Saito, and K. Nakamoto, *J. Chem. Phys.* **57**, 4194 (1972).
6. M. Abe, T. Kitagawa, and Y. Kyogoku, *J. Chem. Phys.* **69**, 4526 (1978).
7. X.-Y. Li, R. S. Czernuszewicz, J. R. Kincaid, Y. O. Su, and T. G. Spiro, *J. Phys. Chem.* **91**, 31 (1990).
8. X.-Y. Li, R. S. Czernuszewicz, J. R. Kincaid, P. Stein, and T. G. Spiro, *J. Phys. Chem.* **94**, 47 (1990).
9. T. G. Spiro and X.-Y. Li, "Resonance Raman spectroscopy of metalloporphyrins," *in* "Biological Applications of Raman Spectroscopy" (T. G. Spiro, ed.), Vol. 3. John Wiley, New York, 1988.
10. For example, see L. Stryer, "Biochemistry." W. H. Freeman, New York, 1975.
11. T. Miyazawa, T. Shimanouchi, and S. Mizushima, *J. Chem. Phys.* **29**, 611 (1958).
12. N.-T. Yu, B. H. Jo, and D. C. O'Shea, *Arch. Biochem. Biophys.* **156**, 71 (1973).
13. R. P. Rava and T. G. Spiro, *J. Am. Chem. Soc.* **106**, 4062 (1984); *Biochemistry* **24**, 1861 (1985).
14. S. A. Asher, *Anal. Chem.* **65**, 59A and 201A (1993).
15. S. Krimm, "Peptides and proteins," *in* "Biological Applications of Raman Spectroscopy" (T. G. Spiro, ed.), Vol. 1. John Wiley, New York, 1987.
16. B. S. Hudson and L. C. Mayne, "Peptides and protein side chains," *in* "Biological Applications of Raman Spectroscopy" (T. G. Spiro, ed.), Vol. 2. John Wiley, New York, 1987.
17. A. T. Tu, "Peptide backbone conformation and microenvironment of protein side chains," *in* "Advances in Spectroscopy" (R. J. H. Clark and R. H. Hester, eds.), Vol. 13. John Wiley, New York, 1986.
18. I. Harada and H. Takeuchi, "Raman and ultraviolet resonance Raman spectra of proteins and related compounds," *in* "Advances in Spectroscopy" (R. J. H. Clark and R. E. Hester, eds.), Vol. 13. John Wiley, New York, 1986.
19. W. L. Peticolas, W. L. Kubasek, G. A. Thomas, and M. Tsuboi, "Nucleic acids," *in* "Biological Applications of Raman Spectroscopy" (T. G. Spiro, ed.), Vol. 1, p. 83. John Wiley, New York, 1987.
20. Y. Nishimura and M. Tsuboi, "Local conformations and polymorphism of DNA duplexes as revealed by their Raman spectra," *in* "Advances in Spectroscopy" (R. J. H. Clark and

R. E. Hester, eds.), Vol. 13, 1986; Y. Nishimura, M. Tusboi, T. Sato, and K. Aoki, *J. Mol. Structure* **146**, 123 (1986).

21. W. L. Kubasek, B. Hudson, and W. L. Peticolas, *Proc. Natl. Acad. Sci.* **82**, 2369 (1985); M. Tsuboi, Y. Nishimura, A. Hirakawa, and W. L. Peticolas, "Resonance Raman spectroscopy and normal modes of nucleic acid bases," *in* "Biological Applications of Raman Spectroscopy" (T. G. Spiro, ed.), Vol. 2. John Wiley, New York, 1987.

22. W. L. Peticolas and M. Tsuboi, "The Raman spectroscopy of nucleic acids," *in* "Infrared and Raman Spectroscopy of Biological Molecules" (T. M. Theophanides, ed.), p. 153. Reidel Pub., Dortrecht, The Netherlands, 1979.

23. W.-D. Wagner and K. Nakamoto, *J. Am. Chem. Soc.* **111**, 1590 (1989); *ibid.* **110**, 4044 (1988).

24. For example, see D. F. Shriver and C. B. Cooper, "Vibrational spectra of metal–metal bonded transition metal compounds," *in* "Advances in Infrared and Raman Spectroscopy" (R. J. H. Clark and R. E. Hester, eds.), Vol. 6. John Wiley, New York, 1979.

25. G. Peters and W. Preetz, *Z. Naturforsch.* **34b**, 1767 (1979).

26. F. A. Cotton and G. Wilkinson, "Advanced Inorganic Chemistry," Third Ed., p. 552. John Wiley, New York, 1972.

27. R. F. Dallinger, *J. Am. Chem. Soc.* **107**, 7202 (1985).

28. R. F. Dallinger, V. M. Miskowski, H. B. Gray, and W. H. Woodruff, *J. Am. Chem. Soc.* **103**, 1595 (1981).

29. D. E. Morris and W. H. Woodruff, "Vibrational spectra and the structure of electronically excited molecules in solution" *in* "Advances in Spectroscopy" (R. J. H. Clark and R. E. Hester, eds.), Vol. 14, p. 185. John Wiley, New York, 1987.

30. J. Laane, *Pure and Appl. Chem.* **59**, 1307 (1987).

31. I. M. Irwin, J. M. Cooke, and J. Laane, *J. Am. Chem. Soc.* **99**, 3273 (1977).

32. J. Laane, "Application of Raman spectroscopy to structural and conformational problems," *in* "Advances in Multi-photon Processes and Spectroscopy" (S. H. Lin, ed.), Vol. 1. World Scientific Press, Singapore, 1984.

33. C. J. Wurrey, J. R. Durig, and L. A. Carreira, "Gas-phase Raman spectroscopy of anharmonic vibrations," *in* "Vibrational Spectra and Structure" (J. R. Durig, ed.), Vol. 5, p. 121. Elsevier, Amsterdam, 1976.

34. W. G. Fateley and F. A. Miller, *Spectrochim. Acta.* **17**, 857 (1961).

35. J. R. Durig, W. E. Bucy, L. A. Carreira, and C. J. Wurrey, *J. Chem. Phys.* **60**, 1754 (1974).

36. J. A. Shelnutt, *J. Phys. Chem.* **87**, 605 (1983).

37. J. R. Kincaid, L. M. Proniewicz, K. Bajdor, A. Bruha, and K. Nakamoto, *J. Am. Chem. Soc.* **107**, 6775 (1985).

38. T. M. Cotton, J.-H. Kim, and R. E. Holt, "Surface-enhanced resonance Raman scattering (SERRS) spectroscopy," *in* "Advances in Biophysical Chemistry," Vol. 2, p. 115. JAI Press, 1992.

39. T. M. Cotton, J.-H. Kim, and G. S. Chumanov, *J. Raman Spectrosc.* **22**, 729 (1991).

40. E. Koglin and J. M. Sequarts, "Surface enhanced Raman scattering of biomolecules," *in* "Topics in Current Chemistry," Vol. 134. Springer-Verlag, Berlin, 1986.

41. R. F. Paisley and M. D. Morris, "Surface enhanced Raman spectroscopy of small molecules," *Progr. Analyt. Spectrosc.* **11**, 111 (1988).

42. T. M. Cotton, S. G. Schultz, and R. P. Van Duyne, *J. Am. Chem. Soc.* **102**, 7960 (1980).

43. T. M. Cotton, R. A. Uphaus, and D. Mobius, *J. Phys. Chem.* **90**, 6071 (1986).

44. E. Koglin, H. H. Lewinsky, and J. M. Sequarts, *Sur. Sci.* **158**, 370 (1985).

45. T. Watanabe, O. Kawanami, H. Katoh, K. Honda, Y. Nishimura, and M. Tsuboi, *Sur. Sci.* **158**, 341 (1985).

46. For example, see T. G. Spiro, "The resonance Raman spectroscopy of metalloporphyrins and heme proteins," *in* "Iron Porphyrins" (A. B. P. Lever and H. B. Gray, eds.), Part II. Addison-Wesley, 1983.

47. T. G. Spiro and T. C. Strekas, *J. Am. Chem. Soc.* **96**, 338 (1974).
48. K. Nakamoto, *Coord. Chem. Rev.* **100**, 363 (1990).
49. L. Duff, E. H. Appleman, D. F. Shriver, and I. M. Klotz, *Biochem. Biophys. Res. Comm.* **90**, 1098 (1979).
50. T. Kitagawa, M. R. Ondrias, D. L. Rousseau, M. Ikeda-Saito, and T. Yonetani, *Nature* **298**, 869 (1982).
51. T. Kitagawa, K. Nagai, and M. Tsubaki, *FEBS Lett.* **104**, 376 (1979).
52. K. Nagai, T. Kagimoto, A. Hayashi, F. Taketa, and T. Kitagawa, *Biochemistry* **22**, 1305 (1983).
53. J. Terner, J. D. Stong, T. G. Spiro, M. Nagumo, M. Nicol, and M. A. El-Sayed, *Proc. Nat. Acad. Sci. USA* **78**, 1313 (1981).
54. L. S. Alexander and H. M. Goff, *J. Chem. Educ.* **59**, 179 (1982).
55. P. M. Champion, B. R. Stallard, G. C. Wagner, and I. C. Gunsalus, *J. Am. Chem. Soc.* **104**, 5469 (1982).
56. Y. Ozaki, T. Kitagawa, Y. Kyogoku, Y. Imai, C. Hashimoto-Yutsudo, and R. Sato, *Biochemistry* **17**, 5826 (1978).
57. O. Bangcharoenpaurpong, A. K. Rigos, and P. M. Champion, *J. Biol. Chem.* **261**, 8089 (1986).
58. S. Hashimoto, Y. Tatsuno, and T. Kitagawa, *Proc. Japan Acad.* **60B**, 345 (1984); *Proc. Natl. Acad. Sci. USA* **83**, 2417 (1986).
59. J. Terner, A. J. Sitter, and M. Reczek, *Biochim. Biophys. Acta* **828**, 73 (1985); *J. Biol. Chem.* **260**, 7515 (1985).
60. T. Kitagawa, "Resonance Raman spectra of reaction intermediates of heme enzymes," in "Advances in Spectroscopy" (R. J. H. Clark and R. E. Hester, eds.), Vol. 13. John Wiley, New York, 1986.
61. D. M. Kurtz, Jr., D. F. Shriver, and I. M. Klotz, *Coord. Chem. Rev.* **24**, 145 (1977).
62. D. M. Kurtz, Jr., D. F. Shriver, and I. M. Klotz, *J. Am. Chem. Soc.* **98**, 5033 (1976).
63. R. E. Stenkamp, L. C. Sieker, L. H. Jensen, J. D. McCallum, and J. Sanders-Loehr, *Proc. Natl. Acad. Sci. USA* **82**, 713 (1985).
64. A. K. Shiemke, T. M. Loehr, and J. Sanders-Loehr, *J. Am. Chem. Soc.* **108**, 2437 (1986).
65. W. H. Woodruff, R. B. Dyer, and J. R. Schoonover, "Resonance Raman spectroscopy of blue copper proteins," in "Biological Applications of Raman Spectroscopy" (T. G. Spiro, ed.), Vol. 3. John Wiley, New York, 1988.
66. T. B. Freedman, J. Sanders-Loehr, and T. M. Loehr, *J. Am. Chem. Soc.* **98**, 2809 (1976).
67. T. J. Thamann, J. Sanders-Loehr, and T. M. Loehr, *J. Am. Chem. Soc.* **99**, 4187 (1977).
68. J. A. Larrabee and T. G. Spiro, *J. Am. Chem. Soc.* **102**, 4217 (1980).
69. J. Ling, K. D. Sharma, J. Sanders-Loehr, T. M. Loehr, R. S. Czernuszewicz, R. Fraczkiewicz, L. Nestor, and T. G. Spiro, to be published.
70. Y. Nonaka, M. Tsuboi, and K. Nakamoto, *J. Raman Spectrosc.* **21**, 133 (1990).
71. D. S. Lu, Y. Nonaka, M. Tsuboi, and K. Nakamoto, *J. Raman Spectrosc.* **21**, 321 (1990).
72. G. D. Strahan, D. S. Lu, M. Tsuboi, and K. Nakamoto, *J. Phys. Chem.* **96**, 6450 (1992).
73. M. Manfait and T. Theophanides, "Drug–nucleic acid interactions," in "Advances in Infrared and Raman Spectroscopy" (R. J. H. Clark and R. E. Hester, eds.), Vol. 13. John Wiley, 1986.
74. D. Gill, R. G. Kilponen, and L. Rimai, *Nature* **227**, 743 (1970).
75. K. Larsson and L. Hellgren, *Experientia* **30**, 48 (1974).
76. K. A. Britton, R. A. Dalterio, W. H. Nelson, D. Britt, and J. F. Sperry, *Appl. Spectrosc.* **42**, 782 (1988).
77. Y. Ozaki and K. Iriyama, "Potential of Raman spectroscopy in medical science," in "Laser Light Scattering Spectroscopy of Biological Objects" (J. Stepanek, P. Anzenbacher, and B. Sedlacek, eds.). Elsevier, Amsterdam, 1987.

78. Y. Ozaki, A. Mizuno, K. Itoh, M. Yoshiura, T. Iwamoto, and K. Iriyama, *Biochemistry* **22**, 6254 (1983).
79. Y. Ozaki, A. Mizuno, K. Itoh, and K. Iriyama, *Tech. Biol. Med.* **5**, 269 (1984).
80. N.-T. Yu, M. Bando, and J. F. R. Kuck, Jr., *Invest. Ophthalmol. Vis. Sci.* **26**, 97 (1985).
81. N.-T. Yu, D. C. DeNagel, D. J.-Y. Ho, and J. F. R. Kuck, "Ocular lenses," *in* "Biological Applications of Raman Spectroscopy" (T. G. Spiro, ed.), Vol. 1, p. 47. John Wiley, 1987.
82. R. LeFevre, J. Barbillat, J.-P. Cuif, P. Dhamelincourt, and J. Laureyns, *C. R. Acad. Sci. Paris* **288**, D-19 (1979).
83. H. Ishida, R. Kamoto, S. Uchida, A. Ishitani, Y. Ozaki, K. Iriyama, E. Tsukie, K. Shibata, F. Ishihara, and H. Kameda, *Appl. Spectrosc.* **41**, 407 (1987).
84. G. A. Schick and D. F. Bocian, " Hydroporphyrins and chlorophylls," *Biochim. Biophys. Acta* **895**, 127 (1987).
85. M. Tasumi and M. Fujiwara, "Vibrational spectra of chlorophylls," *in* "Advances in Spectroscopy" (R. J. H. Clark and R. E. Hester, eds.), Vol. 14. John Wiley, New York, 1987.
86. J. T. McFarland, "Flavins," *in* "Biological Applications of Raman Spectroscopy" (T. G. Spiro, ed.), Vol. 2. John Wiley, New York, 1987.
87. M. D. Morris and R. J. Bienstock, "Resonance Raman spectroscopy of flavins and flavoproteins," *in* "Advances in Spectroscopy" (R. J. H. Clark and R. E. Hester, eds.), Vol. 13. John Wiley, New York, 1986.
88. M. Abe, "Normal coordinate analysis of large molecules of biological interest: Metalloporphyrins and lumiflavin," *in* "Advances in Spectroscopy" (R. J. H. Clark and R. E. Hester, eds.), Vol. 13. John Wiley, New York, 1986.
89. T. G. Spiro, J. Hare, V. Yachandra, A. Gewirth, M. K. Johnson, and E. Remsen, "Resonance Raman spectra of iron–sulfur proteins and analogs," *in* "Metal Ions in Biology" (T. G. Spiro, ed.), Vol. 4. John Wiley, New York, 1982.
90. T. G. Spiro, R. S. Czernuszewicz, and S. Han, "Iron–sulfur proteins and analog complexes," *in* "Biological Applications of Raman Spectroscopy" (T. G. Spiro, ed.), Vol. 3. John Wiley, New York, 1988.
91. L. Que, Jr., "Metal–tyrosinate proteins," *in* "Biological Applications of Raman Spectroscopy)) (T. G. Spiro, ed.), Vol. 3. John Wiley, New York, 1988.
92. R. A. Mathies, S. O. Smith, and I. Palings, "Determination of retinal chromophore structure in rhodopsins," *in* "Biological Applications of Raman Spectroscopy" (T. G. Spiro, ed.), Vol. 2. John Wiley, New York, 1987.
93. M. Stockburger, T. Alshuth, D. Qesterheit, and W. Gartner, "Resonance Raman spectroscopy of bacterio-rhodopsin: Structure and function" *in* "Advances in Spectroscopy" (R. J. H. Clark and R. E. Hester, eds.), Vol. 13. John Wiley, New York, 1986.
94. B. Curry, I. Palings, A. D. Broek, J. A. Pardoen, J. Lugtenburg, and R. Mathies, "Vibrational analysis of the retinal isomers," *in* "Advances in Infrared and Raman Spectroscopy" (R. J. H. Clark and R. E. Hester, eds.), Vol. 12. John Wiley, New York, 1985.
95. M. Lutz, "Resonance Raman studies in photosynthesis," *in* "Advances in Infrared and Raman Spectroscopy" (R. J. H. Clark and R. E. Hester, eds.), Vol. 11. John Wiley, New York, 1984.
96. G. J. Thomas, Jr., "Viruses and nucleoproteins," *in* "Biological Applications of Raman Spectroscopy" (T. G. Spiro, ed.), Vol. 1. John Wiley, New York, 1987.
97. G. J. Thomas, Jr., "Applications of Raman spectroscopy in structural studies of viruses, nucleoproteins and their constituents," *in* "Advances in Spectroscopy" (R. J. H. Clark and R. E. Hester, eds.), Vol. 13. John Wiley, New York, 1986.
98. P. Yager and B. P. Gaber, "Membranes," *in* "Biological Applications of Raman Spectroscopy" (T. G. Spiro, ed.), Vol. 1. John Wiley, New York, 1987.
99. I. W. Levin, "Vibrational spectroscopy in membrane assemblies," *in* "Advances in Infrared

302 Chapter 4. Applications

and Raman Spectroscopy" (R. J. H. Clark and R. E. Hester, eds.), Vol. 11. John Wiley, New York, 1984.

100. P. R. Carey, "Resonance Raman labels and enzyme–substrate reactions" *in* "Biological Applications of Raman Spectroscopy" (T. G. Spiro, ed.), Vol. 2. John Wiley, New York, 1987.

101. P. R. Carey and V. R. Salares, "Raman and resonance Raman studies of biological systems," *in* "Advances in Infrared and Raman Spectroscopy" (R. J. H. Clark and R. E. Hester, eds.), Vol. 7. John Wiley, New York, 1980.

102. P. R. Carey, "Biochemical applications of Raman and resonance Raman spectroscopy." Academic Press, New York, 1982.

103. Y. Ozaki, "Medical applications of Raman spectroscopy," *Appl. Spectrosc. Rev.* **24**, 259 (1988).

104. P. K. Khulbe, A. Agaral, G. S. Raghuvanski, H. D. Bist, H. Hashimoto, T. Kitagawa, T. S. Little, and J. R. Durig, *J. Raman Spectrosc.* **20**, 283 (1989).

105. G. A. Kouvouklis, A. Jayaramann, G. P. Espinosa, and A. S. Cooper, *J. Raman Spectrosc.* **22**, 57 (1991).

106. M. W. Urban and B. C. Cornilsen, "Measurement of non-stoichiometry in $BaTiO_3$ using Raman spectroscopy," *in* "Advances in Material Characterization II" (R. L. Snyder, R. A. Condrate and P. F. Johnson, eds.), pp. 89–96. Plenum Press, New York, 1985.

107. M. A. Krebs and R. A. Condrate, *J. Am. Ceramic Soc.* **65**, C144 (1982).

108. G. Chen, R. G. Haire, and J. R. Peterson, *Appl. Spectrosc.* **46**, 1495 (1992).

109. F. D. Hardcastle and I. E. Wachs, *J. Phys. Chem.* **95**, 5031 (1991).

110. T. J. Dennis, J. P. Hare, H. W. Kroto, R. Taylor, D. R. M. Walton, and P. J. Hendra, *Spec. Acta* **47A**, 1289 (1991).

111. R. C. Haddon, A. F. Hebard, M. J. Rosseinsky, D. W. Murphy, S. J. Duclos, K. B. Lyons, B. Miller, J. M. Rosamilia, R. M. Fleming, A. R. Kortan, S. H. Glarum, A. V. Makhija, A. J. Muller, R. H. Eick, Z. M. Zahurak, R. Tycko, G. Dabbagh, and F. A. Thiel, *Nature* **350**, 320 (1991).

112. R. C. Haddon, A. F. Hebard, M. J. Rosseinsky, D. W. Murphy, S. H. Glarum, T. T. M. Palstra, A. P. Ramirez, S. J. Duclos, R. M. Fleming, T. Siegrist, and R. Tycko, *in* "Fullerenes: Synthesis, Properties, and Chemistry of Large Carbon Clusters" (G. S. Hammond and V. J. Kuck, eds.), ACS Symposium Series, Vol. 481, p. 78. ACS, Washington, D.C., 1992.

113. J. R. Ferraro and J. M. Williams, *Appl. Spectrosc.* **44**, 200 (1990).

114. M. E. Kozlov, K. I. Pokhadnia, and A. A. Yurchenko, *Spec. Acta* **43A**, 323 (1987).

115. S. Sugai and G. Saito, *Solid State Commun.* **58**, 759 (1986).

116. J. R. Ferraro, H. H. Wang, M.-W. Whangbo, and P. Stout, *Appl. Spectrosc.* **46**, 1520 (1992).

117. J. R. Ferraro and V. A. Maroni, *Appl. Spectrosc.* **44**, 351, 1990.

118. V. A. Maroni and John R. Ferraro, "The use of vibrational spectroscopy in the characterization of high-critical temperature ceramic superconductors," *in* "Practical Fourier Transform Infrared Spectroscopy—Industrial Laboratory Chemical Analysis" J. R. Ferraro and K. Krishman, eds.). Academic Press, San Diego (1990), and references therein.

119. Y. M. Guo and R. A. Condrate, *Spectrosc. Letters* **24**, 1185 (1991).

120. K. I. Mullen, D. X. Wang, L. G. Crane, and K. T. Carreon, *Spectrosc.* **7**(5), 24 (1992).

121. K. Xi, S. K. Sharma, G. T. Taylor, and D. W. Muenow, *Appl. Spectrosc.* **46**, 819 (1992).

122. G. Ellis, M. Claybourne, and S. E. Richards, *Spec. Acta* **46A**, 227 (1990).

123. Y. Ozaki, R. Cho, K. Ikegawa, S. Muraishi, and K. Kawauchi, *Appl. Spectrosc.* **46**, 1503 (1992).

124. C. M. Hodges and J. Akhavan, *Spec. Acta* **46A**, 303 (1990); C. M. Hodges, P. J. Hendra, H. A. Willis, and T. Farley, *J. Raman Spectroscopy* **20**, 745 (1989).

125. D. Bourgeois and S. P. Church, *Spec. Acta* **46A**, 295 (1990).
126. C. A. Melendres, "Laser Raman spectroscopy: Principles and applications to corrosion studies," *in* "Electrochemical and Optical Techniques for the Study and Monitoring of Metallic Corrosion" (M. G. S. Ferreira and C. A. Melendres, eds.), pp. 355–388, and references therein. Kluwer Academic Publishers, The Netherlands, 1991.
127. J. J. McMahon, W. Ruther, and C. A. Melendres, *J. Electrochem. Soc.* **135**, 557 (1988).
128. C. Wang, T. J. Vickers, J. B. Schlenoff, and C. D. Mann, *Appl. Spectrosc.* **46**, 1729 (1992).
129. A. A. Garrison, D. J. Trimble, E. C. Muly, M. J. Roberts, and S. W. Kercel, *Amer. Laboratory* Feb. 1, 1990, p. 19.
130. P. J. Hendra, G. Ellis, and D. J. Cutler, *J. Raman Spectroscopy* **19**, 413 (1988).
131. D. D. Archibald, L. T. Lin, and D. E. Honigs, *Appl. Spectrosc.* **42**, 1558 (1988).
132. M. L. Meyrick and S. M. Angel, *Appl. Spectrosc.* **44**, 565 (1990).
133. C. D. Allred and R. L. McCreery, *Appl. Spectrosc.* **44**, 1229 (1990).
134. S. D. Schwab and R. L. McCreery, *Appl. Spectrosc.* **41**, 126 (1987).
135. S. D. Schwab and R. L. McCreery, and F. T. Gamble, *Annal. Chem.* **58**, 2486 (1986).
136. B. Yang and M. D. Morris, *Appl. Spectrosc.* **45**, 512 (1991).
137. S. M. Angel, T. J. Kulp, and T. M. Vess, *Appl. Spectrosc.* **46**, 1085 (1992).
138. M. Mehicic and J. G. Grasselli, "Raman spectroscopy of catalysts," *in* "Analytical Raman Spectroscopy," (J. G. Grasselli and B. J. Bulkin, eds.) Chapter 10. J. Wiley and Sons, New York, 1991.
139. J. G. Grasselli, M. K. Snavely, and B. Bulkin, "Chemical Applications of Raman Spectroscopy." J. Wiley and Sons, New York, 1981.
140. J. G. Grasselli, M. Mehicic, and J. R. Mooney, *Fresenius Z. Anal. Chem.* **324**, 537 (1986).
141. M. J. Smith, G. Kemeny, and F. Walder, *Nicolet FT-Raman Application Notes, AN-9142* (1992).

APPENDICES

Appendix 1

Point Groups and Their Character Tables

Taken from K. Nakamoto, "Infrared and Raman Spectra of Inorganic and Coordination Compounds", Wiley and Sons, New York, 1978. Reprinted with permission of John Wiley and Sons, Inc.

C_s	E	$\sigma(xy)$		
A'	$+1$	$+1$	T_x, T_y, R_z	$\alpha_{xx}, \alpha_{yy}, \alpha_{zz}, \alpha_{xy}$
A''	$+1$	-1	T_z, R_x, R_y	α_{yz}, α_{xz}

C_2	E	$C_2(z)$		
A	$+1$	$+1$	T_z, R_z	$\alpha_{xx}, \alpha_{yy}, \alpha_{zz}, \alpha_{xy}$
B	$+1$	-1	T_x, T_y, R_x, R_y	α_{yz}, α_{xz}

C_i	E	i		
A_g	$+1$	$+1$	R_x, R_y, R_z	all components of α
A_u	$+1$	-1	T_x, T_y, T_z	

C_{2v}	E	$C_2(z)$	$\sigma_v(xz)$	$\sigma_v(yz)$		
A_1	$+1$	$+1$	$+1$	$+1$	T_z	$\alpha_{xx}, \alpha_{yy}, \alpha_{zz}$
A_2	$+1$	$+1$	-1	-1	R_z	α_{xy}
B_1	$+1$	-1	$+1$	-1	T_x, R_y	α_{xz}
B_2	$+1$	-1	-1	$+1$	T_y, R_x	α_{yz}

C_{3v}	E	$2C_3(z)$	$3\sigma_v$		
A_1	$+1$	$+1$	$+1$	T_z	$\alpha_{xx}+\alpha_{yy},\ \alpha_{zz}$
A_2	$+1$	$+1$	-1	R_z	
E	$+2$	-1	0	$(T_x,\ T_y),\ (R_x,\ R_y)$	$(\alpha_{xx}-\alpha_{yy},\ \alpha_{xy}),\ (\alpha_{yz},\ \alpha_{xz})$

C_{4v}	E	$2C_4(z)$	$C_4^2\equiv C_2''$	$2\sigma_v$	$2\sigma_d$		
A_1	$+1$	$+1$	$+1$	$+1$	$+1$	T_z	$\alpha_{xx}+\alpha_{yy},\ \alpha_{zz}$
A_2	$+1$	$+1$	$+1$	-1	-1	R_z	
B_1	$+1$	-1	$+1$	$+1$	-1		$\alpha_{xx}-\alpha_{yy}$
B_2	$+1$	-1	$+1$	-1	$+1$		α_{xy}
E	$+2$	0	-2	0	0	$(T_x,\ T_y),\ (R_x,\ R_y)$	$(\alpha_{yz},\ \alpha_{xz})$

C_p^n (or S_p^n) denotes that the C_p (or S_p) operation is carried out successively n times.

$C_{\infty v}$	E	$2C_\infty^\phi$	$2C_\infty^{2\phi}$	$2C_\infty^{3\phi}$	$\cdots$	$\infty\sigma_v$		
Σ^+	$+1$	$+1$	$+1$	$+1$	$\cdots$	$+1$	T_z	$\alpha_{xx}+\alpha_{yy},\ \alpha_{zz}$
Σ^-	$+1$	$+1$	$+1$	$+1$	$\cdots$	-1	R_z	
Π	$+2$	$2\cos\phi$	$2\cos 2\phi$	$2\cos 3\phi$	$\cdots$	0	$(T_x,\ T_y),\ (R_x,\ R_y)$	$(\alpha_{yz},\ \alpha_{xz})$
Δ	$+2$	$2\cos 2\phi$	$2\cos 2\cdot2\phi$	$2\cos 3\cdot2\phi$	$\cdots$	0		$(\alpha_{xx}-\alpha_{yy},\ \alpha_{xy})$
Φ	$+2$	$2\cos 3\phi$	$2\cos 2\cdot3\phi$	$2\cos 3\cdot3\phi$	$\cdots$	0		
$\cdots$	$\cdots$	$\cdots$	$\cdots$	$\cdots$	$\cdots$	$\cdots$		

C_{2h}	E	$C_2(z)$	$\sigma_h(xy)$	i		
A_g	$+1$	$+1$	$+1$	$+1$	R_z	$\alpha_{xx},\ \alpha_{yy},\ \alpha_{zz},\ \alpha_{xy}$
A_u	$+1$	$+1$	-1	-1	T_z	
B_g	$+1$	-1	-1	$+1$	$R_x,\ R_y$	$\alpha_{yz},\ \alpha_{xz}$
B_u	$+1$	-1	$+1$	-1	$T_x,\ T_y$	

D_3	E	$2C_3(z)$	$3C_2$		
A_1	$+1$	$+1$	$+1$		$\alpha_{xx}+\alpha_{yy},\ \alpha_{zz}$
A_2	$+1$	$+1$	-1	$T_z,\ R_z$	
E	$+2$	-1	0	$(T_x,\ T_y),\ (R_x,\ R_y)$	$(\alpha_{xx}-\alpha_{yy},\ \alpha_{xy}),\ (\alpha_{yz},\ \alpha_{xz})$

$D_{2d} \equiv V_d$	E	$2S_4(z)$	$S_4^2 \equiv C_2''$	$2C_2$	$2\sigma_d$		
A_1	$+1$	$+1$	$+1$	$+1$	$+1$		$\alpha_{xx}+\alpha_{yy},\ \alpha_{zz}$
A_2	$+1$	$+1$	$+1$	-1	-1	R_z	
B_1	$+1$	-1	$+1$	$+1$	-1		$\alpha_{xx}-\alpha_{yy}$
B_2	$+1$	-1	$+1$	-1	$+1$	T_z	α_{xy}
E	$+2$	0	-2	0	0	$(T_x, T_y),\ (R_x, R_y)$	$(\alpha_{yz}, \alpha_{xz})$

D_{3d}	E	$2S_6(z)$	$2S_6^2 \equiv 2C_3$	$S_6^3 \equiv S_2 \equiv i$	$3C_2$	$3\sigma_d$		
A_{1g}	$+1$	$+1$	$+1$	$+1$	$+1$	$+1$		$\alpha_{xx}+\alpha_{yy},\ \alpha_{zz}$
A_{1u}	$+1$	-1	$+1$	-1	$+1$	-1		
A_{2g}	$+1$	$+1$	$+1$	$+1$	-1	-1	R_z	
A_{2u}	$+1$	-1	$+1$	-1	-1	$+1$	T_z	
E_g	$+2$	-1	-1	$+2$	0	0	(R_x, R_y)	$(\alpha_{xx}-\alpha_{yy}, \alpha_{xy}),\ (\alpha_{yz}, \alpha_{xz})$
E_u	$+2$	$+1$	-1	-2	0	0	(T_x, T_y)	

D_{4d}	E	$2S_8(z)$	$2S_8^2 \equiv 2C_4$	$2S_8^3$	$S_8^4 \equiv C_2''$	$4C_2$	$4\sigma_d$		
A_1	$+1$	$+1$	$+1$	$+1$	$+1$	$+1$	$+1$		$\alpha_{xx}+\alpha_{yy},\ \alpha_{zz}$
A_2	$+1$	$+1$	$+1$	$+1$	$+1$	-1	-1	R_z	
B_1	$+1$	-1	$+1$	-1	$+1$	$+1$	-1		
B_2	$+1$	-1	$+1$	-1	$+1$	-1	$+1$	T_z	
E_1	$+2$	$+\sqrt{2}$	0	$-\sqrt{2}$	-2	0	0	(T_x, T_y)	
E_2	$+2$	0	-2	0	$+2$	0	0		$(\alpha_{xx}-\alpha_{yy}, \alpha_{xy})$
E_3	$+2$	$-\sqrt{2}$	0	$+\sqrt{2}$	-2	0	0	(R_x, R_y)	$(\alpha_{yz}, \alpha_{xz})$

$D_{2h} \equiv V_h$	E	$\sigma(xy)$	$\sigma(xz)$	$\sigma(yz)$	i	$C_2(z)$	$C_2(y)$	$C_2(x)$		
A_g	$+1$	$+1$	$+1$	$+1$	$+1$	$+1$	$+1$	$+1$		$\alpha_{xx}, \alpha_{yy}, \alpha_{zz}$
A_u	$+1$	-1	-1	-1	-1	$+1$	$+1$	$+1$		
B_{1g}	$+1$	$+1$	-1	-1	$+1$	$+1$	-1	-1	R_z	α_{xy}
B_{1u}	$+1$	-1	$+1$	$+1$	-1	$+1$	-1	-1	T_z	
B_{2g}	$+1$	-1	$+1$	-1	$+1$	-1	$+1$	-1	R_y	α_{xz}
B_{2u}	$+1$	$+1$	-1	$+1$	-1	-1	$+1$	-1	T_y	
B_{3g}	$+1$	-1	-1	$+1$	$+1$	-1	-1	$+1$	R_x	α_{yz}
B_{3u}	$+1$	$+1$	$+1$	-1	-1	-1	-1	$+1$	T_x	

D_{3h}	E	$2C_3(z)$	$3C_2$	σ_h	$2S_3$	$3\sigma_v$		
A_1'	$+1$	$+1$	$+1$	$+1$	$+1$	$+1$		$\alpha_{xx}+\alpha_{yy},\ \alpha_{zz}$
A_1''	$+1$	$+1$	$+1$	-1	-1	-1		
A_2'	$+1$	$+1$	-1	$+1$	$+1$	-1	R_z	
A_2''	$+1$	$+1$	-1	-1	-1	$+1$	T_z	
E'	$+2$	-1	0	$+2$	-1	0	(T_x, T_y)	$(\alpha_{xx}-\alpha_{yy}, \alpha_{xy})$
E''	$+2$	-1	0	-2	$+1$	0	(R_x, R_y)	$(\alpha_{yz}, \alpha_{xz})$

$\mathbf{D}_{4h}$	E	$2C_4(z)$	$C_4^2 \equiv C_2''$	$2C_2$	$2C_2'$	σ_h	$2\sigma_v$	$2\sigma_d$	$2S_4$	$S_2 \equiv i$		
A_{1g}	$+1$	$+1$	$+1$	$+1$	$+1$	$+1$	$+1$	$+1$	$+1$	$+1$		$\alpha_{xx}+\alpha_{yy},\ \alpha_{zz}$
A_{1u}	$+1$	$+1$	$+1$	$+1$	$+1$	-1	-1	-1	-1	-1		
A_{2g}	$+1$	$+1$	$+1$	-1	-1	$+1$	-1	-1	$+1$	$+1$	R_z	
A_{2u}	$+1$	$+1$	$+1$	-1	-1	-1	$+1$	$+1$	-1	-1	T_z	
B_{1g}	$+1$	-1	$+1$	$+1$	-1	$+1$	$+1$	-1	-1	$+1$		$\alpha_{xx}-\alpha_{yy}$
B_{1u}	$+1$	-1	$+1$	$+1$	-1	-1	-1	$+1$	$+1$	-1		
B_{2g}	$+1$	-1	$+1$	-1	$+1$	$+1$	-1	$+1$	-1	$+1$		α_{xy}
B_{2u}	$+1$	-1	$+1$	-1	$+1$	-1	$+1$	-1	$+1$	-1		
E_g	$+2$	0	-2	0	0	-2	0	0	0	$+2$	(R_x, R_y)	$(\alpha_{yz}, \alpha_{xz})$
E_u	$+2$	0	-2	0	0	$+2$	0	0	0	-2	(T_x, T_y)	

$\mathbf{D}_{5h}$	E	$2C_5(z)$	$2C_5^2$	σ_h	$5C_2$	$5\sigma_v$	$2S_5$	$2S_5^3$		
A_1'	$+1$	$+1$	$+1$	$+1$	$+1$	$+1$	$+1$	$+1$		$\alpha_{xx}+\alpha_{yy},\ \alpha_{zz}$
A_1''	$+1$	$+1$	$+1$	-1	$+1$	-1	-1	-1		
A_2'	$+1$	$+1$	$+1$	$+1$	-1	-1	$+1$	$+1$	R_z	
A_2''	$+1$	$+1$	$+1$	-1	-1	$+1$	-1	-1	T_z	
E_1'	$+2$	$2\cos 72°$	$2\cos 144°$	$+2$	0	0	$+2\cos 72°$	$+2\cos 144°$	(T_x, T_y)	
E_1''	$+2$	$2\cos 72°$	$2\cos 144°$	-2	0	0	$-2\cos 72°$	$-2\cos 144°$	(R_x, R_y)	$(\alpha_{yz}, \alpha_{xz})$
E_2'	$+2$	$2\cos 144°$	$2\cos 72°$	$+2$	0	0	$+2\cos 144°$	$+2\cos 72°$		$(\alpha_{xx}-\alpha_{yy},\ \alpha_{xy})$
E_2''	$+2$	$2\cos 144°$	$2\cos 72°$	-2	0	0	$-2\cos 144°$	$-2\cos 72°$		

$\mathbf{D}_{6h}$	E	$2C_6(z)$	$2C_6^2\equiv 2C_3$	$C_6^3\equiv C_2''$	$3C_2$	$3C_2'$	σ_h	$3\sigma_v$	$3\sigma_d$	$2S_6$	$2S_3$	$S_6^3\equiv S_2\equiv i$		
A_{1g}	$+1$	$+1$	$+1$	$+1$	$+1$	$+1$	$+1$	$+1$	$+1$	$+1$	$+1$	$+1$		$\alpha_{xx}+\alpha_{yy},\ \alpha_{zz}$
A_{1u}	$+1$	$+1$	$+1$	$+1$	$+1$	$+1$	-1	-1	-1	-1	-1	-1		
A_{2g}	$+1$	$+1$	$+1$	$+1$	-1	-1	$+1$	-1	-1	$+1$	$+1$	$+1$	R_z	
A_{2u}	$+1$	$+1$	$+1$	$+1$	-1	-1	-1	$+1$	$+1$	-1	-1	-1	T_z	
B_{1g}	$+1$	-1	$+1$	-1	$+1$	-1	-1	-1	$+1$	$+1$	-1	$+1$		
B_{1u}	$+1$	-1	$+1$	-1	$+1$	-1	$+1$	$+1$	-1	-1	$+1$	-1		
B_{2g}	$+1$	-1	$+1$	-1	-1	$+1$	-1	$+1$	-1	$+1$	-1	$+1$		
B_{2u}	$+1$	-1	$+1$	-1	-1	$+1$	$+1$	-1	$+1$	-1	$+1$	-1		
E_{1g}	$+2$	$+1$	-1	-2	0	0	-2	0	0	-1	$+1$	$+2$	(R_x, R_y)	$(\alpha_{yz},\ \alpha_{xz})$
E_{1u}	$+2$	$+1$	-1	-2	0	0	$+2$	0	0	$+1$	-1	-2	(T_x, T_y)	
E_{2g}	$+2$	-1	-1	$+2$	0	0	$+2$	0	0	-1	-1	$+2$		$(\alpha_{xx}-\alpha_{yy},\ \alpha_{xy})$
E_{2u}	$+2$	-1	-1	$+2$	0	0	-2	0	0	$+1$	$+1$	-2		

$\mathbf{D}_{\infty h}$	E	$2C_\infty^{\phi}$	$2C_\infty^{2\phi}$	$2C_\infty^{3\phi}$	$\cdots$	σ_h	∞C_2	$\infty \sigma_v$	$2S_\infty^{\phi}$	$2S_\infty^{2\phi}$	$\cdots$	$S_2\equiv i$		
Σ_g^{+}	$+1$	$+1$	$+1$	$+1$	$\cdots$	$+1$	$+1$	$+1$	$+1$	$+1$	$\cdots$	$+1$		$\alpha_{xx}+\alpha_{yy},\ \alpha_{zz}$
Σ_u^{+}	$+1$	$+1$	$+1$	$+1$	$\cdots$	-1	-1	$+1$	-1	-1	$\cdots$	-1	T_z	
Σ_g^{-}	$+1$	$+1$	$+1$	$+1$	$\cdots$	$+1$	-1	-1	$+1$	$+1$	$\cdots$	$+1$	R_z	
Σ_u^{-}	$+1$	$+1$	$+1$	$+1$	$\cdots$	-1	$+1$	-1	-1	-1	$\cdots$	-1		
Π_g	$+2$	$2\cos\phi$	$2\cos 2\phi$	$2\cos 3\phi$	$\cdots$	-2	0	0	$-2\cos\phi$	$-2\cos 2\phi$	$\cdots$	$+2$	(R_x, R_y)	$(\alpha_{yz},\ \alpha_{xz})$
Π_u	$+2$	$2\cos\phi$	$2\cos 2\phi$	$2\cos 3\phi$	$\cdots$	$+2$	0	0	$+2\cos\phi$	$+2\cos 2\phi$	$\cdots$	-2	(T_x, T_y)	
Δ_u	$+2$	$2\cos 2\phi$	$2\cos 4\phi$	$2\cos 6\phi$	$\cdots$	-2	0	0	$-2\cos 2\phi$	$-2\cos 4\phi$	$\cdots$	-2		
Δ_g	$+2$	$2\cos 2\phi$	$2\cos 4\phi$	$2\cos 6\phi$	$\cdots$	$+2$	0	0	$+2\cos 2\phi$	$+2\cos 4\phi$	$\cdots$	$+2$		$(\alpha_{xx}-\alpha_{yy},\ \alpha_{xy})$
Φ_g	$+2$	$2\cos 3\phi$	$2\cos 6\phi$	$2\cos 9\phi$	$\cdots$	-2	0	0	$-2\cos 3\phi$	$-2\cos 6\phi$	$\cdots$	$+2$		
Φ_u	$+2$	$2\cos 3\phi$	$2\cos 6\phi$	$2\cos 9\phi$	$\cdots$	$+2$	0	0	$+2\cos 3\phi$	$+2\cos 6\phi$	$\cdots$	-2		
$\vdots$														

T_d	E	$8C_3$	$6\sigma_d$	$6S_4$	$3S_4^2\equiv 3C_2$		
A_1	$+1$	$+1$	$+1$	$+1$	$+1$		$\alpha_{xx}+\alpha_{yy}+\alpha_{zz}$
A_2	$+1$	$+1$	-1	-1	$+1$		
E	$+2$	-1	0	0	$+2$		$(\alpha_{xx}+\alpha_{yy}-2\alpha_{zz},\ \alpha_{xx}-\alpha_{yy})$
F_1	$+3$	0	-1	$+1$	-1	(R_x, R_y, R_z)	
F_2	$+3$	0	$+1$	-1	-1	(T_x, T_y, T_z)	$(\alpha_{xy}, \alpha_{yz}, \alpha_{xz})$

O_h	E	$8C_3$	$6C_2$	$6C_4$	$3C_4^2\equiv 3C_2''$	$S_2\equiv i$	$6S_4$	$8S_6$	$3\sigma_h$	$6\sigma_d$		
A_{1g}	$+1$	$+1$	$+1$	$+1$	$+1$	$+1$	$+1$	$+1$	$+1$	$+1$		$\alpha_{xx}+\alpha_{yy}+\alpha_{zz}$
A_{1u}	$+1$	$+1$	$+1$	$+1$	$+1$	-1	-1	-1	-1	-1		
A_{2g}	$+1$	$+1$	-1	-1	$+1$	$+1$	-1	$+1$	$+1$	-1		
A_{2u}	$+1$	$+1$	-1	-1	$+1$	-1	$+1$	-1	-1	$+1$		
E_g	$+2$	-1	0	0	$+2$	$+2$	0	-1	$+2$	0		$(\alpha_{xx}+\alpha_{yy}-2\alpha_{zz},\ \alpha_{xx}-\alpha_{yy})$
E_u	$+2$	-1	0	0	$+2$	-2	0	$+1$	-2	0		
F_{1g}	$+3$	0	-1	$+1$	-1	$+3$	$+1$	0	-1	-1	(R_x, R_y, R_z)	
F_{1u}	$+3$	0	-1	$+1$	-1	-3	-1	0	$+1$	$+1$	(T_x, T_y, T_z)	
F_{2g}	$+3$	0	$+1$	-1	-1	$+3$	-1	0	-1	$+1$		$(\alpha_{xy}, \alpha_{yz}, \alpha_{xz})$
F_{2u}	$+3$	0	$+1$	-1	-1	-3	$+1$	0	$+1$	-1		

I_h	E	$12C_5$	$12C_5^2$	$20C_3$	$15C_2$	i	$12S_{10}$	$12S_{10}^3$	$20S_6$	15σ		
A_g	1	1	1	1	1	1	1	1	1	1		$\alpha_{xx}+\alpha_{yy}+\alpha_{zz}$
F_{1g}	3	$\frac{1}{2}(1+\sqrt{5})$	$\frac{1}{2}(1-\sqrt{5})$	0	-1	3	$\frac{1}{2}(1-\sqrt{5})$	$\frac{1}{2}(1+\sqrt{5})$	0	-1	(R_x, R_y, R_z)	
F_{2g}	3	$\frac{1}{2}(1-\sqrt{5})$	$\frac{1}{2}(1+\sqrt{5})$	0	-1	3	$\frac{1}{2}(1+\sqrt{5})$	$\frac{1}{2}(1-\sqrt{5})$	0	-1		
G_g	4	-1	-1	1	0	4	-1	-1	1	0		
H_g	5	0	0	-1	1	5	0	0	-1	1		$(2\alpha_{zz}-\alpha_{xx}-\alpha_{yy},$ $\alpha_{xx}-\alpha_{yy},$ $\alpha_{xy}, \alpha_{yz}, \alpha_{xz})$
A_u	1	1	1	1	1	-1	-1	-1	-1	-1		
F_{1u}	3	$\frac{1}{2}(1+\sqrt{5})$	$\frac{1}{2}(1-\sqrt{5})$	0	-1	-3	$-\frac{1}{2}(1-\sqrt{5})$	$-\frac{1}{2}(1+\sqrt{5})$	0	1	(T_x, T_y, T_z)	
F_{2u}	3	$\frac{1}{2}(1-\sqrt{5})$	$\frac{1}{2}(1+\sqrt{5})$	0	-1	-3	$-\frac{1}{2}(1+\sqrt{5})$	$-\frac{1}{2}(1-\sqrt{5})$	0	1		
G_u	4	-1	-1	1	0	-4	1	1	-1	0		
H_u	5	0	0	-1	1	-5	0	0	1	-1		

Appendix 2

General Formulas for Calculating the Number of Normal Vibrations in Each Symmetry Species

These tables were quoted from G. Hertzberg, "Molecular Spectra and Molecular Structure," Vol. II: "Infrared and Raman Spectra of Polyatomic Molecules." Van Nostrand, Princeton, New Jersey, 1945.

A. Point Groups Including Only Nondegenerate Vibrations

Point Group	Total Number of Atoms	Species	Number of Vibrations[a]
C_2	$2m + m_0$	A	$3m + m_0 - 2$
		B	$3m + 2m_0 - 4$
C_s	$2m + m_0$	A'	$3m + 2m_0 - 3$
		A''	$3m + m_0 - 3$
$C_i \equiv S_2$	$2m + m_0$	A_g	$3m - 3$
		A_u	$3m + 3m_0 - 3$
C_{2v}	$4m + 2m_{xz}$ $+ 2m_{yz} + m_0$	A_1	$3m + 2m_{xz} + 2m_{yz} + m_0 - 1$
		A_2	$3m + m_{xz} + m_{yz} - 1$
		B_1	$3m + 2m_{xz} + m_{yz} + m_0 - 2$
		B_2	$3m + m_{xz} + 2m_{yz} + m_0 - 2$
C_{2h}	$4m + 2m_h$ $+ 2m_2 + m_0$	A_g	$3m + 2m_h + m_2 - 1$
		A_u	$3m + m_h + m_2 + m_0 - 1$
		B_g	$3m + m_h + 2m_2 - 2$
		B_u	$3m + 2m_h + 2m_2 + 2m_0 - 2$

313

A. *continued*

Point Group	Total Number of Atoms	Species	Number of Vibations[a]
$D_{2h} \equiv V_h$	$8m + 4m_{xy} + 4m_{xz}$ $+ 4m_{yz} + 2m_{2x}$ $+ 2m_{2y} + 2m_{2z} + m_0$	A_g	$3m + 2m_{xy} + 2m_{xz} + 2m_{yz} + m_{2x} + m_{2y} + m_{2z}$
		A_u	$3m + m_{xy} + m_{xz} + m_{yz}$
		B_{1g}	$3m + 2m_{xy} + m_{xz} + m_{yz} + m_{2x} + m_{2y} - 1$
		B_{1u}	$3m + m_{xy} + 2m_{xz} + 2m_{yz} + m_{2x} + m_{2y} + m_{2z} + m_0 - 1$
		B_{2g}	$3m + m_{xy} + 2m_{xz} + m_{yz} + m_{2x} + m_{2z} - 1$
		B_{2u}	$3m + 2m_{xy} + m_{xz} + 2m_{yz} + m_{2x} + m_{2y} + m_{2z} + m_0 - 1$
		B_{3g}	$3m + m_{xy} + m_{xz} + 2m_{yz} + m_{2y} + m_{2x} - 1$
		B_{3u}	$3m + 2m_{xy} + 2m_{xz} + m_{yz} + m_{2x} + m_{2y} + m_{2z} + m_0 - 1$

[a] Note that m is always the number of sets of equivalent nuclei not on any element of symmetry; m_0 is the number of nuclei lying on all symmetry elements present; m_{xy}, m_{xz}, m_{yz} are the numbers of sets of nuclei lying on the xy, xz, yz plane, respectively, but not on any axes going through these planes; m_2 is the number of sets of nuclei on a twofold axis but not at the point of intersection with another element of symmetry; m_{2x}, m_{2y}, m_{2z} are the numbers of sets of nuclei lying on the x, y, z axis if they are twofold axes, but not on all of them; m_h is the number of sets of nuclei on a plane σ_h but not on the axis perpendicular to this plane.

B. Point Groups Including Degenerate Vibrations

Point Group	Total Number of Atoms	Species	Number ov Vibations[a]
D_3	$6m + 3m_2$ $+ 2m_3 + m_0$	A_1	$3m + m_2 + m_3$
		A_2	$3m + 2m_2 + m_3 + m_0 - 2$
		E	$6m + 3m_2 + 2m_3 + m_0 - 2$
C_{3v}	$6m + 3m_v + m_0$	A_1	$3m + 2m_v + m_0 - 1$
		A_2	$3m + m_v - 1$
		E	$6m + 3m_v + m_0 - 2$
C_{4v}	$8m + 4m_v$ $+ 4m_d + m_0$	A_1	$3m + 2m_v + 2m_d + m_0 - 1$
		A_2	$3m + m_v + m_d - 1$
		B_1	$3m + 2m_v + m_d$
		B_2	$3m + m_v + 2m_d$
		E	$6m + 3m_v + 3m_d + m_0 - 2$
$C_{\infty v}$	m_0	Σ^+	$m_0 - 1$
		Σ^-	0
		Π	$m_0 - 2$
		$\Delta, \Phi, \ldots$	0

B. *continued*

Point Group	Total Number of Atoms	Species	Number of Vibrations[a]
$D_{2d} \equiv V_d$	$8m + 4m_d + 4m_2$ $+ 2m_4 + m_0$	A_1	$3m + 2m_d + m_2 + m_4$
		A_2	$3m + m_d + 2m_2 - 1$
		B_1	$3m + m_d + m_2$
		B_2	$3m + 2m_d + 2m_2 + m_4 + m_0 - 1$
		E	$6m + 3m_d + 3m_2 + 2m_4 + m_0 - 2$
D_{3d}	$12m + 6m_d$ $+ 6m_2 + 2m_6 + m_0$	A_{1g}	$3m + 2m_d + m_2 + m_6$
		A_{1u}	$3m + m_4 + m_2$
		A_{2g}	$3m + m_d + 2m_2 - 1$
		A_{2u}	$3m + 2m_d + 2m_2 + m_6 + m_0 - 1$
		E_g	$6m + 3m_d + 3m_2 + m_6 - 1$
		E_u	$6m + 3m_d + 3m_2 + m_6 + m_0 - 1$
D_{4d}	$16m + 8m_d$ $+ 8m_2 + 2m_8 + m_0$	A_1	$3m + 2m_d + m_2 + m_8$
		A_2	$3m + m_d + 2m_2 - 1$
		B_1	$3m + m_d + m_2$
		B_2	$3m + 2m_d + 2m_2 + m_8 + m_0 - 1$
		E_1	$6m + 3m_d + 3m_2 + m_8 + m_0 - 1$
		E_2	$6m + 3m_d + 3m_2$
		E_3	$6m + 3m_d + 3m_2 + m_8 - 1$
D_{3h}	$12m + 6m_v + 6m_h$ $+ 3m_2 + 2m_3 + m_0$	A_1'	$3m + 2m_v + 2m_h + m_2 + m_3$
		A_1''	$3m + m_v + m_h$
		A_2'	$3m + m_v + 2m_h + m_2 - 1$
		A_2''	$3m + 2m_v + m_h + m_2 + m_3 + m_0 - 1$
		E'	$6m + 3m_v + 4m_h + 2m_2 + m_3 + m_0 - 1$
		E''	$6m + 3m_v + 2m_h + m_2 + m_3 - 1$
D_{4h}	$16m + 8m_v + 8m_d$ $+ 8m_h + 4m_2 + 4m_2'$ $+ 2m_4 + m_0$	A_{1g}	$3m + 2m_v + 2m_d + 2m_h + m_2 + m_2' + m_4$
		A_{1u}	$3m + m_v + m_d + m_h$
		A_{2g}	$3m + m_v + m_d + 2m_h + m_2 + m_2' - 1$
		A_{2u}	$3m + 2m_v + 2m_d + m_h + m_2 + m_2' + m_4 + m_0 - 1$
		B_{1g}	$3m + 2m_v + m_d + 2m_h + m_2 + m_2'$
		B_{1u}	$3m + m_v + 2m_d + m_h + m_2'$
		B_{2g}	$3m + m_v + 2m_d + 2m_h + m_2 + m_2'$
		B_{2u}	$3m + 2m_v + m_d + m_h + m_2$
		E_g	$6m + 3m_v + 3m_d + 2m_h + m_2 + m_2' + m_4 - 1$
		E_u	$6m + 3m_v + 3m_d + 4m_h + 2m_2 + 2m_2' + m_4 + m_0 - 1$

B. *continued*

Point Group	Total Number of Atoms	Species	Number of vibrations[a]
$\mathbf{D}_{5h}$	$20m + 10m_v + 10m_h$ $+ 5m_2 + 2m_5 + m_0$	A'_1	$3m + 2m_v + 2m_h + m_2 + m_5$
		A''_1	$3m + m_v + m_h$
		A'_2	$3m + m_v + 2m_h + m_2 - 1$
		A''_2	$3m + 2m_v + m_h + m_2 + m_5 + m_0 - 1$
		E'_1	$6m + 3m_v + 4m_h + 2m_2 + m_5 + m_0 - 1$
		E''_1	$6m + 3m_v + 2m_h + m_2 + m_5 - 1$
		E'_2	$6m + 3m_v + 4m_h + 2m_2$
		E''_2	$6m + 3m_v + 2m_h + m_2$
$\mathbf{D}_{6h}$	$24m + 12m_v + 12m_d$ $+ 12m_h + 6m_2 + 6m'_2$ $+ 2m_6 + m_0$	A_{1g}	$3m + 2m_v + 2m_d + 2m_h + m_2 + m'_2 + m_6$
		A_{1u}	$3m + m_v + m_d + m_h$
		A_{2g}	$3m + m_v + m_d + 2m_h + m_2 + m'_2 - 1$
		A_{2u}	$3m + 2m_v + 2m_d + m_h + m_2 + m'_2 + m_6 + m_0 - 1$
		B_{1g}	$3m + m_v + 2m_d + m_h + m'_2$
		B_{1u}	$3m + 2m_v + m_d + 2m_h + m_2 + m'_2$
		B_{2g}	$3m + m_v + 2m_d + m_h + m_2$
		B_{2u}	$3m + m_v + 2m_d + 2m_h + m_2 + m'_2$
		E_{1g}	$6m + 3m_v + 3m_d + 2m_h + m_2 + m'_2 + m_6 - 1$
		E_{1u}	$6m + 3m_v + 3m_d + 4m_h + 2m_2 + 2m'_2 + m_6 + m_0 - 1$
		E_{2g}	$6m + 3m_v + 3m_d + 4m_h + 2m_2 + 2m'_2$
		E_{2u}	$6m + 3m_v + 3m_d + 2m_h + m_2 + m'_2$
$\mathbf{D}_{\infty h}$	$2m_x + m_0$	Σ_g^+	m_x
		Σ_u^+	$m_x + m_0 - 1$
		Σ_g^-, Σ_u^-	0
		Π_g	$m_x - 1$
		Π_u	$m_x + m_0 - 1$
		Δ_g, Δ_u	0
		$\Phi_g, \Phi_u, \ldots$	0
$\mathbf{T}_d$	$24m + 12m_d$ $+ 6m_2 + 4m_3 + m_0$	A_1	$3m + 2m_d + m_2 + m_3$
		A_2	$3m + m_d$
		E	$6m + 3m_d + m_2 + m_3$
		F_1	$9m + 4m_d + 2m_2 + m_3 - 1$
		F_2	$9m + 5m_d + 3m_2 + 2m_3 + m_0 - 1$

B. *continued*

Point Group	Total Number of Atoms	Species	Number of Vibations[a]
O_h	$48m + 24m_h + 24m_d$ $+ 12m_2 + 8m_3$ $+ 6m_4 + m_0$	A_{1g}	$3m + 2m_h + 2m_d + m_2 + m_3 + m_4$
		A_{1u}	$3m + m_h + m_d$
		A_{2g}	$3m + 2m_h + m_d + m_2$
		A_{2u}	$3m + m_h + 2m_d + m_2 + m_3$
		E_g	$6m + 4m_h + 3m_d + 2m_2 + m_3 + m_4$
		E_u	$6m + 2m_h + 3m_d + m_2 + m_3$
		F_{1g}	$9m + 4m_h + 4m_d + 2m_2 + m_3 + m_4 - 1$
		F_{1u}	$9m + 5m_h + 5m_d + 3m_2 + 2m_3 + 2m_4 + m_0 - 1$
		F_{2g}	$9m + 4m_h + 5m_d + 2m_2 + 2m_3 + m_4$
		F_{2u}	$9m + 5m_h + 4m_d + 2m_2 + m_3 + m_4$

[a]Note that m is the number of sets of nuclei not on any element of symmetry; m_0 is the number of nuclei on all elements of symmetry; $m_2, m_3, m_4, \ldots$ are the numbers of sets of nuclei on a twofold, threefold, fourfold, ... axis but not on any other element of symmetry that does not wholly coincide with that axis; m_2' is the number of sets of nuclei on the twofold axis called C_2' in the preceding character tables; m_v, m_d, m_h are the numbers of sets of nuclei on planes $\sigma_v, \sigma_d, \sigma_h$, respectively, but not on any other element of symmetry.

Appendix 3

Direct Products of Irreducible Representations

This material was reproduced with permission from E. B. Wilson, J. C. Decius and P. C. Cross, "Molecular Vibrations," McGraw-Hill, New York, 1955.

$$A \times A = A, \quad B \times B = A, \quad A \times B = B, \quad A \times E = E, \quad B \times E = E, \quad A \times F = F,$$
$$B \times F = F, \quad g \times g = g, \quad u \times u = g, \quad u \times g = u \quad '\times' = ', \quad ''\times'' = ', \quad '\times'' = '',$$
$$A \times E_1 = E_1, \quad A \times E_2 = E_2, \quad B \times E_1 = E_2, \quad B \times E_2 = E_1.$$

Subscripts on A or B:

$1 \times 1 = 1$, $2 \times 2 = 1$, $1 \times 2 = 2$, except for $\mathbf{D_2} = \mathbf{V}$ and $\mathbf{D_{2h}} = \mathbf{V}_h$, where
$1 \times 2 = 3$, $2 \times 3 = 1$, $1 \times 3 = 2$.

Doubly degenerate representations:

For $\mathbf{C_3, C_{3h}, C_{3v}, D_3, D_{3h}, D_{3d}, C_6, C_{6h}, C_{6v}, D_6, D_{6h}, S_6, O, O_h, T, T_d, T_h}$:

$$E_1 \times E_1 = E_2 \times E_2 = A_1 + A_2 + E_2,$$
$$E_1 \times E_2 = B_1 + B_2 + E_1,$$

For $\mathbf{C_4, C_{4v}, C_{4h}, D_{2d}, D_4, D_{4h}, S_4}$: $E \times E = A_1 + A_2 + B_1 + B_2$
For groups in above lists which have symbols A, B, or E without subscripts, read $A_1 = A_2 = A$, etc.

Triply degenerate representations:

For $\mathbf{T_d, O, O_h}$: $E \times F_1 = E \times F_2 = F_1 + F_2$
$$F_1 \times F_1 = F_2 \times F_2 = A_1 + E + F_1 + F_2$$
$$F_1 \times F_2 = A_2 + E + F_1 + F_2$$

For $\mathbf{T, T_h}$: Drop subscripts 1 and 2 from A and F

Linear molecules ($\mathbf{C_{\infty v}}$ and $\mathbf{D_{\infty h}}$):

$$\Sigma^+ \times \Sigma^+ = \Sigma^- \times \Sigma^- = \Sigma^+; \qquad \Sigma^+ \times \Sigma^- = \Sigma^-$$
$$\Sigma^+ \times \Pi = \Sigma^- \times \Pi = \Pi; \qquad \Sigma^+ \times \Delta = \Sigma^- \times \Delta = \Delta; \qquad \text{etc.}$$
$$\Pi \times \Pi = \Sigma^+ + \Sigma^- + \Delta$$
$$\Delta \times \Delta = \Sigma^+ + \Sigma^- + \Gamma$$

$\Pi \times \Delta = \Pi + \Phi$

Appendix 4

Site Symmetries for the 230 Space Groups

Taken with permission from J. R. Ferraro and J. S. Ziomek, "Introductory Group Theory and Its Application to Molecular Structure." Plenum Press, New York, 1975.

Space Group[a]		Site Symmetries[b]
1 $P1$	C_1^1	$C_1(1)$
2 $P\bar{1}$	C_i^1	$8C_i(1); C_1(2)$
3 $P2$	C_2^1	$4C_2(1); C_1(2)$
4 $P2_1$	C_2^2	$C_1(2)$
5 B_2 or C_2	C_2^3	$2C_2(1); C_1(2)$
6 Pm	C_s^1	$2C_s(1); C_1(2)$
7 Pb or Pc	C_s^2	$C_1(2)$
8 Bm or Cm	C_s^3	$C_s(1); C_1(2)$
9 Bb or Cc	C_s^4	$C_1(2)$
10 $P2/m$	C_{2h}^1	$8C_{2h}(1); 4C_2(2); 2C_s(2); C_1(4)$
11 $P2_1/m$	C_{2h}^2	$4C_i(2); C_s(2); C_1(4)$
12 $B2/m$ or $C2/m$	C_{2h}^3	$4C_{2h}(1); 2C_i(2); 2C_2(2); C_s(2); C_1(4)$
13 $P2/b$ or $P2/c$	C_{2h}^4	$4C_i(2); 2C_2(2); C_1(4)$
14 $P2_1/b$ or $P2_1/c$	C_{2h}^5	$4C_i(2); C_1(4)$
15 $B2/b$ or $C2/c$	C_{2h}^6	$4C_i(2); C_2(2); C_1(4)$
16 $P222$	D_2^1	$8D_2(1); 12C_2(2); C_1(4)$
17 $P222_1$	D_2^2	$4C_2(2); C_1(4)$
18 $P2_12_12$	D_2^3	$2C_2(2); C_1(4)$
19 $P2_12_12_1$	D_2^4	$C_1(4)$
20 $C222_1$	D_2^5	$2C_2(2); C_1(4)$
21 $C222$	D_2^6	$4D_2(1); 7C_2(2); C_1(4)$
22 $F222$	D_2^7	$4D_2(1); 6C_2(2); C_1(4)$
23 $I222$	D_2^8	$4D_2(1); 6C_2(2); C_1(4)$
24 $I2_12_12_1$	D_2^9	$3C_2(2); C_1(4)$
25 $Pmm2$	C_{2v}^1	$4C_{2v}(1); 4C_s(2); C_1(4)$
26 $Pmc2_1$	C_{2v}^2	$2C_s(2); C_1(4)$
27 $Pcc2$	C_{2v}^3	$4C_2(2); C_1(4)$
28 $Pma2$	C_{2v}^4	$2C_2(2); C_s(2); C_1(4)$

Continued

Appendix 4 (continued)

Space Group^a		Site Symmetries^b
29 Pca2$_1$	C_{2v}^5	$C_1(4)$
30 Pnc2	C_{2v}^6	$2C_2(2)$; $C_1(4)$
31 Pmn2$_1$	C_{2v}^7	$C_s(2)$; $C_1(4)$
32 Pba2	C_{2v}^8	$2C_2(2)$; $C_1(4)$
33 Pna2$_1$	C_{2v}^9	$C_1(4)$
34 Pnn2	C_{2v}^{10}	$2C_2(2)$; $C_1(4)$
35 Cmm2	C_{2v}^{11}	$2C_{2v}(1)$; $C_2(2)$; $2C_s(2)$; $C_1(4)$
36 Cmc2$_1$.	C_{2v}^{12}	$C_s(2)$; $C_1(4)$
37 Ccc2	C_{2v}^{13}	$3C_2(2)$; $C_1(4)$
38 Amm2	C_{2v}^{14}	$2C_{2v}(1)$; $3C_s(2)$; $C_1(4)$
39 Abm2	C_{2v}^{15}	$2C_2(2)$; $C_s(2)$; $C_1(4)$
40 Ama2	C_{2v}^{16}	$C_2(2)$; $C_s(2)$; $C_1(4)$
41 Aba2	C_{2v}^{17}	$C_2(2)$; $C_1(4)$
42 Fmm2	C_{2v}^{18}	$C_{2v}(1)$; $C_2(2)$; $2C_s(2)$; $C_1(4)$
43 Fdd2	C_{2v}^{19}	$C_2(2)$; $C_1(4)$
44 Imm2	C_{2v}^{20}	$2C_{2v}(1)$; $2C_s(2)$; $C_1(4)$
45 Iba2	C_{2v}^{21}	$2C_2(2)$; $C_1(4)$
46 Ima2	C_{2v}^{22}	$C_2(2)$; $C_s(2)$; $C_1(4)$
47 Pmmm	D_{2h}^1	$8D_{2h}(1)$; $12C_{2v}(2)$; $6C_s(4)$; $C_1(8)$
48 Pnnn	D_{2h}^2	$4D_2(2)$; $2C_i(4)$; $6C_2(4)$; $C_1(8)$
49 Pccm	D_{2h}^3	$4C_{2h}(2)$; $4D_2(2)$; $8C_2(4)$; $C_s(4)$; $C_1(8)$
50 Pban	D_{2h}^4	$4D_2(2)$; $2C_i(4)$; $6C_2(4)$; $C_1(8)$
51 Pmma	D_{2h}^5	$4C_{2h}(2)$; $2C_{2v}(2)$; $2C_2(4)$; $3C_s(4)$; $C_1(8)$
52 Pnna	D_{2h}^6	$2C_i(4)$; $2C_2(4)$; $C_1(8)$
53 Pmna	D_{2h}^7	$4C_{2h}(2)$; $3C_2(4)$; $C_s(4)$; $C_1(8)$
54 Pcca	D_{2h}^8	$2C_i(4)$; $3C_2(4)$; $C_1(8)$
55 Pbam	D_{2h}^9	$4C_{2h}(2)$; $2C_2(4)$; $2C_s(4)$; $C_1(8)$
56 Pccn	D_{2h}^{10}	$2C_i(4)$; $2C_2(4)$; $C_1(8)$
57 Pbcm	D_{2h}^{11}	$2C_i(4)$; $C_2(4)$; $C_s(4)$; $C_1(8)$
58 Pnnm	D_{2h}^{12}	$4C_{2h}(2)$; $2C_2(4)$; $C_s(4)$; $C_1(8)$
59 Pmmm	D_{2h}^{13}	$2C_{2v}(2)$; $2C_i(4)$; $2C_s(4)$; $C_1(8)$
60 Pbcn	D_{2h}^{14}	$2C_i(4)$; $C_2(4)$; $C_1(8)$
61 Pbca	D_{2h}^{15}	$2C_i(4)$; $C_1(8)$
62 Pnma	D_{2h}^{16}	$2C_i(4)$; $C_s(4)$; $C_1(8)$
63 Cmcm	D_{2h}^{17}	$2C_{2h}(2)$; $C_{2v}(2)$; $C_i(4)$; $C_2(4)$; $2C_s(4)$; $C_1(8)$
64 Cmca	D_{2h}^{18}	$2C_{2h}(2)$; $C_i(4)$; $2C_2(4)$; $C_s(4)$; $C_i(8)$
65 Cmmm	D_{2h}^{19}	$4D_{2h}(1)$; $2C_{2h}(2)$; $6C_{2v}(2)$; $C_2(4)$; $4C_s(4)$; $C_1(8)$
66 Cccm	D_{2h}^{20}	$2D_2(2)$; $4C_{2h}(2)$; $5C_2(4)$; $C_s(4)$; $C_1(8)$
67 Cmma	D_{2h}^{21}	$2D_2(2)$; $4C_{2h}(2)$; $C_{2v}(2)$; $5C_2(4)$; $2C_s(4)$; $C_1(8)$
68 Ccca	D_{2h}^{22}	$2D_2(2)$; $2C_i(4)$; $4C_2(4)$; $C_1(8)$
69 Fmmm	D_{2h}^{23}	$2D_{2h}(1)$; $3C_{2h}(2)$; $D_2(2)$; $3C_{2v}(2)$; $3C_2(4)$; $3C_s(4)$; $C_1(8)$
70 Fddd	D_{2h}^{24}	$2D_2(2)$; $2C_i(4)$; $3C_2(4)$; $C_1(8)$
71 Immm	D_{2h}^{25}	$4D_{2h}(1)$; $6C_{2v}(2)$; $C_i(4)$; $3C_s(4)$; $C_1(8)$
72 Ibam	D_{2h}^{26}	$2D_2(2)$; $2C_{2h}(2)$; $C_i(4)$; $4C_2(4)$; $C_s(4)$; $C_1(8)$
73 Ibca	D_{2h}^{27}	$2C_i(4)$; $3C_2(4)$; $C_1(8)$
74 Imma	D_{2h}^{28}	$4C_{2h}(2)$; $C_{2v}(2)$; $2C_2(4)$; $2C_s(4)$; $C_1(8)$

Appendix 4 (*continued*)

Space Group[a]		Site Symmetries[b]
75 $P4$	C_4^1	$2C_4(1)$; $C_2(2)$; $C_1(4)$
76 $P4_1$	C_4^2	$C_1(4)$
77 $P4_2$	C_4^3	$3C_2(2)$; $C_1(4)$
78 $P4_3$	C_4^4	$C_1(4)$
79 $I4$	C_4^5	$C_4(1)$; $C_2(2)$; $C_1(4)$
80 $I4_1$	C_4^6	$C_2(2)$; $C_1(4)$
81 $P\bar{4}$	S_4^1	$4S_4(1)$; $3C_2(2)$; $C_1(4)$
82 $I\bar{4}$	S_4^2	$4S_4(1)$; $2C_2(2)$; $C_1(4)$
83 $P4/m$	C_{4h}^1	$4C_{4h}(1)$; $2C_{2h}(2)$; $2C_4(2)$; $C_2(4)$; $2C_s(4)$; $C_1(8)$
84 $P4_2/m$	C_{4h}^2	$4C_{2h}(2)$; $2S_4(2)$; $3C_2(4)$; $C_s(4)$; $C_1(8)$
85 $P4/n$	C_{4h}^3	$2S_4(2)$; $C_4(2)$; $2C_i(4)$; $C_2(4)$; $C_1(8)$
86 $P4_2/n$	C_{4h}^4	$2S_4(2)$; $2C_i(4)$; $2C_2(4)$; $C_1(8)$
87 $I4/m$	C_{4h}^5	$2C_{4h}(1)$; $C_{2h}(2)$; $S_4(2)$; $C_4(2)$; $C_i(4)$; $C_2(4)$; $C_s(4)$; $C_1(8)$
88 $I4_1/a$	C_{4h}^6	$2S_4(2)$; $2C_i(4)$; $C_2(4)$; $C_1(8)$
89 $P422$	D_4^1	$4D_4(1)$; $2D_2(2)$; $2C_4(2)$; $7C_2(4)$; $C_1(8)$
90 $P42_12$	D_4^2	$2D_2(2)$; $C_4(2)$; $3C_2(4)$; $C_1(8)$
91 $P4_122$	D_4^3	$3C_2(4)$; $C_1(8)$
92 $P4_12_12$	D_4^4	$C_2(4)$; $C_1(8)$
93 $P4_222$	D_4^5	$6D_2(2)$; $9C_2(4)$; $C_1(8)$
94 $P4_22_12$	D_4^6	$2D_2(2)$; $4C_2(4)$; $C_1(8)$
95 $P4_322$	D_4^7	$3C_2(4)$; $C_1(8)$
96 $P4_32_12$	D_4^8	$C_2(4)$; $C_1(8)$
97 $I422$	D_4^9	$2D_4(1)$; $2D_2(2)$; $C_4(2)$; $5C_2(4)$; $C_1(8)$
98 $I4_122$	D_4^{10}	$2D_2(2)$; $4C_2(4)$; $C_1(8)$
99 $P4mm$	C_{4v}^1	$2C_{4v}(1)$; $C_{2v}(2)$; $3C_s(4)$; $C_1(8)$
100 $P4bm$	C_{4v}^2	$C_4(2)$; $C_{2v}(2)$; $C_s(4)$; $C_1(8)$
101 $P4_2cm$	C_{4v}^3	$2C_{2v}(2)$; $C_2(4)$; $C_s(4)$; $C_1(8)$
102 $P4_2nm$	C_{2v}^4	$C_{2v}(2)$; $C_2(4)$; $C_s(4)$; $C_1(8)$
103 $P4cc$	C_{4v}^5	$2C_4(2)$; $C_2(4)$; $C_1(8)$
104 $P4nc$	C_{4v}^6	$C_4(2)$; $C_2(4)$; $C_1(8)$
105 $P4_2mc$	C_{4v}^7	$3C_{2v}(2)$; $2C_s(4)$; $C_1(8)$
106 $P4_2bc$	C_{4v}^8	$2C_2(4)$; $C_1(8)$
107 $I4mm$	C_{4v}^9	$C_{4v}(1)$; $C_{2v}(2)$; $2C_s(4)$; $C_1(8)$
108 $I4cm$	C_{4v}^{10}	$C_4(2)$; $C_{2v}(2)$; $C_s(4)$; $C_1(8)$
109 $I4_1md$	C_{4v}^{11}	$C_{2v}(2)$; $C_s(4)$; $C_1(8)$
110 $I4_1cd$	C_{4v}^{12}	$C_2(4)$; $C_1(8)$
111 $P\bar{4}2m$	D_{2d}^1	$4D_{2d}(1)$; $2D_2(2)$; $2C_{2v}(2)$; $5C_2(4)$; $C_s(4)$; $C_1(8)$
112 $P\bar{4}2c$	D_{2d}^2	$4D_2(2)$; $2S_4(2)$; $7C_2(4)$; $C_1(8)$
113 $P\bar{4}2_1m$	D_{2d}^3	$2S_4(2)$; $C_{2v}(2)$; $C_2(4)$; $C_s(4)$; $C_1(8)$
114 $P\bar{4}2_1c$	D_{2d}^4	$2S_4(2)$; $2C_2(4)$; $C_1(8)$
115 $P\bar{4}m_2$	D_{2d}^5	$4D_{2d}(1)$; $3C_{2v}(2)$; $2C_2(4)$; $2C_s(4)$; $C_1(8)$
116 $P\bar{4}c2$	D_{2d}^6	$2D_2(2)$; $2S_4(2)$; $5C_2(4)$; $C_1(8)$
117 $P\bar{4}b_2$	D_{2d}^7	$2S_4(2)$; $2D_2(2)$; $4C_2(4)$; $C_1(8)$
118 $P\bar{4}n_2$	D_{2d}^8	$2S_4(2)$; $2D_2(2)$; $4C_2(4)$; $C_1(8)$
119 $I\bar{4}m_2$	D_{2d}^9	$4D_{2d}(1)$; $2C_{2v}(2)$; $2C_2(4)$; $C_s(4)$; $C_1(8)$
120 $I\bar{4}c2$	D_{2d}^{10}	$D_2(2)$; $2S_4(2)$; $D_2(2)$; $4C_2(4)$; $C_1(8)$

Continued

Appendix 4 (*continued*)

Space Group[a]		Site Symmetries[b]
121 $I\bar{4}2m$	D_{2d}^{11}	$2D_{2d}(1)$; $D_2(2)$; $S_4(2)$; $C_{2v}(2)$; $3C_2(4)$; $C_s(4)$; $C_1(8)$
122 $I\bar{4}2d$	D_{2d}^{12}	$2S_4(4)$; $2C_2(4)$; $C_1(8)$
123 $P4/mmm$	D_{4h}^{1}	$4D_{4h}(1)$; $2D_{2h}(2)$; $2C_{4v}(2)$; $7C_{2v}(4)$; $5C_s(8)$; $C_1(16)$
124 $P4/mcc$	D_{4h}^{2}	$D_4(2)$; $C_{4h}(2)$; $D_4(2)$; $C_{4h}(2)$; $C_{2h}(4)$; $D_2(4)$; $2C_4(4)$; $4C_2(8)$; $C_s(8)$; $C_1(16)$
125 $P4/nbm$	D_{4h}^{3}	$2D_4(2)$; $2D_{2d}(2)$; $2C_{2h}(4)$; $C_4(4)$; $C_{2v}(4)$; $4C_2(8)$; $C_s(8)$; $C_1(16)$
126 $P4/nnc$	D_{2h}^{4}	$2D_4(2)$; $D_2(4)$; $S_4(4)$; $C_4(4)$; $C_i(4)$; $4C_2(8)$; $C_1(8)$
127 $P4/mbm$	D_{4h}^{5}	$2C_{4h}(2)$; $2D_{2h}(2)$; $C_4(4)$; $3C_{2v}(4)$; $3C_s(8)$; $C_1(16)$
128 $P4/mnc$	D_{4h}^{6}	$2C_{4h}(2)$; $C_{2h}(4)$; $D_2(4)$; $C_4(4)$; $2C_2(8)$; $C_s(8)$; $C_1(16)$
129 $P4/nmm$	D_{4h}^{7}	$2D_{2d}(2)$; $C_{4v}(2)$; $2C_{2h}(4)$; $C_{2v}(4)$; $2C_2(8)$; $2C_s(8)$; $C_1(16)$
130 $P4/ncc$	D_{4h}^{8}	$D_2(4)$; $S_4(4)$; $C_4(4)$; $C_i(8)$; $2C_2(8)$; $C_1(16)$
131 $P4_2/mmc$	D_{4h}^{9}	$4D_{2h}(2)$; $2D_{2d}(2)$; $7C_{2v}(4)$; $C_2(8)$; $3C_s(8)$; $C_1(16)$
132 $P4_2/mcm$	D_{4h}^{10}	$2D_{2h}(2)$; $2D_{2d}(2)$; $D_2(4)$; $C_{2h}(4)$; $4C_{2v}(4)$; $3C_2(8)$; $2C_s(8)$; $C_1(16)$
133 $P4_2/nbc$	D_{4h}^{11}	$3D_2(4)$; $S_4(4)$; $C_i(8)$; $5C_2(8)$; $C_1(16)$
134 $P4_2/nnm$	D_{4h}^{12}	$2D_{2d}(2)$; $2D_2(4)$; $2C_{2h}(4)$; $C_{2v}(4)$; $5C_2(8)$; $C_s(8)$; $C_1(16)$
135 $P4_2/mbc$	D_{4h}^{13}	$C_{2h}(4)$; $S_4(4)$; $C_{2h}(4)$; $D_2(4)$; $3C_2(8)$; $C_s(8)$; $C_1(16)$
136 $P4_2/mnm$	D_{4h}^{14}	$2D_{2h}(2)$; $C_{2h}(4)$; $S_4(4)$; $3C_{2v}(4)$; $C_2(8)$; $2C_s(8)$; $C_1(16)$
137 $P4_2/nmc$	D_{4h}^{15}	$2D_{2d}(2)$; $2C_{2v}(4)$; $C_i(8)$; $C_2(8)$; $C_s(8)$; $C_1(16)$
138 $P4_2/ncm$	D_{4h}^{16}	$D_2(4)$; $S_4(4)$; $2C_{2h}(4)$; $C_{2v}(4)$; $3C_2(8)$; $C_s(8)$; $C_1(16)$
139 $I4/mmm$	D_{4h}^{17}	$2D_{4h}(1)$; $D_{2h}(2)$; $D_{2d}(2)$; $C_{4v}(2)$; $C_{2h}(4)$; $4C_{2v}(4)$; $C_2(8)$; $3C_s(8)$; $C_1(16)$
140 $I4/mcm$	D_{2h}^{18}	$D_4(2)$; $D_{2d}(2)$; $C_{4h}(2)$; $D_{2h}(2)$; $C_{2h}(4)$; $C_4(4)$; $2C_{2v}(4)$; $2C_2(8)$; $2C_s(8)$; $C_1(16)$
141 $I4_1/amd$	D_{4h}^{19}	$2D_{2d}(2)$; $2C_{2h}(4)$; $C_{2v}(4)$; $2C_2(8)$; $C_s(8)$; $C_1(16)$
142 $I4_1/acd$	D_{4h}^{20}	$S_4(4)$; $D_2(4)$; $C_i(8)$; $3C_2(8)$; $C_1(16)$
143 $P3$	C_3^{1}	$3C_3(1)$; $C_1(3)$
144 $P3_1$	C_3^{2}	$C_1(3)$
145 $P3_2$	C_3^{3}	$C_1(3)$
146 $R3$	C_3^{4}	$C_3(1)$; $C_1(3)$
147 $P\bar{3}$	C_{3i}^{1}	$2C_3(1)$; $2C_3(2)$; $2C_i(3)$; $C_1(6)$
148 $R\bar{3}$	C_{3i}^{2}	$2C_3(1)$; $C_3(2)$; $2C_i(3)$; $C_1(6)$
149 $P312$	D_3^{1}	$6D_3(1)$; $3C_3(2)$; $2C_2(3)$; $C_1(6)$
150 $P321$	D_3^{2}	$2D_3(1)$; $2C_3(2)$; $2C_2(3)$; $C_1(6)$
151 $P3_112$	D_3^{3}	$2C_2(3)$; $C_1(6)$
152 $P3_121$	D_3^{4}	$2C_2(3)$; $C_1(6)$
153 $P3_212$	D_3^{5}	$2C_2(2)$; $C_1(6)$
154 $P3_221$	D_3^{6}	$2C_2(3)$; $C_1(6)$
155 $R32$	D_3^{7}	$2D_3(1)$; $C_3(2)$; $2C_2(3)$; $C_1(6)$
156 $P3m1$	C_{3v}^{1}	$3C_{3v}(1)$; $C_s(3)$; $C_1(6)$
157 $P31m$	C_{3v}^{2}	$C_{3v}(1)$; $C_3(2)$; $C_s(3)$; $C_1(6)$
158 $P3c1$	C_{3v}^{3}	$3C_3(2)$; $C_1(6)$
159 $P31c$	C_{3v}^{4}	$2C_3(2)$; $C_1(6)$
160 $R3m$	C_{3v}^{5}	$C_{3v}(1)$; $C_s(3)$; $C_1(6)$
161 $R3c$	C_{3v}^{6}	$C_3(2)$; $C_1(6)$

Appendix 4 (*continued*)

Space Group[a]		Site Symmetries[b]
162 $P\bar{3}1m$	D_{3d}^1	$2D_{3d}(1)$; $2D_3(2)$; $C_{3v}(2)$; $2C_{2h}(3)$; $C_3(4)$; $2C_2(6)$; $C_s(6)$; $C_1(12)$
163 $P\bar{3}1c$	D_{3d}^2	$D_3(2)$; $C_{3i}(2)$; $2D_3(2)$; $2C_3(4)$; $C_i(6)$; $C_2(6)$; $C_1(12)$
164 $P\bar{3}m1$	D_{3d}^3	$2D_{3d}(1)$; $2C_{3v}(2)$; $2C_{2h}(3)$; $2C_2(6)$; $C_s(6)$; $C_1(12)$
165 $P\bar{3}c1$	D_{3d}^4	$D_3(2)$; $C_{3i}(2)$; $2C_3(4)$; $C_i(6)$; $C_1(6)$; $C_2(12)$
166 $R\bar{3}m$	D_{3d}^5	$2D_{3d}(1)$; $C_{3v}(2)$; $2C_{2h}(3)$; $2C_2(6)$; $C_s(6)$; $C_1(12)$
167 $R\bar{3}c$	D_{3d}^6	$D_3(2)$; $C_{3i}(2)$; $C_3(4)$; $C_i(6)$; $C_2(6)$; $C_1(12)$
168 $P6$	C_6^1	$C_6(1)$; $C_3(2)$; $C_2(3)$; $C_1(6)$
169 $P6_1$	C_6^2	$C_1(6)$
170 $P6_5$	C_6^3	$C_1(6)$
171 $P6_2$	C_6^4	$2C_2(3)$; $C_1(6)$
172 $P6_4$	C_6^5	$2C_2(3)$; $C_1(6)$
173 $P6_3$	C_6^6	$2C_3(2)$; $C_1(6)$
174 $P\bar{6}$	C_{3h}^1	$6C_{3h}(1)$; $3C_3(3)$; $2C_s(3)$; $C_1(6)$
175 $P6/m$	C_{6h}^1	$2C_{6h}(1)$; $2C_{3h}(2)$; $C_6(2)$; $2C_{2h}(3)$; $C_3(4)$; $C_2(6)$; $2C_s(6)$; $C_1(12)$
176 $P6_3/m$	C_{6h}^2	$3C_{3h}(2)$; $2C_{3i}(2)$; $2C_3(4)$; $C_i(6)$; $C_s(6)$; $C_1(12)$
177 $P622$	D_6^1	$2D_6(1)$; $2D_3(2)$; $C_6(2)$; $2D_2(3)$; $C_4(4)$; $5C_2(6)$; $C_1(12)$
178 $P6_122$	D_6^2	$2C_2(6)$; $C_1(12)$
179 $P6_522$	D_6^3	$2C_2(6)$; $C_1(12)$
180 $P6_222$	D_6^4	$4D_2(3)$; $6C_2(6)$; $C_1(12)$
181 $P6_422$	D_6^5	$4D_2(3)$; $6C_2(6)$; $C_1(12)$
182 $P6_333$	D_6^6	$4D_3(2)$; $2C_3(4)$; $2C_2(6)$; $C_1(12)$
183 $P6mm$	C_{6v}^1	$C_{6v}(1)$; $C_{3v}(2)$; $C_{2v}(3)$; $2C_s(6)$; $C_1(12)$
184 $P6cc$	C_{6v}^2	$C_6(2)$; $C_3(4)$; $C_2(6)$; $C_1(12)$
185 $P6_3cm$	C_{6v}^3	$C_{3v}(2)$; $C_3(4)$; $C_s(6)$; $C_1(12)$
186 $P6_3mc$	C_{6v}^4	$2C_{3v}(2)$; $C_s(6)$; $C_1(12)$
187 $P\bar{6}m2$	D_{3h}^1	$6D_{3h}(1)$; $3C_{3v}(2)$; $2C_{2v}(3)$; $3C_s(6)$; $C_1(12)$
188 $P\bar{6}c2$	D_{3h}^2	$3D_3(2)$; $C_{3h}(2)$; $C_{3h}(2)$; $C_{3h}(2)$; $3C_3(4)$; $C_2(6)$; $C_s(6)$; $C_1(12)$
189 $P\bar{6}2m$	D_{3h}^3	$2D_{3h}(1)$; $2C_{3h}(2)$; $C_{3v}(2)$; $2C_{2v}(3)$; $C_3(4)$; $3C_s(6)$; $C_1(12)$
190 $P\bar{6}2c$	D_{3h}^4	$D_3(2)$; $3C_{3h}(2)$; $2C_3(4)$; $C_2(6)$; $C_s(6)$; $C_1(12)$
191 $P6/mmm$	D_{6h}^1	$2D_{6h}(1)$; $2D_{3h}(2)$; $C_{6v}(2)$; $2D_{2h}(3)$; $C_{3v}(4)$; $5C_{2v}(6)$; $4C_s(12)$; $C_1(24)$
192 $P6/mcc$	D_{6h}^2	$D_6(2)$; $C_{6h}(2)$; $D_3(4)$; $C_{3h}(4)$; $C_6(4)$; $D_2(6)$; $C_{2h}(6)$; $C_3(8)$; $3C_2(12)$; $C_s(12)$; $C_1(24)$
193 $P6_3/mcm$	D_{6h}^3	$D_{3h}(2)$; $D_{3d}(2)$; $C_{3h}(4)$; $D_3(4)$; $C_6(4)$; $C_{2h}(6)$; $C_{2v}(6)$; $C_3(8)$; $C_2(12)$; $2C_2(12)$; $C_1(24)$
194 $P6_3/mmc$	D_{6h}^4	$D_{3d}(2)$; $3D_{3h}(2)$; $2C_{3v}(4)$; $C_{2h}(6)$; $C_{2v}(6)$; $C_2(12)$; $2C_s(12)$; $C_1(24)$
195 $P23$	T^1	$2T(1)$; $2D_2(3)$; $C_3(4)$; $4C_2(6)$; $C_1(12)$
196 $F23$	T^2	$4T(1)$; $C_3(4)$; $2C_2(6)$; $C_1(12)$
197 $I23$	T^3	$T(1)$; $D_2(2)$; $C_3(4)$; $2C_2(6)$; $C_1(12)$
198 $P2_13$	T^4	$C_3(4)$; $C_1(12)$
199 $I2_13$	T^5	$C_3(4)$; $C_2(6)$; $C_1(12)$
200 $Pm3$	T_h^1	$2T_h(1)$; $2D_{2h}(3)$; $4C_{2v}(6)$; $C_3(8)$; $2C_s(12)$; $C_1(24)$
201 $Pn3$	T_h^3	$T(2)$; $2C_{3i}(4)$; $D_2(6)$; $C_3(8)$; $2C_2(12)$; $C_1(24)$

Continued

Appendix 4 (*continued*)

Space Group[a]		Site Symmetries[b]
202 $Fm3$	T_h^3	$2T_h(1)$; $T(2)$; $C_{2h}(6)$; $C_{2v}(6)$; $C_3(8)$; $C_2(12)$; $C_s(12)$; $C_1(24)$
203 $Fd3$	T_h^4	$2T(2)$; $2C_{3i}(4)$; $C_3(8)$; $C_2(12)$; $C_1(24)$
204 $Im3$	T_h^5	$T_h(1)$; $D_{2h}(3)$; $C_{3i}(4)$; $2C_{2v}(6)$; $C_3(8)$; $C_s(12)$; $C_1(24)$
205 $Pa3$	T_h^6	$2C_{3i}(4)$; $C_3(8)$; $C_1(24)$
206 $Ia3$	T_h^7	$2C_{3i}(4)$; $C_3(8)$; $C_2(12)$; $C_1(24)$
207 $P432$	O^1	$2O(1)$; $2D_4(3)$; $2C_4(6)$; $C_3(8)$; $3C_2(12)$; $C_1(24)$
208 $P4_232$	O^2	$T(2)$; $2D_3(4)$; $3D_2(6)$; $C_3(8)$; $5C_2(12)$; $C_1(24)$
209 $F432$	O^3	$2O(1)$; $T(2)$; $D_2(6)$; $C_4(6)$; $C_3(8)$; $3C_2(12)$; $C_1(24)$
210 $F4_132$	O^4	$2T(2)$; $2D_3(4)$; $C_3(8)$; $2C_2(12)$; $C_1(24)$
211 $I432$	O^5	$O(1)$; $D_4(3)$; $D_3(4)$; $D_2(6)$; $C_4(6)$; $C_3(8)$; $3C_2(12)$; $C_1(24)$
212 $P4_332$	O^6	$2D_3(4)$; $C_3(8)$; $C_2(12)$; $C_1(24)$
213 $P4_132$	O^7	$2D_3(4)$; $C_3(8)$; $C_2(12)$; $C_1(24)$
214 $I4_132$	O^8	$2D_3(4)$; $2D_2(6)$; $C_3(8)$; $3C_2(12)$; $C_1(24)$
215 $P\bar{4}3m$	T_d^1	$2T_d(1)$; $2D_{2d}(3)$; $C_{3v}(4)$; $2C_{2v}(6)$; $C_2(12)$; $C_s(12)$; $C_1(24)$
216 $F\bar{4}3m$	T_d^2	$4T_d(1)$; $C_{3v}(4)$; $2C_{2v}(6)$; $C_s(12)$; $C_1(24)$
217 $I\bar{4}3m$	T_d^3	$T_d(1)$; $D_{2d}(3)$; $C_{3v}(4)$; $S_4(6)$; $C_{2v}(6)$; $C_2(12)$; $C_s(12)$; $C_1(24)$
218 $P\bar{4}3n$	T_d^4	$T(2)$; $D_2(6)$; $2S_4(6)$; $C_3(8)$; $3C_2(12)$; $C_1(24)$
219 $F\bar{4}3c$	T_d^5	$2T(2)$; $2S_4(6)$; $C_3(8)$; $2C_2(12)$; $C_1(24)$
220 $I\bar{4}3d$	T_d^6	$2S_4(6)$; $C_3(8)$; $C_2(12)$; $C_1(24)$
221 $Pm3m$	O_h^1	$2O_h(1)$; $2D_{4h}(3)$; $2C_{4v}(6)$; $C_{3v}(8)$; $3C_{2v}(12)$; $3C_s(24)$; $C_1(48)$
222 $Pn3n$	O_h^2	$O(2)$; $D_4(6)$; $C_{3i}(8)$; $S_4(12)$; $C_4(12)$; $C_3(16)$; $2C_2(24)$; $C_1(48)$
223 $Pm3n$	O_h^3	$T_h(2)$; $D_{2h}(6)$; $2D_{2d}(6)$; $D_3(8)$; $3C_{2v}(12)$; $C_3(16)$; $C_2(24)$; $C_s(24)$; $C_1(48)$
224 $Pn3m$	O_h^4	$T_d(2)$; $2D_{3d}(4)$; $D_{2d}(6)$; $C_{3v}(8)$; $D_2(12)$; $C_{2v}(12)$; $3C_2(24)$; $C_1(48)$
225 $Fm3m$	O_h^5	$2O_h(1)$; $T_d(2)$; $D_{2h}(6)$; $C_{4v}(6)$; $C_{3v}(8)$; $3C_{2v}(12)$; $2C_s(24)$; $C_1(48)$
226 $Fd3c$	O_h^6	$O(2)$; $T_h(2)$; $D_{2d}(6)$; $C_{4h}(6)$; $C_{2v}(12)$; $C_4(12)$; $C_3(16)$; $C_2(24)$; $C_s(24)$; $C_1(48)$
227 $Fd3m$	O_h^7	$2T_d(2)$; $2D_{3d}(4)$; $C_{3v}(8)$; $C_{2v}(12)$; $C_s(24)$; $C_2(24)$; $C_1(48)$
228 $Fd3c$	O_h^8	$T(4)$; $D_3(8)$; $C_{3i}(8)$; $S_4(12)$; $C_3(16)$; $2C_2(24)$; $C_1(48)$
229 $Im3m$	O_h^9	$O_h(1)$; $D_{4h}(3)$; $D_{3d}(4)$; $D_{2d}(6)$; $C_{4v}(6)$; $C_{3v}(8)$; $2C_{2v}(12)$; $C_2(24)$; $2C_s(24)$; $C_1(48)$
230 $Ia3d$	O_h^{10}	$C_{3i}(8)$; $D_3(8)$; $D_2(12)$; $S_4(12)$; $C_3(16)$; $2C_2(24)$; $C_1(48)$

Note the following equivalent nomenclatures: $C_i \equiv S_2$, $C_s \equiv C_{1h}$, $D_2 \equiv V$, $D_{2h} \equiv V_h$, $D_{2d} \equiv V_d$, and $C_{3i} \equiv S_6$.

[a] N. F. M. Henry and K. Lonsdale (Eds.), "International Tables for X-Ray Crystallography," Vol. 1, Kynoch Press, Birmingham, U.K., 1965.

[b] R. S. Halford, *J. Chem. Phys.* **14**, 8 (1946).

Appendix 5

Determination of the Proper Correlation Using Wyckoff's Tables

Taken from W. G. Fateley, F. R. Dollish, N. T. McDevitt and F. F. Bentley, "Infrared and Raman Selection Rules for Molecular and Lattice Vibrations: The Correlation Method," Wiley-Interscience, New York, 1972, courtesy of Wiley-Interscience.

Space Group Number			Site Correlation					
			$C_2(z)$	$C_2(y)$	$C_2(x)$	$\sigma(xy)$	$\sigma(zx)$	$\sigma(yz)$
D_2^i	16	D_2^1	q, r, s, t	m, n, o, p	i, j, k, l			
	17	D_2^2		c, d	a, b			
	18	D_2^3	a, b					
	20	D_2^5		b	a			
	21	D_2^6	i, j, k	g, h	e, f			
	22	D_2^7	g, h	f, i	e, j			
	23	D_2^8	i, j	g, h	e, f			
	24	D_2^9	c	b	a			
C_{2v}^i	25	C_{2v}^1					e, f	g, h
	26	C_{2v}^2						a, b
	28	C_{2v}^4						c
	31	C_{2v}^7						a
	35	C_{2v}^{11}					d	e
	36	C_{2v}^{12}						a
	38	C_{2v}^{14}					c	d, e
	39	C_{2v}^{15}					c	
	40	C_{2v}^{16}						b
	42	C_{2v}^{18}					d	c
	44	C_{2v}^{20}					c	d
	46	C_{2v}^{22}						b

Continued

325

Appendix 5

Space Group Number		Site Correlation					
		$C_2(z)$	$C_2(y)$	$C_2(x)$	$\sigma(xy)$	$\sigma(zx)$	$\sigma(yz)$
D_{2h}^i 47	D_{2h}^1	q, r, s, t	m, n, o, p	i, j, k, l	y, z	w, x	u, v
48	D_{2h}^2	k, l	i, j	g, h			
49	D_{2h}^3	a, b, c, d, m, n, o, p	k, l	i, j	q		
50	D_{2h}^4	k, l	i, j	g, h			
51	D_{2h}^5	e, f	a, b, c, d, g, h			i, j	k
52	D_{2h}^6	c		d			
53	D_{2h}^7		g	a, b, c, d, e, f			h
54	D_{2h}^8	d, e	c				
55	D_{2h}^9	a, b, c, d, e, f			g, h		
56	D_{2h}^{10}	c, d					
57	D_{2h}^{11}			c	d		
58	D_{2h}^{12}	a, b, c, d, e, f			g		
59	D_{2h}^{13}	a, b				f	e
60	D_{2h}^{14}		c				
62	D_{2h}^{16}					c	
63	D_{2h}^{17}		c	a, b, e	g		f
64	D_{2h}^{18}		e	a, b, d			f
65	D_{2h}^{19}	e, f, k, l, m	i, j	g, h	p, q	o	n
66	D_{2h}^{20}	c, d, e, f, i, j, k	h	g	l		
67	D_{2h}^{21}	g, l	e, f, j, k	c, d, h, i		n	m
68	D_{2h}^{22}	g, h	f	e			
69	D_{2h}^{23}	e, i, j	d, h, k	c, g, l	o	n	m
70	D_{2h}^{24}	g	f	e			
71	D_{2h}^{25}	i, j	g, h	e, f	n	m	l
72	D_{2h}^{26}	c, d, h, i	g	f	j		
73	D_{2h}^{27}	e	d	c			
74	D_{2h}^{28}	e	c, d, g	a, b, f		i	h

	Space Group Number	C_2	C_2'	C_2''	C_2 / σ_v	C_2 / σ_d	σ_h	σ_v	σ_d
D_4^i	89 D_4^1	i	e, f, l, m, n, o	j, k					
	90 D_4^2	d		e, f, a, b					
	91 D_4^3		a, b	c					
	92 D_4^4			a					
	93 D_4^5	g, h, i	a, b, c, d, j, k, l, m	e, f, n, o					
	94 D_4^6	c, d		a, b, e, f					
	95 D_4^7		a, b	c					
	96 D_4^8			a					
	97 D_4^9	f	c, h, i	d, g, j					
	98 D_4^{10}	c	f	a, b, d, e					
C_{4v}^i	99 C_{4v}^1				c, e, f	d			
	100 C_{4v}^2					b, c			
	101 C_{4v}^3					a, b, d			
	102 C_{4v}^4					a, c			
	105 C_{4v}^7				a, b, c, d, e				
	107 C_{4v}^9				b, d				
	108 C_{4v}^{10}					c			
	109 C_{4v}^{11}				a, b	b, c			
D_{2d}^i	111 D_{2d}^1	m	i, j, k, l						
	112 D_{2d}^2	k, l, m	g, h, i, j						
	113 D_{2d}^3	d							
	114 D_{2d}^4	c, d							
	115 D_{2d}^5		h, i						
	116 D_{2d}^6	g, h, i	e, f						
	117 D_{2d}^7	e, f	g, h						
	118 D_{2d}^8	e, h	f, g						
	119 D_{2d}^9		g, h						

Space Group Number	C_2	C_2	C_2''	C_2 / σ_v	C_2 / σ_d	σ_h	σ_v	σ_d
D_{2d}^i								
120 D_{2d}^{10}	f, g	e, h						r
121 D_{2d}^{11}	h	f, g				m		
122 D_{2d}^{12}	c	d						
D_{4h}^i								
123 D_{4h}^{1}		e, f, l, m, n, o	j, k	i		p, q	s, t	
124 D_{4h}^{2}	i, e	f, k, l	j			m		m
125 D_{4h}^{3}		c, d, k, l	e, f, i, j		h			m
126 D_{4h}^{4}	g	c, i, j	h					
127 D_{4h}^{5}			c, d, g, h		f	i, j		k
128 D_{4h}^{6}	c, f		d, g			h		
129 D_{4h}^{7}			a, b, d, e, g, h	j			i	j
130 D_{4h}^{8}	e		a, f					
131 D_{4h}^{9}		a, b, c, d, j, k, l, m	e, f, n	g, h, i		q	o, p	o
132 D_{4h}^{10}	f, k	b, d, e, l, m	a, c, i, j		g, h	n		
133 D_{4h}^{11}	f, g	a, b, h, i	c, j					
134 D_{4h}^{12}	h	a, b, c, i, j	d, e, f, k, l		g	h		m
135 D_{4h}^{13}	a, c, e, f		d, g					
136 D_{4h}^{14}	c, h		a, b, f, g		e	i		j
137 D_{4h}^{15}			a, b, f	c, d			g	
138 D_{4h}^{16}	f		a, c, d, g, h		e			i
139 D_{4h}^{17}			d, f, h, k	g		l	n	m
140 D_{4h}^{18}		c, i, j	d, e, h, i		g			l
141 D_{4h}^{19}		b, j	a, b, g			k	h	
142 D_{4h}^{20}	d	c, d, f	b, f	e				

Space Group Number		σ_h	σ_v
D_{2h}^i			
187	D_{3h}^1	l, m	n
188	D_{3h}^2	k	
189	D_{3h}^3	j, k	i
190	D_{3h}^4	h	

Space Group Number		σ_v	σ_d
C_{6v}^i			
183	C_{6v}^1	b, e	d
185	C_{6v}^3		a, c
186	C_{6v}^4	a, b, c	

Space Group Number		C_2	C_2'	C_2''
D_6^i				
177	D_6^1	i	j, k	c, d, l, m
178	D_6^2		a	b
179	D_6^3		a	b
180	D_6^4	e, f	g, h	i, j
181	D_6^5	e, f	g, h	i, j
182	D_6^6		a, g	b, c, d, h

Space Group Number		C_2	C_2'	C_2''	σ_h	σ_d	σ_v
D_{4h}^i							
191	D_{6h}^1	i	j, k	l, m, c, d	p, q	h, o	n
192	D_{6h}^2	g, i	j	c, k	l		k
193	D_{6h}^3		a, g	b, d, f, i	j	k, e, f	
194	D_{6h}^4		a, g, i	b, c, d, h	j	k, e, f	

Space Group Number			C_2	C'_2	$3C_2$	$C_2, 2C'_2$
O^i	207	O^1	h	i, j		
	208	O^2	h, i, j	k, l	d	e, f
	209	O^3	i	g, h		d
	210	O^4	f	g		
	211	O^5	g	h, i		d
	212	O^6		d		
	213	O^7		d		
	214	O^8	f	g, h		c, d

Space Group Number			C_2	$3C_2$	$C_2, 2C'_2$	C'_2, σ_h	C_2, σ_h	C_2, σ_d	σ_h	σ_d
O^i_h	221	O^1_h				i, j	h		k, l	m
	222	O^2_h	g			h				
	223	O^3_h		b		c, d, j	f, g, h		k	
	224	O^4_h	d, h		f	i, j		g		k
	225	O^5_h			d	h, i		g	j	k
	226	O^6_h				c, h	e		i	
	227	O^7_h				h		f		g
	228	O^8_h	f			g				
	229	O^9_h				i, d, h	g		j	k
	230	O^{10}_h	f		c	g				

Appendix 6

Correlation Tables

We wish to express our gratitude for permission to reproduce these tables from the book of W. G. Fateley *et al.* for the use of their comprehensive tables (1). Reprinted with permission of John Wiley and Sons, Inc., New York.

C_4	C_2
A	A
B	A
E	$2B$

C_6	C_3	C_2	C_1
A	A	A	A
B	A	B	A
E_1	E	$2B$	$2A$
E_2	E	$2A$	$2A$

D_2	C_2^z	C_2^y	C_2^x
A	A	A	A
B_1	A	B	B
B_2	B	A	B
B_3	B	B	A

D_3	C_3	C_2
A_1	A	A
A_2	A	B
E	E	$A+B$

D_4	C_2' D_2	C_2'' D_2	C_4	C_2	C_2' C_2	C_2'' C_2
A_1	A	A	A	A	A	A
A_2	B_1	B_1	A	A	B	B
B_1	A	B_1	B	A	A	B
B_2	B_1	A	B	A	B	A
E	B_2+B_3	B_2+B_3	E	$2B$	$A+B$	$A+B$

D_5	C_5	C_2
A_1	A	A
A_2	A	B
E_1	E_1	$A+B$
E_2	E_2	$A+B$

D_6	C_6	C_2' D_3	C_2'' D_3	D_2	C_3	C_2	C_2' C_2	C_2'' C_2
A_1	A	A_1	A_1	A	A	A	A	A
A_2	A	A_2	A_2	B_1	A	A	B	B
B_1	B	A_1	A_2	B_2	A	B	A	B
B_2	B	A_2	A_1	B_3	A	B	B	A
E_1	E_1	E	E	B_2+B_3	E	$2B$	$A+B$	$A+B$
E_2	E_2	E	E	$A+B_1$	E	$2A$	$A+B$	$A+B$

C_{2v}	C_2	C_s $\sigma(zx)$	C_s $\sigma(yz)$
A_1	A	A'	A'
A_2	A	A''	A''
B_1	B	A'	A''
B_2	B	A''	A'

C_{3v}	C_3	C_s
A_1	A	A'
A_2	A	A''
E	E	$A'+A''$

C_{4v}	C_4	C_{2v} σ_v	C_{2v} σ_d	C_2	C_s σ_v	C_s σ_d
A_1	A	A_1	A_1	A	A'	A'
A_2	A	A_2	A_2	A	A''	A''
B_1	B	A_1	A_2	A	A'	A''
B_2	B	A_2	A_1	A	A''	A'
E	E	B_1+B_2	B_1+B_2	$2B$	$A'+A''$	$A'+A''$

C_{5v}	C_5	C_s
A_1	A	A'
A_2	A	A''
E_1	E_1	$A'+A''$
E_2	E_2	$A'+A''$

C_{6v}	C_6	C_{3v} σ_v	C_{3v} σ_d	C_{2v} $\sigma_v \rightarrow \sigma(zx)$	C_3	C_2	C_s σ_v	C_s σ_d
A_1	A	A_1	A_1	A_1	A	A	A'	A'
A_2	A	A_2	A_2	A_2	A	A	A''	A''
B_1	B	A_1	A_2	B_1	A	B	A'	A''
B_2	B	A_2	A_1	B_2	A	B	A''	A'
E_1	E_1	E	E	B_1+B_2	E	$2B$	$A'+A''$	$A'+A''$
E_2	E_2	E	E	A_1+A_2	E	$2A$	$A'+A''$	$A'+A''$

C_{2h}	C_2	C_s	C_i
A_g	A	A'	A_g
B_g	B	A''	A_g
A_u	A	A''	A_u
B_u	B	A'	A_u

C_{3h}	C_3	C_s	C_1
A'	A	A'	A
E'	E	$2A'$	$2A$
A''	A	A''	A
E''	E	$2A''$	$2A$

$C_{3i} \equiv S_6$	C_3	C_i	C_1
A_g	A	A_g	A
E_g	E	$2A^*$	$2A^*$
A_u	A	A_u	A
E_u	E	$2A^*$	$2A^*$

* The coefficient 2 is not used in this condition.

$C_{\infty v}$	C_{6v}	C_{4v}	C_{3v}	C_{2v}
$A_1 \equiv \Sigma^+$	A_1	A_1	A_1	A_1
$A_2 \equiv \Sigma^-$	A_2	A_2	A_2	A_2
$E_1 \equiv \Pi$	E_1	E	E	$B_1 + B_2$
$E_2 \equiv \Delta$	E_2	$B_1 + B_2$	E	$A_1 + A_2$
$E_3 \equiv \Phi$	$B_2 + B_1$	E	$A_1 + A_2$	$B_1 + B_2$
$E_4 \equiv \Gamma$	E_2	$A_1 + A_2$	E	$A_1 + A_2$
...				

$D_{\infty h}$	D_{6h}	C_{6v}	C_{3v}	D_{4h}	C_{4v}	C_{2v}	$C_{\infty v}$
Σ_g^+	A_{1g}	A_1	A_1	A_{1g}	A_1	A_1	$\Sigma^+ \equiv A_1$
Σ_g^-	A_{2g}	A_2	A_2	A_{2g}	A_2	A_2	$\Sigma^- \equiv A_2$
Π_g	E_{1g}	E_1	E	E_g	E	$B_1 + B_2$	$\Pi \equiv E_1$
Δ_g	E_{2g}	E_2	E	$B_{1g} + B_{2g}$	$B_1 + B_2$	$A_1 + A_2$	$\Delta \equiv E_2$
...							
Σ_u^+	A_{2u}	A_1	A_1	A_{2u}	A_1	A_1	$\Sigma^+ \equiv A_1$
Σ_u^-	A_{1u}	A_2	A_2	A_{1u}	A_2	A_2	$\Sigma^- \equiv A_2$
Π_u	E_{1u}	E_1	E	E_u	E	$B_1 + B_2$	$\Pi \equiv E_1$
Δ_u	E_{2u}	E_2	E	$B_{1u} + B_{2u}$	$B_1 + B_2$	$A_1 + A_2$	$\Delta \equiv E_2$
...							

C_{4h}	C_4	S_4	C_{2h}	C_2	C_s	C_i	C_1
A_g	A	A	A_g	A	A'	A_g	A
B_g	B	B	A_g	A	A'	A_g	A
E_g	E	E	$2B_g$	$2B$	$2A''$	$2A_g$	$2A$
A_u	A	B	A_u	A	A''	A_u	A
B_u	B	A	A_u	A	A''	A_u	A
E_u	E	E	$2B_u$	$2B$	$2A'$	$2A_u$	$2A$

C_{5h}	C_5	C_s	C_1
A'	A	A'	A
E_1'	E_1	$2A'$	$2A$
E_2'	E_2	$2A'$	$2A$
A''	A	A''	A
E_1''	E_1	$2A''$	$2A$
E_2''	E_2	$2A''$	$2A$

C_{6h}	C_6	C_{3h}	S_6	C_{2h}	C_3	C_2	C_s	C_i	C_1
A_g	A	A'	A_g	A_g	A	A	A'	A_g	A
B_g	B	A''	A_g	B_g	A	B	A''	A_g	A
E_{1g}	E_1	E''	E_g	$2B_g$	E	$2B$	$2A''$	$2A_g$	$2A$
E_{2g}	E_2	E'	E_g	$2A_g$	E	$2A$	$2A'$	$2A_g$	$2A$
A_u	A	A''	A_u	A_u	A	A	A''	A_u	A
B_u	B	A'	A_u	B_u	A	B	A'	A_u	A
E_{1u}	E_1	E'	E_u	$2B_u$	E	$2B$	$2A'$	$2A_u$	$2A$
E_{2u}	E_2	E''	E_u	$2A_u$	E	$2A$	$2A''$	$2A_u$	$2A$

D_{2h}	D_2	$C_2(z)$ C_{2v}	$C_2(y)$ C_{2v}	$C_2(x)$ C_{2v}	$C_2(z)$ C_{2h}	$C_2(y)$ C_{2h}	$C_2(x)$ C_{2h}
A_g	A	A_1	A_1	A_1	A_g	A_g	A_g
B_{1g}	B_1	A_2	B_2	B_1	A_g	B_g	B_g
B_{2g}	B_2	B_1	A_2	B_2	B_g	A_g	B_g
B_{3g}	B_3	B_2	B_1	A_2	B_g	B_g	A_g
A_u	A	A_2	A_2	A_2	A_u	A_u	A_u
B_{1u}	B_1	A_1	B_1	B_2	A_u	B_u	B_u
B_{2u}	B_2	B_2	A_1	B_1	B_u	A_u	B_u
B_{3u}	B_3	B_1	B_2	A_1	B_u	B_u	A_u

D_{2h} (cont.)	$C_2(z)$ C_2	$C_2(y)$ C_2	$C_2(x)$ C_2	$\sigma(xy)$ C_s	$\sigma(zx)$ C_s	$\sigma(yz)$ C_s	C_i
A_g	A	A	A	A'	A'	A'	A_g
B_{1g}	A	B	B	A'	A''	A''	A_g
B_{2g}	B	A	B	A''	A'	A''	A_g
B_{3g}	B	B	A	A''	A''	A'	A_g
A_u	A	A	A	A''	A''	A''	A_u
B_{1u}	A	B	B	A''	A'	A'	A_u
B_{2u}	B	A	B	A'	A''	A'	A_u
B_{3u}	B	B	A	A'	A'	$A_{,,}$	A_u

D_{3h}	C_{3h}	D_3	C_{3v}	$\sigma_h \rightarrow \sigma_v(zy)$ C_{2v}	C_3	C_2	σ_h C_s	σ_v C_s
A_1'	A'	A_1	A_1	A_1	A	A	A'	A'
A_2'	A'	A_2	A_2	B_2	A	B	A'	A''
E'	E'	E	E	$A_1 + B_2$	E	$A + B$	$2A'$	$A' + A''$
A_1''	A''	A_1	A_2	A_2	A	A	A''	A''
A_2''	A''	A_2	A_1	B_1	A	B	A''	A'
E''	E''	E	E	$A_2 + B_1$	E	$A + B$	$2A''$	$A' + A''$

D_{4h}	D_4	C'_2 D_{2d}	C''_2 D_{2d}	C_{4v}	C_{4h}	C'_2 D_{2h}	C''_2 D_{2h}	C_4	S_4
A_{1g}	A_1	A_1	A_1	A_1	A_g	A_g	A_g	A	A
A_{2g}	A_2	A_2	A_2	A_2	A_g	B_{1g}	B_{1g}	A	A
B_{1g}	B_1	B_1	B_2	B_1	B_g	A_g	B_{1g}	B	B
B_{2g}	B_2	B_2	B_1	B_2	B_g	B_{1g}	A_g	B	B
E_g	E	E	E	E	E_g	$B_{2g}+B_{3g}$	$B_{2g}+B_{3g}$	E	E
A_{1u}	A_1	B_1	B_1	A_2	A_u	A_u	A_u	A	B
A_{2u}	A_2	B_2	B_2	A_1	A_u	B_{1u}	B_{1u}	A	B
B_{1u}	B_1	A_1	A_2	B_2	B_u	A_u	B_{1u}	B	A
B_{2u}	B_2	A_2	A_1	B_1	B_u	B_{1u}	A_u	B	A
E_u	E	E	E	E	E_u	$B_{2u}+B_{3u}$	$B_{2u}+B_{3u}$	E	E

D_{4h} (cont.)	C'_2 D_2	C''_2 D_2	C_2, σ_v C_{2v}	C_2, σ_d C_{2v}	C'_2 C_{2v}	C''_2 C_{2v}
A_{1g}	A	A	A_1	A_1	A_1	A_1
A_{2g}	B_1	B_1	A_2	A_2	B_1	B_1
B_{1g}	A	B_1	A_1	A_2	A_1	B_1
B_{2g}	B_1	A	A_2	A_1	B_1	A_1
E_g	B_2+B_3	B_2+B_3	B_1+B_2	B_1+B_2	A_2+B_2	A_2+B_2
A_{1u}	A	A	A_2	A_2	A_2	A_2
A_{2u}	B_1	B_1	A_1	A_1	B_2	B_2
B_{1u}	A	B_1	A_2	A_1	A_2	B_2
B_{2u}	B_1	A	A_1	A_2	B_2	A_2
E_u	B_2+B_3	B_2+B_3	B_1+B_2	B_1+B_2	A_1+B_1	A_1+B_1

D_{4h} (cont.)	C_2 C_{2h}	C'_2 C_{2h}	C''_2 C_{2h}	C_2 C_2	C'_2 C_2	C''_2 C_2	σ_h C_s	σ_v C_s	σ_d C_s	C_i
A_{1g}	A_g	A_g	A_g	A	A	A	A'	A'	A'	A_g
A_{2g}	A_g	B_g	B_g	A	B	B	A'	A''	A''	A_g
B_{1g}	A_g	A_g	B_g	A	A	B	A'	A''	A''	A_g
B_{2g}	A_g	B_g	A_g	A	B	A	A'	A''	A'	A_g
E_g	$2B_g$	A_g+B_g	A_g+B_g	$2B$	$A+B$	$A+B$	$2A''$	$A'+A''$	$A'+A''$	$2A_g$
A_{1u}	A_u	A_u	A_u	A	A	A	A''	A''	A''	A_u
A_{2u}	A_u	B_u	B_u	A	B	B	A''	A'	A'	A_u
B_{1u}	A_u	A_u	B_u	A	A	B	A''	A''	A'	A_u
B_{2u}	A_u	B_u	A_u	A	B	A	A''	A'	A''	A_u
E_u	$2B_u$	A_u+B_u	A_u+B_u	$2B$	$A+B$	$A+B$	$2A'$	$A'+A''$	$A'+A''$	$2A_u$

D_{5h}	D_5	C_{5v}	C_{5h}	C_5	$\sigma_h\to\sigma(zx)$ C_{2v}	C_2	σ_h C_5	σ_v C_s
A'_1	A_1	A_1	A'	A	A_1	A	A'	A'
A'_2	A_2	A_2	A'	A	B_1	B	A'	A''
E'_1	E_1	E_1	E'_1	E_1	A_1+B_1	$A+B$	$2A'$	$A'+A''$
E'_2	E_2	E_2	E'_2	E_1	A_1+B_1	$A+B$	$2A'$	$A'+A''$
A''_1	A_1	A_2	A''	A	A_2	A	A''	A''
A''_2	A_2	A_1	A''	A	B_2	B	A''	A'
E''_1	E_1	E_1	E''_1	E_1	A_2+B_2	$A+B$	$2A''$	$A'+A''$
E''_2	E_2	E_2	E''_2	E_2	A_2+B_2	$A+B$	$2A''$	$A'+A''$

D_{6h}	D_6	C'_2 D_{3h}	C''_2 D_{3h}	C_{6v}	C_{6h}	C''_2 D_{3d}	C'_2 D_{3d}	$\sigma_h\to\sigma(xy)$ $\sigma_v\to\sigma(yz)$ D_{2h}
A_{1g}	A_1	A'_1	A'_1	A_1	A_g	A_{1g}	A_{1g}	A_g
A_{2g}	A_2	A'_2	A'_2	A_2	A_g	A_{2g}	A_{2g}	B_{1g}
B_{1g}	B_1	A''_1	A''_2	B_2	B_g	A_{2g}	A_{1g}	B_{2g}
B_{2g}	B_2	A''_2	A''_1	B_1	B_g	A_{1g}	A_{2g}	B_{2g}
E_{1g}	E_1	E''	E''	E_1	E_{1g}	E_g	E_g	$B_{2g}+B_{1g}$
E_{2g}	E_2	E'	E'	E_2	E_{2g}	E_g	E_g	A_g+B_{1g}
A_{1u}	A_1	A''_1	A''_1	A_2	A_u	A_{1u}	A_{1u}	A_u
A_{2u}	A_2	A''_2	A''_2	A_1	A_u	A_{2u}	A_{2u}	B_{1u}
B_{1u}	B_1	A'_1	A'_2	B_1	B_u	A_{2u}	A_{1u}	B_{2u}
B_{2u}	B_2	A'_2	A'_1	B_2	B_u	A_{1u}	A_{2u}	B_{3u}
E_{1u}	E_1	E'	E'	E_1	E_{1u}	E_u	E_u	$B_{2u}+B_{3u}$
E_{2u}	E_2	E''	E''	E_2	E_{2u}	E_u	E_u	A_u+B_u

D_{6h} (cont.)	C_6	C_{3h}	C'_2 D_3	C''_2 D_3	σ_v C_{3v}	σ_d C_{3v}	S_6	D_2
A_{1g}	A	A'	A_1	A_1	A_1	A_1	A_g	A
A_{2g}	A	A'	A_2	A_2	A_2	A_2	A_g	B_1
B_{1g}	B	A''	A_1	A_2	A_2	A_1	A_g	B_2
B_{2g}	B	A''	A_2	A_1	A_1	A_2	A_g	B_3
E_{1g}	E_1	E''	E	E	E	E	E_g	B_2+B_3
E_{2g}	E_2	E'	E	E	E	E	E_g	$A+B_1$
A_{1u}	A	A''	A_1	A_1	A_2	A_2	A_u	A
A_{2u}	A	A''	A_2	A_2	A_1	A_1	A_u	B_1
B_{1u}	B	A'	A_1	A_2	A_1	A_2	A_u	B_2
B_{2u}	B	A'	A_2	A_1	A_2	A_1	A_u	B_3
E_{1u}	E_1	E'	E	E	E	E	E_u	B_2+B_3
E_{2u}	E_2	E''	E	E	E	E	E_u	$A+B_1$

D_{6h} (cont.)	C_2 / C_{2v}	C'_2 / C_{2v}	C''_2 / C_{2v}	C_2 / C_{2h}	C'_2 / C_{2h}	C''_2 / C_{2h}	C_3	C_2 / C_2
A_{1g}	A_1	A_1	A_1	A_g	A_g	A_u	A	A
A_{2g}	A_2	B_1	B_1	A_g	B_g	B_g	A	A
B_{1g}	B_1	A_2	B_2	B_g	A_g	B_g	A	B
B_{2g}	B_2	B_2	A_2	B_g	B_g	A_g	A	B
E_{1g}	B_1+B_2	A_2+B_2	A_2+B_2	$2B_g$	A_g+B_g	A_g+B_g	E	$2B$
E_{2g}	A_1+A_2	A_1+B_1	A_1+B_1	$2A_g$	A_g+B_g	A_g+B_g	E	$2A$
A_{1u}	A_2	A_2	A_2	A_u	A_u	A_u	A	A
A_{2u}	A_1	B_1	B_2	A_u	B_u	B_u	A	A
B_{1u}	B_2	A_1	B_1	B_u	A_u	B_u	A	B
B_{2u}	B_1	B_2	A_1	B_u	B_u	A_u	A	B
E_{1u}	B_2+B_1	A_1+B_2	A_1+B_1	$2B_u$	A_u+B_u	A_u+B_u	E	$2B$
E_{2u}	A_2+A_1	A_2+B_1	A_2+B_2	$2A_u$	A_u+B_u	A_u+B_u	E	$2A$

D_{6h} (cont.)	C'_2 / C_2	C''_2 / C_2	σ_h / C_s	σ_d / C_s	σ_v / C_s	C_i
A_{1g}	A	A	A'	A'	A'	A_g
A_{2g}	B	B	A'	A''	A''	A_g
B_{1g}	A	B	A''	A'	A''	A_g
B_{2g}	B	A	A''	A''	A'	A_g
E_{1g}	$A+B$	$A+B$	$2A''$	$A'+A''$	$A'+A''$	$2A_g$
E_{2g}	$A+B$	$A+B$	$2A'$	$A'+A''$	$A'+A''$	$2A_g$
A_{1u}	A	A	A''	A''	A''	A_u
A_{2u}	B	B	A''	A'	A'	A_u
B_{1u}	A	B	A'	A''	A'	A_u
B_{2u}	B	A	A'	A'	A''	A_u
E_{1u}	$A+B$	$A+B$	$2A'$	$A'+A''$	$A'+A''$	$2A_u$
E_{2u}	$A+B$	$A+B$	$2A''$	$A'+A''$	$A'+A''$	$2A_u$

D_{2d}	S_4	$C_2 \to C_2(z)$ / D_2	C_{2v}	C_2 / C_2	C'_2 / C_2	C_s
A_1	A	A	A_1	A	A	A'
A_2	A	B_1	A_2	A	B	A''
B_1	B	A	A_2	A	A	A''
B_2	B	B_1	A_1	A	B	A'
E	E	B_2+B_3	B_1+B_2	$2B$	$A+B$	$A'+A''$

D_{3d}	D_3	C_{3v}	S_6	C_3	C_{2h}	C_2	C_s	C_i
A_{1g}	A_1	A_1	A_g	A	A_g	A	A'	A_g
A_{2g}	A_2	A_2	A_g	A	B_g	B	A''	A_g
E_g	E	E	E_g	E	A_g+B_g	$A+B$	$A'+A''$	$2A_g$
A_{1u}	A_1	A_2	A_u	A	A_u	A	A''	A_u
A_{2u}	A_2	A_1	A_u	A	B_u	B	A'	A_u
E_u	E	E	E_u	E	A_u+B_u	$A+B$	$A'+A''$	$2A_u$

D_{4d}	D_4	C_{4v}	S_8	C_4	C_{2v}	C_2 C_2	C'_2 C_2	C_s
A_1	A_1	A_1	A	A	A_1	A	A	A'
A_2	A_2	A_2	A	A	A_2	A	B	A''
B_1	A_1	A_2	B	A	A_2	A	A	A''
B_2	A_2	A_1	B	A	A_1	A	B	A'
E_1	E	E	E_1	E	B_1+B_2	$2B$	$A+B$	$A'+A''$
E_2	B_1+B_2	B_1+B_2	E_2	$2B$	A_1+A_2	$2A$	$A+B$	$A'+A''$
E_3	E	E	E_3	E	B_1+B_2	$2B$	$A+B$	$A'+A''$

D_{5d}	D_5	C_{5v}	C_5	C_2	C_s	C_i
A_{1g}	A_1	A_1	A	A	A'	A_g
A_{2g}	A_2	A_2	A	B	A''	A_g
E_{1g}	E_1	E_1	E_1	$A+B$	$A'+A''$	$2A_g$
E_{2g}	E_2	E_2	E_2	$A+B$	$A'+A''$	$2A_g$
A_{1u}	A_1	A_2	A	A	A''	A_u
A_{2u}	A_2	A_1	A	B	A'	A_u
E_{1u}	E_1	E_1	E_1	$A+B$	$A'+A''$	$2A_u$
E_{2u}	E_2	E_2	E_2	$A+B$	$A'+A''$	$2A_u$

D_{6d}	D_6	C_{6v}	C_6	D_{2d}	D_3	C_{3v}
A_1	A_1	A_1	A	A_1	A_1	A_1
A_2	A_2	A_2	A	A_2	A_2	A_2
B_1	A_1	A_2	A	B_1	A_1	A_2
B_2	A_2	A_1	A	B_2	A_2	A_1
E_1	E_1	E_1	E_1	E	E	E
E_2	E_2	E_2	E_2	B_1+B_2	E	E
E_3	B_1+B_2	B_1+B_2	$2B$	E	A_1+A_2	A_1+A_2
E_4	E_2	E_2	E_2	A_1+A_2	E	E
E_5	E_1	E_1	E_1	E	E	E

D_{6d} (cont.)	D_2	C_{2v}	S_4	C_3	C_2 C_2	C_2' C_2	C_s
A_1	A	A_1	A	A	A	A	A'
A_2	B_1	A_2	A	A	B	B	A''
B_1	A	A_2	B	A	A	A	A''
B_2	B_1	A_1	B	A	A	B	A'
E_1	$B_2 + B_3$	$B_1 + B_2$	E	E	$2B$	$A + B$	$A' + A''$
E_2	$A + B_1$	$A_1 + A_2$	$2B$	E	$2A$	$A + B$	$A' + A''$
E_3	$B_2 + B_3$	$B_1 + B_2$	E	$2A$	$2B$	$A + B$	$A' + A''$
E_4	$A + B_1$	$A_1 + A_2$	$2A$	E	$2A$	$A + B$	$A' + A''$
E_5	$B_2 + B_1$	$B_1 + B_2$	E	E	$2B$	$A + B$	$A' + A''$

S_4	C_2	C_1
A	A	A
B	A	A
E	$2B$	$2A$

S_8	C_4	C_2	C_1
A	A	A	A
B	A	A	A
E_1	E	$2B$	$2A$
E_2	$2B$	$2A$	$2A$
E_3	E	$2B$	$2A$

T	D_2	C_3	C_2	C_1
A	A	A	A	A
E	$2A$	E	$2A$	$2A$
F	$B_1 + B_2 + B_3$	$A + E$	$A + 2B$	$3A$

T_h	T	D_{2h}	S_6	D_2
A_g	A	A_g	A_g	A
E_g	E	$2A_g$	E_g	$2A$
F_g	F	$B_{1g} + B_{2g} + B_{3g}$	$A_g + E_g$	$B_1 + B_2 + B_3$
A_u	A	A_u	A_u	A
E_u	E	$2A_u$	E_u	$2A$
F_u	F	$B_{1u} + B_{2u} + B_{3u}$	$A_u + E_u$	$B_1 + B_2 + B_3$

Note. For S_6 correlation see $C_{3i} \equiv S_6$ on page 332.

T_h (cont.)	C_{2v}	C_{2h}	C_3	C_2	C_s	C_i	C_1
A_g	A_1	A_g	A	A	A'	A_g	A
E_g	$2A_1$	$2A_2$	E	$2A$	$2A'$	$2A_g$	$2A$
F_g	$A_2+B_1+B_2$	A_g+2B_g	$A+E$	$A+2B$	$A'+2A''$	$3A_g$	$3A$
A_u	A_2	A_u	A	A	A''	A_u	A
E_u	$2A_2$	$2A_u$	E	$2A$	$2A''$	$2A_u$	$2A$
F_u	$A_1+B_1+B_2$	A_u+2B_u	$A+E$	$A+2B$	$2A'+A''$	$3A_u$	$3A$

T_d	T	D_{2d}	C_{3v}	S_4	D_2	C_{2v}
A_1	A	A_1	A_1	A	A	A_1
A_2	A	B_1	A_2	B	A	A_2
E	E	A_1+B_1	E	$A+B$	$2A$	A_1+A_2
F_1	F	A_2+E	A_2+E	$A+E$	$B_1+B_2+B_3$	$A_2+B_1+B_2$
F_2	F	B_2+E	A_1+E	$B+E$	$B_1+B_2+B_3$	$B_1+B_2+B_3$

T_d (cont.)	C_3	C_2	C_s
A_1	A	A	A'
A_2	A	A	A''
E	E	$2A$	$A'+A''$
F_1	$A+E$	$A+2B$	$A'+2A''$
F_2	$A+E$	$A+2B$	$2A'+A''$

O	T	D_4	D_3	C_4	$3C_2$ D_2	$C_2, 2C_2'$ D_2
A_1	A	A_1	A_1	A	A	A
A_2	A	B_1	A_2	B	A	B_1
E	E	A_1+B_1	E	$A+B$	$2A$	$A+B_1$
F_1	F	A_2+E	A_2+E	$A+E$	$B_1+B_2+B_3$	$B_1+B_2+B_3$
F_2	F	B_2+E	A_1+E	$B+E$	$B_1+B_2+B_3$	$A+B_2+B_3$

O (cont.)	C_3	C_2	C_2
A_1	A	A	A
A_2	A	A	B
E	E	$2A$	$A+B$
F_1	$A+E$	$A+2B$	$A+2B$
F_2	$A+E$	$A+2B$	$2A+B$

$O_h{}^a$	O	T_d	T_h	T	D_{4h}	D_{3d}	D_{4d}	C_{3v}	D_3	$D_{3i} \equiv S_6$
A_{1g}	A_1	A_1	A_g	A	A_{1g}	A_{1g}	A_{1g}	A_1	A_1	A_g
A_{2g}	A_2	A_2	A_g	A	B_{1g}	A_{2g}	B_{1g}	A_2	A_2	A_g
E_g	E	E	E_g	E	$A_{1g}+B_{1g}$	E_g	$A_{1g}+B_{1g}$	E	E	E_g
F_{1g}	F_1	F_1	F_g	F	$A_{2g}+E_g$	$A_{2g}+E_g$	$A_{2g}+E_g$	A_2+E	A_2+E	A_g+E_g
F_{2g}	F_2	F_2	F_g	F	$B_{2g}+E_g$	$A_{1g}+E_q$	$B_{2g}+E_g$	A_1+E	A_1+E	A_g+E_g
A_{1u}	A_1	A_2	A_u	A	A_{1u}	A_{1u}	A_{1u}	A_2	A_1	A_u
A_{2u}	A_2	A_1	A_u	A	B_{1u}	A_{2u}	B_{1u}	A_1	A_2	A_u
E_u	E	E	E_u	E	$A_{1u}+B_{1u}$	E_u	$A_{1u}+B_{1u}$	E	E	E_u
F_{1u}	F_1	F_2	F_u	F	$A_{2u}+E_u$	$A_{2u}+E_u$	$A_{2u}+E_u$	A_1+E	A_2+E	A_u+E_u
F_{2u}	F_2	F_1	F_u	F	$B_{2u}+E_u$	$A_{1u}+E_u$	$B_{2u}+E_u$	A_2+E	A_1+E	A_u+E_u

O_h (cont.)	C_3	C_2, σ_d D_{2d}	C_2', σ_h D_{2d}	C_{4v}	D_4	C_{4h}	S_4	C_4
A_{1g}	A	A_1	A_1	A_1	A_1	A_g	A	A
A_{2g}	A	B_1	B_2	B_1	B_1	B_g	B	B
E_g	E	A_1+B_1	A_1+B_2	A_1+B_1	A_1+B_1	A_g+B_g	$A+B$	$A+B$
F_{1g}	$A+E$	A_2+E	A_2+E	A_2+E	A_2+E	A_g+E_g	$A+E$	$A+E$
F_{2g}	$A+E$	B_2+E	B_1+E	B_2+E	B_2+E	B_g+E_g	$B+E$	$B+E$
A_{1u}	A	B_1	B_1	A_2	A_1	A_u	B	A
A_{2u}	A	A_1	A_2	B_2	B_1	B_u	A	B
E_u	E	A_1+B_1	A_2+B_1	A_2+B_2	A_1+B_1	A_u+B_u	$A+B$	$A+B$
F_{1u}	$A+E$	B_2+E	B_2+E	A_1+E	A_2+E	A_u+E_u	$B+E$	$B+E$
F_{2u}	$A+E$	A_2+E	A_1+E	B_1+E	B_2+E	B_u+E_u	$A+E$	$B+E$

^{a}To find correlations with smaller subgroups, carry out the correlation in two steps: for example, if the correlation of O_h with C_{2v} is desired, use the table to pass from O_h to T_d and then employ the table for T_d to go on to C_{2v}.

O_h (cont.)	$3C_2$ D_{2h}	$C_2, 2C_2'$ D_{2h}	C_2, σ_h C_{2v}	C_2, σ_d C_{2v}
A_{1g}	A_g	A_g	A_1	A_1
A_{2g}	A_g	B_{1g}	A_1	A_2
E_g	$2A_g$	A_g+B_{1g}	$2A_1$	A_1+A_2
F_{1g}	$B_{1g}+B_{2g}+B_{3g}$	$B_{1g}+B_{2g}+B_{3g}$	$A_2+B_1+B_2$	$A_2+B_1+B_2$
F_{2g}	$B_{1g}+B_{2g}+B_{3g}$	$A_{1g}+B_{2g}+B_{3g}$	$A_2+B_1+B_2$	$A_1+B_1+B_2$
A_{1u}	A_u	A_u	A_2	A_2
A_{2u}	A_u	B_{2u}	A_2	A_1
E_u	$2A_u$	A_u+B_u	$2A_2$	A_1+A_2
F_{1u}	$B_{1u}+B_{2u}+B_{3u}$	$B_{1u}+B_{2u}+B_{3u}$	$A_1+B_1+B_2$	$A_1+B_1+B_2$
F_{2u}	$B_{1u}+B_{2u}+B_{3u}$	$A_u+B_{2u}+B_{3u}$	$A_1+B_1+B_2$	$A_2+B_1+B_2$

O_h (cont.)	C'_2, σ_h / C_{2v}	$3C_2$ / D_2	$C_2, 2C'_2$ / D_2	C_2, σ_h / C_{2h}	C'_2, σ_h / C_{2h}
A_{1g}	A_1	A	A	A_g	A_g
A_{2g}	B_1	A	B_1	A_g	B_g
E_g	A_1+B_1	$2A$	$A+B_1$	$2A_g$	A_g+B_g
F_{1g}	$A_2+B_1+B_2$	$B_1+B_2+B_3$	$B_1+B_2+B_3$	A_g+2B_g	A_g+2B_g
F_{2g}	$A_1+A_2+B_2$	$B_1+B_2+B_3$	$A+B_2+B_3$	A_g+2B_g	$2A_g+B_g$
A_{1u}	A_2	A	A	A_u	A_u
A_{2u}	B_2	A	B_1	A_u	B_u
E_u	A_2+B_2	$2A$	$A+B_1$	$2A_u$	A_u+B_u
F_{1u}	$A_1+B_1+B_2$	$B_1+B_2+B_3$	$B_1+B_2+B_3$	A_u+2B_u	A_u+2B_u
F_{2u}	$A_1+A_2+B_1$	$B_1+B_2+B_3$	$A+B_2+B_3$	A_u+2B_u	$2A_u+B_u$

O_h (cont.)	σ_h / C_s	σ_d / C_s	C_2 / C_2	C'_2 / C_2	C_i	C_1
A_{1g}	A'	A'	A	A	A_g	A
A_{2g}	A'	A''	A	B	A_g	A
E_g	$2A'$	$A'+A''$	$2A$	$A+B$	$2A_g$	$2A$
F_{1g}	$A'+A''$	$A'+2A''$	$A+2B$	$A+2B$	$3A_g$	$3A$
F_{2g}	$A'+2A''$	$2A'+A''$	$A+2B$	$2A+B$	$3A_g$	$3A$
A_{1u}	A''	A''	A	A	A_u	A
A_{2u}	A''	A'	A	B	A_u	A
E_u	$2A''$	$A'+A''$	$2A$	$A+B$	$2A_u$	$2A$
F_{1u}	$2A'+A''$	$2A'+A''$	$A+2B$	$A+2B$	$3A_u$	$3A$
F_{2u}	$2A'+A''$	$A'+2A''$	$A+2B$	$2A+B$	$3A_u$	$3A$

I_h	I	C_5	C_3	C_2	C_1
A_g	A	A	A	A	A
A_u	A	A	A	A	A
F_{1g}	F_1	$A+E_1$	$A+E$	$A+2B$	$3A$
F_{1u}	F_1	$A+E_1$	$A+E$	$A+2B$	$3A$
F_{2g}	F_2	$A+E_2$	$A+E$	$A+2B$	$3A$
F_{2u}	F_2	$A+E_2$	$A+E$	$A+2B$	$3A$
G_{1g}	G_1	E_1+E_2	$2A+E$	$2A+2B$	$4A$
G_{1u}	G_1	E_1+E_2	$2A+E$	$2A+2B$	$4A$
H_g	H	$A+E_1+E_2$	$A+2E$	$3A+2B$	$5A$
H_u	H	$A+E_1+E_2$	$A+2E$	$3A+2B$	$5A$

References

1. W. G. Fateley, F. R. Dollish, N. T. McDevitt, and F. F. Bentley, "Infrared and Raman Selection Rules for Molecular and Lattice Vibrations: The Correlation Method." Wiley–Interscience, New York, 1972.

Appendix 7

Principle of Laser Action

The principle of laser action is based on "population inversion." At thermal equilibrium, the ratio of populations at the ground and excited states is determined by the Maxwell–Boltzmann distribution law (Section 1.3). Namely, the population at the excited state decreases exponentially as the transition energy ($h\nu$) increases. To maintain such equilibrium, molecules at the excited state revert to the ground state by emitting photons of $h\nu$ (spontaneous emission). Thus, the population at the ground state is always larger than that in the excited state. Under such a circumstance, a photon of $h\nu$ is more likely to be absorbed by the ground state species than to stimulate emission from the excited state. "Stimulated emission" occurs when the population at the excited state becomes larger than that in the ground state (population inversion). Then, the first few spontaneously emitted photons "stimulate" emissions of others, leading to a cascade of emissions that have the same energy ($h\nu$) and phase as the original photon. The population inversion can be created by combining a laser medium (gas/liquid/solid) with a power supply (flash lamp, electrical discharge, etc.), and the laser beam thus obtained is amplified by trapping it in an optical cavity (resonator).

The structure of a CW gas laser has been shown in Section 2.2. In the He–Ne laser, He (1mm Hg) and Ne (0.1 mm Hg) gases are mixed in the plasma tube. As shown in Fig. 1, the He atoms are excited to the ^1S and ^3S states by electrical discharge to create population inversion. Collisions of these excited state He atoms with Ne atoms produce excited-state Ne atoms, which produce stimulated emission at 632.8 and 1,152.3 nm. The latter is eliminated by using a prism. In an Ar-Ion laser, the population inversion is created by collision with energetic electrons, and the excited-state Ar ion emits a series of lines, including those of 488.0 and 514.5 nm.

In the case of a Nd:YAG (neodymium-doped yttrium aluminum garnet), the Nd^{3+} ion doped in a YAG crystal is excited to the $^4F_{3/2}$ state by flash lamp. Then, the transition from this state to the $^4I_{11/2}$ state produces a laser beam at 1,064 nm. This beam is converted into giant pulses with huge power and narrow width by using the Q-switching, cavity-dumping and modelocking

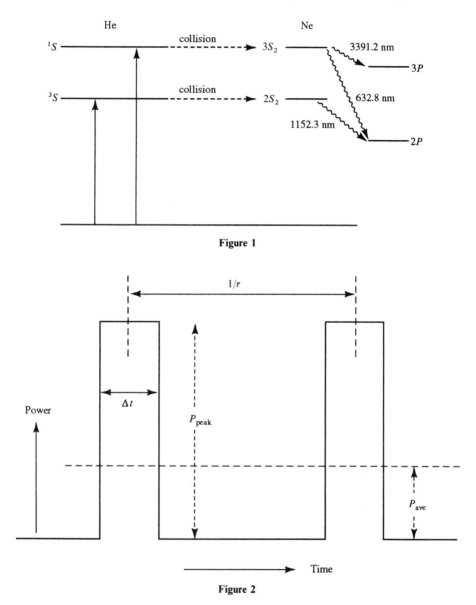

Figure 1

Figure 2

techniques (1). For example, a Quanta-Ray DCR-3 Nd:YAG (Q-switching) laser provides pulses that have 7–9 ns width, ~100 MW peak power and an optimum repetition rate of 20 Hz.

Figure 2 shows the relationship between the peak power and the average

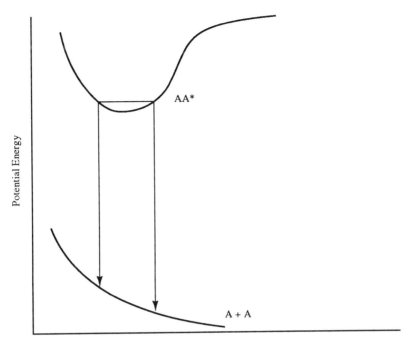

Internuclear Distance

Figure 3

power. It is seen that

$$P_{ave} = P_{peak} \times r \times \Delta t,$$

where r is the repetition rate and Δt is the pulse width. In the case of a CW laser, $r = 1$ and $\Delta t = 1$. Thus,

$$P_{ave} = P_{peak}.$$

The structure of a dye laser is basically the same as that of a gas laser except that a circulating dye solution is used instead of a gas. Large organic dye molecules such as rhodamine 6G exhibit strong absorption bands ($\pi-\pi^x$ transitions) in the visible region. An energy level diagram of such a dye may be represented by Fig. 3-6 in Chapter 3. The absorption band due to $S_0 (v = 0) \rightarrow S_1 (v' = 0, 1, 2, \ldots)$ transitions is broad and continuous because each electronic level is accompanied by a series of vibrational levels that are blurred by rotational and collisional broadening. Molecules excited to various sublevels of S_1 fall to the $v' = 0$ state via radiationless transitions, and then revert to $S_0 (v = 0, 1, 2, \ldots)$ to give fluorescence. Some S_1 state molecules fall to T_1 (triplet) state via intersystem crossing and cause phosphorescence ($T_1 \rightarrow S_0$ transitions), which limits dye performance. If a dye solution is irradiated by a strong laser beam or flash lamp, "population inversion" is created, and stimulated emission occurs from all occupied levels of S_1, resulting in a strong but broad fluorescence band. This emission can be confined into a narrow, selective wavelength region by adding wavelength selective devices to the dye cavity.

The principle of excimer lasers is based on the fact that rare gas atoms such as Kr and Xe can form molecules, "excimers," at their electronic excited states (Fig. 3). If rare gases at high pressures are irradiated by rapid electrical discharge, the resulting "population inversion" produces tunable, high-power pulsed laser beams in the vacuum UV region (Kr_2^*, 146 nm; Xe_2^*, 172 nm). Rare-gas halide excimers provide laser lines in the UV region (KrCl*, 222nm; KrF*, 249 nm; XeCl*, 308 nm; and XeF*, 351 nm).

For more information, the reader should consult reference books and review articles concerning lasers.

References

1. C. Breck Hitz, "Understanding LASER Technology." PennWell Publishing Co., Tulsa, Pklahoma, 1985.
2. For example, J. Hecht, "The Laser Guidebook." McGraw-Hill, New York, 1986.
3. For example, J. C. Wright and M. J. Wirth, *Anal. Chem.* **52**, 1087A (1980).

Appendix 8

Raman Spectra of Typical Solvents

The Raman spectra shown here were reproduced with permission from H. Hamaguchi and A. Hirakawa, "Raman Spectroscopic Methods," Gakukai Shupan Center and Japan Spectroscopy Society, Tokyo, 1988. The following exmperimental conditions were employed:

Source: Coherent CR-2 Ar-ion laser, 488.0 nm, ~ 100 mW.

Spectrometer: Spex Model 1877 triple polychromator with 100 μm slit width (resolution, ~ 5 cm^{-1})

Detector: PAR OMA-III System with Model 1420 intensified diode array detector

Frequency calibrations were made by using Ne emission lines. Accuracy of wavenumbers given in the figures is ± 1 cm^{-1}.

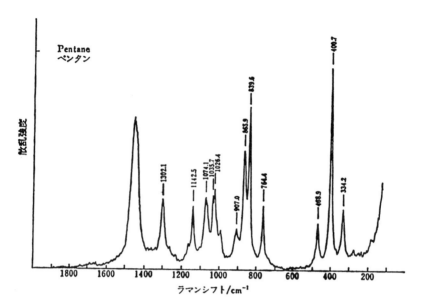

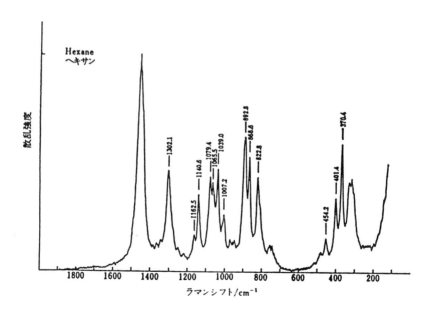

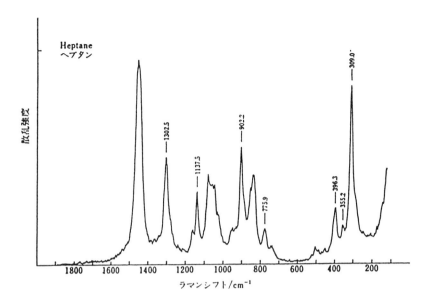

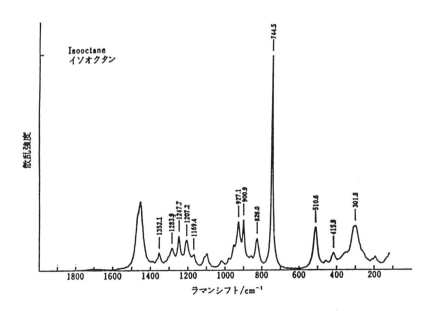

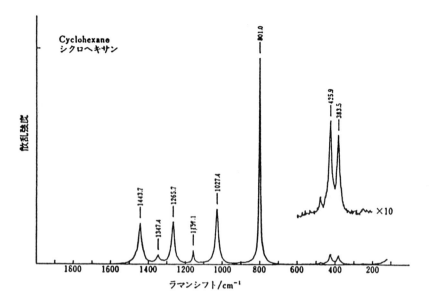

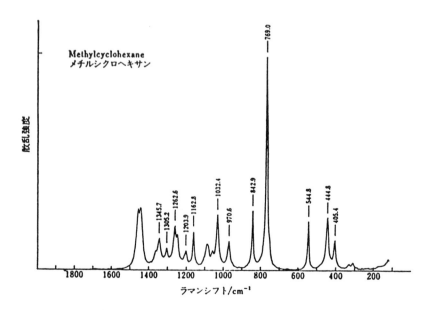

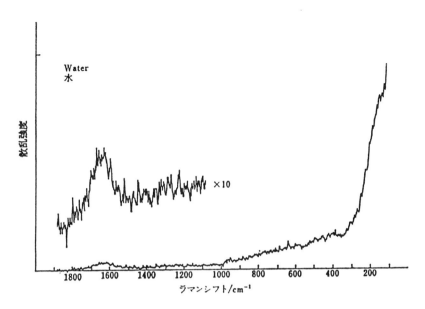

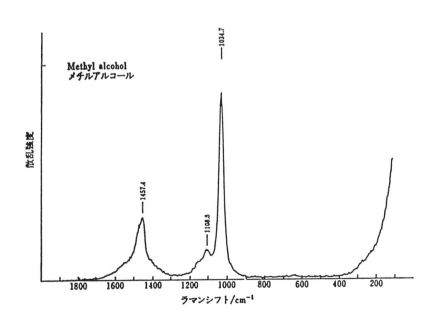

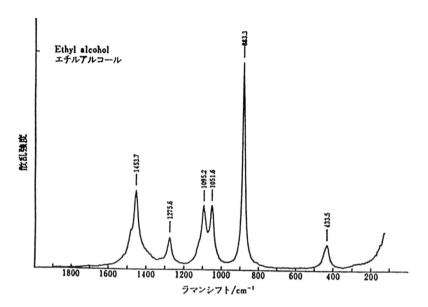

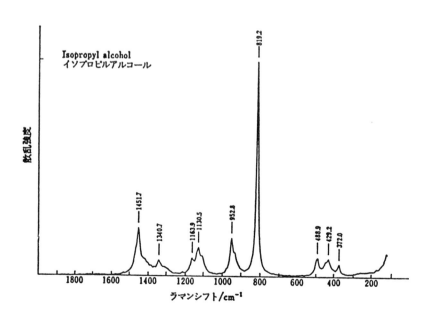

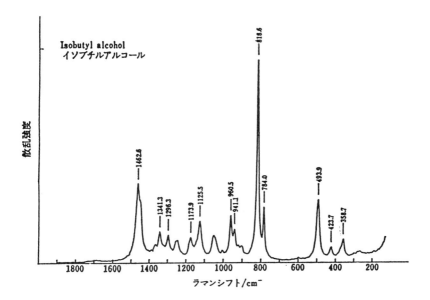

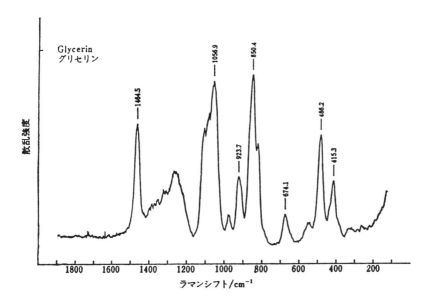

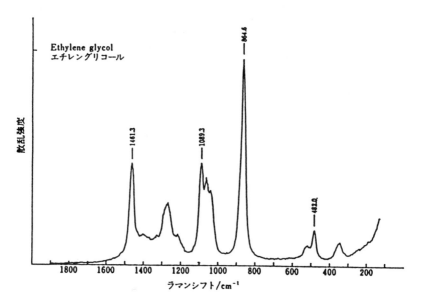

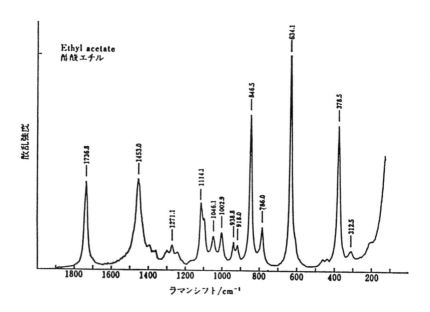

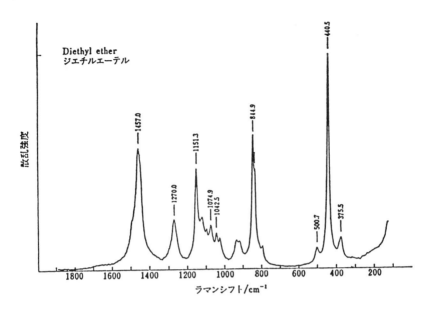

Diethyl ether
ジエチルエーテル

散乱強度

ラマンシフト/cm⁻¹

1457.0, 1270.0, 1151.3, 1074.9, 1042.5, 814.9, 500.7, 440.5, 375.5

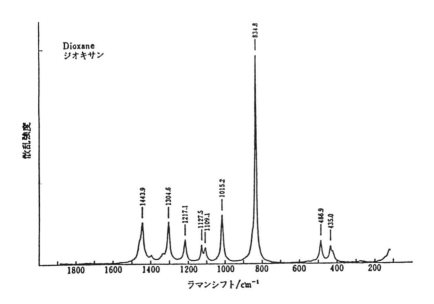

Dioxane
ジオキサン

散乱強度

ラマンシフト/cm⁻¹

1443.9, 1304.6, 1217.1, 1127.5, 1109.1, 1015.2, 834.8, 486.9, 435.0

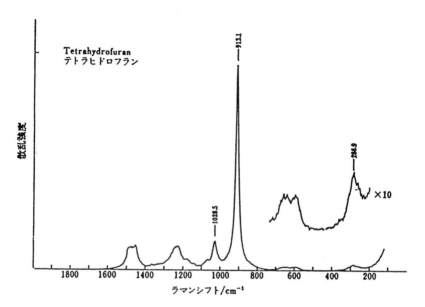

Tetrahydrofuran
テトラヒドロフラン

913.1

1028.5

284.9

×10

ラマンシフト/cm⁻¹

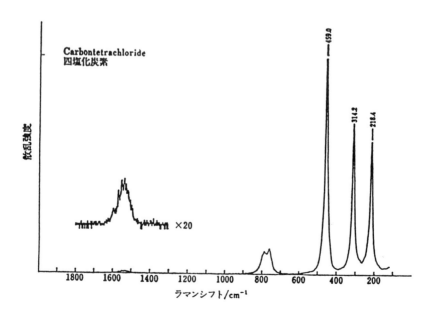

Carbontetrachloride
四塩化炭素

459.0

314.2

218.4

×20

ラマンシフト/cm⁻¹

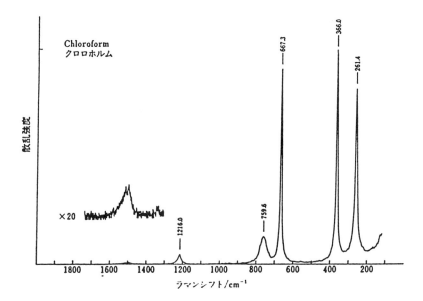

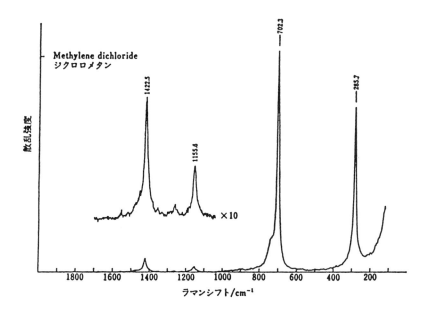

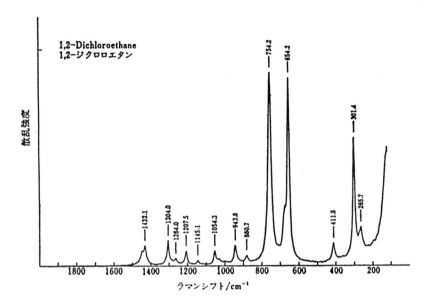

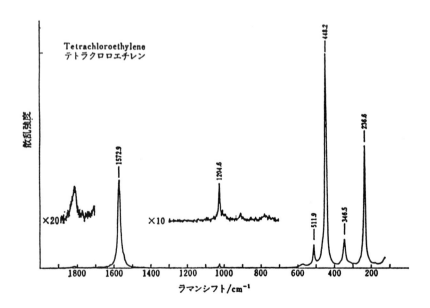

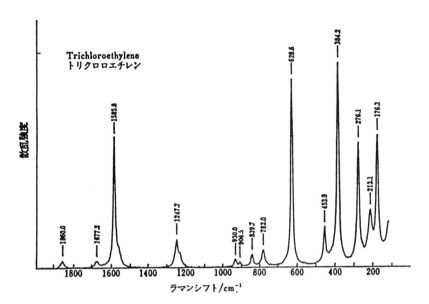

Trichloroethylene
トリクロロエチレン

散乱強度

1860.0 1677.2 1585.8 1247.7 930.0 906.5 839.7 782.0 628.6 453.9 384.2 276.1 211.1 176.3

ラマンシフト/cm⁻¹

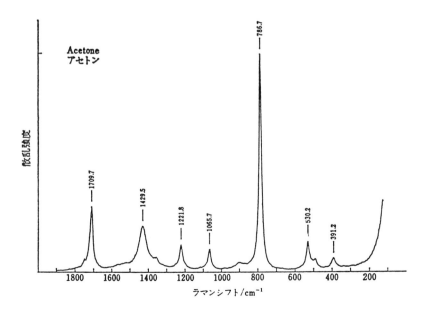

Acetone
アセトン

散乱強度

1709.7 1429.5 1221.8 1065.7 786.7 530.2 391.2

ラマンシフト/cm⁻¹

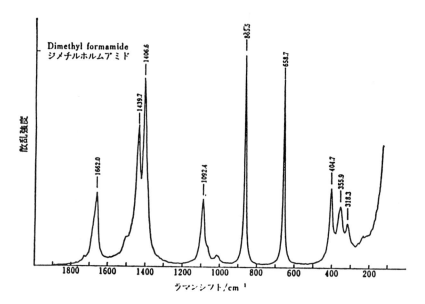

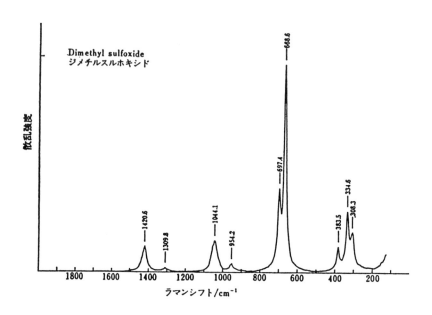

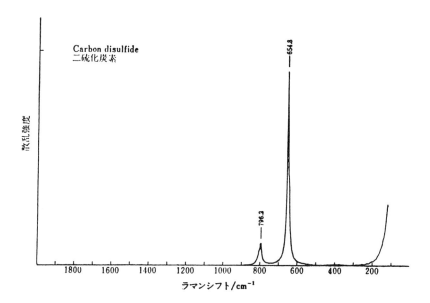

Carbon disulfide
二硫化炭素

654.8

796.2

散乱強度

ラマンシフト/cm⁻¹

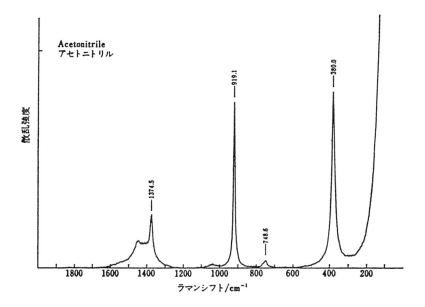

Acetonitrile
アセトニトリル

919.1

380.0

1374.5

748.6

散乱強度

ラマンシフト/cm⁻¹

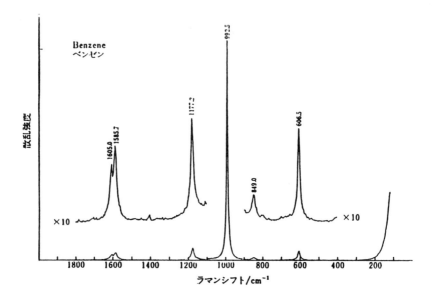

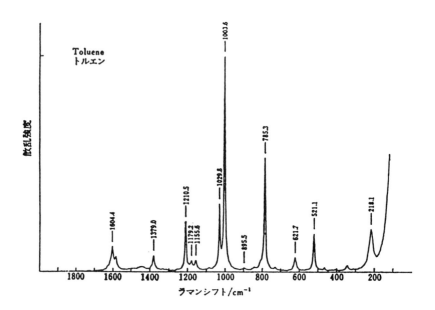

Index

ISBN 0-12-253990-7

9 780122 539909

90040